Protein Folding Protocols

METHODS IN MOLECULAR BIOLOGY™

John M. Walker, SERIES EDITOR

METHODS IN MOLECULAR BIOLOGY™

Protein Folding Protocols

Edited by

Yawen Bai

Laboratory of Biochemistry, National Cancer Institute,
National Institutes of Health, Bethesda, MD

and

Ruth Nussinov

Center for Cancer Research Nanobiology Program SAIC,
National Cancer Institute-Frederick, Frederick, MD
and Department of Human Genetics, Sackler School of Medicine,
Tel Aviv University, Tel Aviv, Israel

HUMANA PRESS ✳ TOTOWA, NEW JERSEY

Cover illustration: Figure 3 from Chapter 15, "Intermediates and Transition States in Protein Folding," by D. Thirumalai and Dmitri K. Klimov.

Production Editor: Erika J. Wasenda

Cover design by Patricia F. Cleary

Library of Congress Cataloging in Publication Data

Protein folding protocols / edited by Yawen Bai and Ruth Nussinov.

p. ; cm. -- (Methods in molecular biology ; 350)

Includes bibliographical references and index.

ISBN 1-58829-622-9 (alk. paper)

1. Protein folding. 2. Proteins--Conformation. 3. Proteins--Analysis.

[DNLM: 1. Protein Folding. 2. Protein Conformation. 3. Proteins--analysis. QU 55 P96635 2006] I. Bai, Yawen. II. Nussinov, Ruth. III. Series: Methods in molecular biology (Clifton, N.J.) ; 350.

QP551.P695822 2006

572'.633--dc22

2006002791

Preface

Protein Folding Protocols presents protocols for studying and characterizing steps and conformational ensembles populating pathways in protein folding from the unfolded to the folded state. It further presents a sample of approaches toward the prediction of protein structure starting from the amino acid sequence, in the absence of overall homologous sequences. Protein folding is a crucial step in the transfer of genetic information from the DNA to the protein. The Genome Project has led to a huge number of available DNA sequences and, therefore, protein sequences. The Structural Genomics initiative largely aims to obtain "new" folds not currently present in the Protein Data Bank. Yet, the number of available structures inevitably lags behind the number of sequences. At the same time, an equally important problem is to find out the types and scope of dissimilar (nonhomologous) protein sequences that adopt a similar fold. Assembling data and comprehension of the sequence space of protein folds should be very useful in computational protein structure prediction. This would enhance the scope of homology modeling, which currently is the method of choice. Thus, experimental and theoretical studies on the relationship between sequence and structure are critical. Figuring out the relationship between sequence and structure would further assist in the prediction of fibril structures observed in protein misfolding diseases, and in figuring out the conformational changes and dynamics resulting from mutations. Protein folding is one of the most important and challenging problems in current molecular and chemical biology. This book reviews some of the recently developed methods for studying protein folding.

The starting point of a folding process is the unfolded state. Eliezer describes how some of the local structural properties of the unfolded state may be characterized using multidimensional nuclear magnetic resonance (NMR). Gebel and Shortle review how the global structures of the unfolded state may be characterized by measuring the residual dipolar couplings, again with NMR. Another state that generally occurs in the process of protein folding is the rate-limiting transition state. Pandit and coworkers describe how it can be characterized by using the psi value analysis. Rao and coworkers employ molecular dynamics simulation to characterize the transition state of small proteins. They propose a technique to estimate the folding probability of those structures that are sampled along near-equilibrium, constant temperature molecular dynamics simulations. Bai and coworkers describe how the native-state hydrogen exchange method could be combined with protein engineering to populate the intermediate state for high-resolution structure determination by multidimensional NMR. Lassalle and Akasaka describe how the high-pressure NMR technique can be used to denature proteins and populate partially unfolded intermediates. Thirumalai and Klimov describe the use of computer simulations and simple models to investigate the formation of the intermediates. They show that the equilibrium intermediates occur "on pathway" and that there is a substantial probability that they be revisited after the native state is reached, in contrast to kinetic intermediates. Haspel and coworkers described how a building block-based protein folding model may be used toward a reduction of the computational complexity of protein folding. The model is based on the cutting of the target protein sequences into building block fragments that are relatively stable and whose conformations in solution are similar to those observed when the fragments are chain connected in the protein molecule from which they were derived.

One of the cutting-edge methods for studying protein folding is the single molecule technique. Ng and coworkers show that atomic force microscopy may be used to study the process of protein folding and unfolding. Schuler describes how the single molecule technique coupled with fluorescence resonance transfer may be used to study the population of proteins under equilibrium conditions and the process of kinetic folding. The study of very fast folding process, in the time-scale of microseconds or less, constitutes another important area in current protein folding studies. Gai and coworkers describe the use of T-jump coupled with infrared spectroscopy to characterize such a fast folding process. Fierz and coworkers outline a new method that uses triplet–triplet energy transfer to measure conformational dynamics in polypeptide chains in the unfolded ensembles, which can set the uplimit of folding times. Streicher and Makhatadze describe recent advances in the analysis of conformational transition in peptides using differential scanning calorimetry. Zhou describes the replica exchange molecular dynamics (REMD) method for enhanced sampling of the conformational space. REMD couples molecular dynamics trajectories with a temperature exchange Monte Carlo process. Replicas are run in parallel at different temperatures. This procedure allows surmounting local barriers encountered in the simulation. REMD has proven to be a very powerful tool, and it is increasingly used in MD simulations for small proteins/peptides. Jernigan and Kloczowski describe a computational result that relates the packing regularities of biological structures to their dynamics. Prediction of protein structures is the central issue of protein folding studies. Casadio and coworkers describe how protein structures might be predicted with and without the existence of homologs. This feat has been achieved with the aid of machine learning-based methods specifically suited for predicting structural features. For protein–protein interaction, a knowledge-based strategy may provide predictions of putative interaction patches on the protein surface.

It behooves us to note that not all proteins fold into stable conformations. Recently, there has been an increasing amount of data that a large percentage of the proteins exist in what is commonly called the natively disordered state. This state describes a spectrum of conformations with variable stabilities. Although on its own the stability of the native state is low (hence the term *disordered*), binding to its partner enhances its stability. Thus, here too, the native state is the functional state. Folding, misfolding, and protein disorder are all related to each other. Progress in the understanding of one assists in comprehension of the other. Above all, all relate to protein function and misfunction.

Altogether, *Protein Folding Protocols* is a comprehensive collection of chapters describing a broad range of techniques to study, predict, and analyze the protein folding process. It covers experiment and theory, bioinformatics approaches, and state-of-the-art simulation protocols for better sampling of the conformational space. Protein folding remains one of the most challenging problems in the biological/natural sciences. Making progress in this area will have tremendous implications, ranging from drug design, functional assignment, comprehension of the nature of regulation, figuring out molecular machines, viral entry into cells, and putting together cellular pathways and their dynamics. The challenge still remains; however, experiment and theory have been making steps toward an eventual practical solution.

Yawen Bai
Ruth Nussinov

Contents

Contributors

KAZUYUKI AKASAKA • *Department of Biotechnological Science, School of Biology-Oriented Science and Technology, Kinki University, Kinokawa City, Wakayama Prefecture, Japan*

YAWEN BAI • *Laboratory of Biochemistry, National Cancer Institute, National Institutes of Health, Bethesda, MD*

AMEDEO CAFLISCH • *Department of Biochemistry, University of Zürich, Zürich, Switzerland*

RITA CASADIO • *Biocomputing Group, CIRB/Department of Biology, University of Bologna, Bologna, Italy*

JANE CLARKE • *Department of Chemistry, University of Cambridge, Medical Research Council Center for Protein Engineering, Cambridge, UK*

ROBIN S. DOTHAGER • *Department of Biochemistry, University of Illinois Champaign-Urbana, Urbana, IL*

DEGUO DU • *Department of Chemistry, University of Pennsylvania, Philadelphia, PA*

DAVID ELIEZER • *Department of Biochemistry and Program in Structural Biology, Weill Medical College of Cornell University, New York, NY*

PIERO FARISELLI • *Biocomputing Group, CIRB/Department of Biology, University of Bologna, Bologna, Italy*

HANQIAO FENG • *Laboratory of Biochemistry, National Cancer Institute, National Institutes of Health, Bethesda, MD*

BEAT FIERZ • *Department of Biophysical Chemistry, Biozentrum, University of Basel, Basel, Switzerland*

FENG GAI • *Department of Chemistry, University of Pennsylvania, Philadelphia, PA*

ERIKA B. GEBEL • *Department of Biological Chemistry, Johns Hopkins University School of Medicine, Baltimore, MD*

NURIT HASPEL • *Department of Exact Sciences, School of Computer Science, Tel Aviv University, Tel Aviv, Israel*

YUVAL INBAR • *Department of Exact Sciences, School of Computer Science, Tel Aviv University, Tel Aviv, Israel*

ROBERT L. JERNIGAN • *Department of Biochemistry, Biophysics, and Molecular Biology, Laurence H. Baker Center for Bioinformatics and Biological Statistics, Iowa State University, Ames, IA*

KARIN JODER • *Department of Biophysical Chemistry, Biozentrum, University of Basel, Basel, Switzerland*

THOMAS KIEFHABER • *Department of Biophysical Chemistry, Biozentrum, University of Basel, Basel, Switzerland*

DMITRI K. KLIMOV • *Bioinformatics and Computational Biology Program, School of Computational Sciences, George Mason University, Manassas, VA*

ANDRZEJ KLOCZKOWSKI • *Department of Biochemistry, Biophysics, and Molecular Biology, Iowa State University, Ames, IA*

BRYAN A. KRANTZ • *Department of Microbiology and Molecular Genetics, Harvard Medical School, Boston, MA*

FLORIAN KRIEGER • *Department of Biophysical Chemistry, Biozentrum, University of Basel, Basel, Switzerland*

MICHAEL W. LASSALLE • *Exploratory Research for Advanced Technology "Actin Filament Dynamics" Project, Japan Science and Technology Corporation, Hyogo, Japan*

GEORGE I. MAKHATADZE • *Department of Biochemistry and Molecular Biology, Penn State College of Medicine, Hershey, PA*

PIER LUIGI MARTELLI • *Biocomputing Group, CIRB/Department of Biology, University of Bologna, Bologna, Italy*

SEAN P. NG • *Department of Chemistry, University of Cambridge, Medical Research Council Center for Protein Engineering, Cambridge, UK*

RUTH NUSSINOV • *Center for Cancer Research Nanobiology Program SAIC, National Cancer Institute-Frederick, Frederick, MD and Department of Human Genetics, Sackler School of Medicine, Tel Aviv University, Tel Aviv, Israel*

ADARSH D. PANDIT • *Department of Biochemistry and Molecular Biology, University of Chicago, Chicago, IL*

LUCY G. RANDLES • *Department of Chemistry, University of Cambridge, Medical Research Council Center for Protein Engineering, Cambridge, UK*

FRANCESCO RAO • *Department of Biochemistry, University of Zürich, Zürich, Switzerland*

BENJAMIN SCHULER • *Department of Biochemistry, University of Zürich, Zürich, Switzerland*

GIOVANNI SETTANNI • *Department of Biochemistry, University of Zürich, Zürich, Switzerland*

DAVID SHORTLE • *Department of Biological Chemistry, Johns Hopkins University School of Medicine, Baltimore, MD*

TOBIN R. SOSNICK • *Department of Biochemistry and Molecular Biology, University of Chicago, Chicago, IL*

WERNER W. STREICHER • *Department of Biochemistry and Molecular Biology, Penn State College of Medicine, Hershey, PA*

GIANLUCA TASCO • *Biocomputing Group, CIRB/Department of Biology, University of Bologna, Bologna, Italy*

D. THIRUMALAI • *Institute for Physical Science and Technology and Department of Chemistry and Biochemistry, University of Maryland, College Park, MD*

CHUNG-JUNG TSAI • *Center for Cancer Research Nanobiology Program SAIC, National Cancer Institute-Frederick, Frederick, MD*

HUI-HSU (GAVIN) TSAI • *Center for Cancer Research Nanobiology Program SAIC, National Cancer Institute-Frederick, Frederick, MD*

GILAD WAINREB • *Department of Human Genetics, Sackler School of Medicine, Tel Aviv University, Tel Aviv, Israel*

HAIM J. WOLFSON • *School of Computer Science, Tel Aviv University, Tel Aviv, Israel*

YAO XU • *Department of Chemistry, University of Pennsylvania, Philadelphia, PA*

RUHONG ZHOU • *Department of Chemistry, Columbia University, New York, NY and Computational Biology Center, IBM Thomas J. Watson Research Center, Yorktown Heights, NY*

ZHENG ZHOU • *Laboratory of Biochemistry, National Cancer Institute, National Institutes of Health, Bethesda, MD*

1

Infrared Temperature-Jump Study of the Folding Dynamics of α-Helices and β-Hairpins

Feng Gai, Deguo Du, and Yao Xu

Summary

The laser-induced temperature-jump (*T*-jump) technique in conjunction with infrared spectroscopy provides a versatile means to study the early events in protein folding. Compared with the commonly used stopped-flow kinetic methods, the *T*-jump initiation technique offers a faster time resolution. It allows the study of protein folding processes occurring on the nanosecond-to-microsecond time-scales. In addition, infrared spectroscopy is a powerful tool for characterizing backbone conformation and dynamics. In this chapter, we mainly discuss the application of this technique to the study of the helix–coil transition dynamics and the mechanism of β-hairpin formation.

Key Words: α-Helix; β-hairpin; dynamics; infrared; protein folding; temperature-jump.

1. Introduction

The folding rates of proteins and peptides span many orders of magnitude. Using the technique of stopped-flow kinetics in conjunction with site-directed mutagenesis and hydrogen-deuterium exchange, it has been possible to study the sequence of folding events of many single-domain proteins on the millisecond time-scale in great detail *(1–3)*. However, the folding of monomeric α-helices and β-hairpins occurs on the nanosecond-to-microsecond time-scale *(4–7)* and has only recently been studied experimentally. Although many computational and theoretical studies have also contributed significantly to our understanding of how these short peptides fold, a comprehensive review of relevant studies is beyond the scope of this chapter.

The α-helix and β-sheet are common secondary structure elements in proteins. Therefore, understanding their folding mechanisms will shed light on the question of how proteins fold. However, because of their fast folding rate, studying the folding dynamics of monomeric helices and β-hairpins requires a fast initiation

From: *Methods in Molecular Biology, vol. 350: Protein Folding Protocols*
Edited by: Y. Bai and R. Nussinov © Humana Press Inc., Totowa, NJ

method. Although many photo or photochemical initiation methods are capable of triggering a folding/unfolding event on the nanosecond or even picosecond time-scale *(8–15)*, the laser-induced temperature-jump (*T*-jump) technique *(4,16–21)* has been the most widely used in the study of the dynamics of the helix–coil transition *(16,22)* and β-hairpin formation *(23,24)*, as well as other fast events in protein folding *(25–30)*. The primary reason for this is that it does not require the additional chemical modification of the polypeptide of interest, even though it is limited to conditions where the peptide undergoes a net folding or unfolding transition. This contrasts with other photochemical triggering methods, which generally involve the removal of a designed conformational constraint by employing a laser pulse. For example, an aryl disulfide bond has been used to constrain an alanine-based peptide to be distorted from its equilibrium form, and after photolysis the dissociation of the disulfide bond generates two thiol radicals, and the peptide then commences to fold into an α-helical conformation *(9,31)*. Similarly, the reversible photoisomerization of azobenzene, linking two cysteine residues in a designed peptide, has also been demonstrated to have the capability to control the helix content in steady state *(12)* and has recently been used to study the early folding kinetics of α-helices and β-hairpins *(32–34)*.

Because several reviews on the *T*-jump method, as well as its application on the study of fast events in protein folding, have already existed in the literature *(4,19–21,35)*, here we only briefly discuss the laser-induced *T*-jump method. This method utilizes very short laser pulses centered around 1.5 or 1.9 μm to heat up a H_2O or D_2O sample solution within a very short period of time, e.g., a few nanoseconds *(36)*. The resulting *T*-jump amplitude depends on the pump volume and the pulse energy, whereas the useful observation time window, in which the temperature remains approximately constant, is governed by the time it takes for the heat to diffuse out of the laser interaction volume, which ranges from a few to tens of milliseconds, depending on the setup *(19)*. Because the stability of a protein, or its free energy, depends on temperature, a rapid increase in temperature thus allows, with the use of an appropriate conformational probe, the measurement of the *T*-jump-induced relaxation kinetics and, therefore, the folding and unfolding rates of the system of interest. For example, for a simple two-state scenario, e.g., U ⇔ N (*U* and *N* represent the unfolded and the folded states, respectively), the observed relaxation rate constant is exactly the sum of the folding (k_f) and unfolding (k_u) rate constants. Thus, the folding and unfolding rate constants can be obtained individually if the equilibrium constant ($K_{eq} = k_f/k_u$) of this two-state reaction at the final temperature is known.

2. *T*-Jump Infrared Technique

The fast *T*-jump initiation technique can be combined with a variety of time-resolved spectroscopic methods, such as fluorescence *(37)*, circular dichroism

(38–40), and vibrational spectroscopy *(41–45)*, to study protein folding and conformational dynamics. In particular, infrared (IR) spectroscopy has been proven to be versatile in such studies. For example, the IR amide bands are valuable markers of protein secondary structures, whereas some of the amino acid side chain vibrations are suitable reporters of local environment *(45,46)*. Among the eight amide vibrational modes, the amide I band, which arises mostly from the amide C = O stretching vibration, is the most intense and is best characterized *(41,47)*. The amide I band of proteins or peptides has long been used as a global conformational reporter. The exact shape and position of the amide I band of a protein, however, are determined by many factors, including various couplings among individual vibrators, hydrogen bonding, solvation, and backbone conformations *(44,45)*. As a result, it is often difficult to quantitatively interpret amide I bands of proteins because they are almost invariably congested with components arising from different structural ensembles. To obtain site-specific structural information, a commonly used technique is isotope editing, wherein one or several amide ^{12}C = Os are substituted with ^{13}C = Os. Because of the isotopic effect, the amide I absorbance of the ^{13}C-labeled carbonyls shifts toward the lower frequency, therefore, (sometimes) permitting site-specific studies *(48,49)*.

Infrared studies of proteins in aqueous solution are complicated by the fact that H_2O absorbs very strongly in the spectral region that overlaps with the amide I band. D_2O, on the other hand, has a relatively low absorbance between 1400 and 1800 cm^{-1} and is therefore commonly employed in IR studies (in D_2O the amide I band is referred to as amide I′). In a *T*-jump IR experiment, however, the *T*-jump pulses will induce a change in the D_2O absorbance. To recover the protein signal, therefore, the solvent signal should be appropriately measured and subtracted from the total absorbance change measured with the protein sample.

The 3-ns and 1.9-μm *T*-jump pulse of our *T*-jump infrared apparatus *(17,35)* is generated via Raman shifting the 1.06-μm output of an injection-seeded Nd:YAG laser in a Raman cell containing a mixture of H_2 and Ar. The latter helps to stabilize the 1.9-μm output and also improve the beam quality. A stable pump source is rather important for *T*-jump experiments because the kinetic events in protein folding are temperature dependent. A continuous wave lead salts infrared diode laser, which is tunable from 1550–1800 cm^{-1}, is used as the probe. Transient absorbance changes of the sample induced by the *T*-jump pulses are detected by a 50 MHz MCT detector, and digitization of the signal is accomplished by a digital oscilloscope. To ensure a uniform *T*-jump distribution within the laser interaction volume, a thin optical path length, typically 52 μm, is used. Furthermore, to provide information for both background subtraction and *T*-jump amplitude determination, a sample cell with dual compartments is used to measure the T-jump-induced signals of both the peptide sample and reference (i.e., buffer) under the same conditions. Characterizing the T-jump amplitude is done by using the T-jump-induced absorbance change of the buffer at the probing frequency ν, $\Delta A(\Delta T, \nu)$ and the following equation: $\Delta A(\Delta T, \nu) = a(\nu)*\Delta T + b(\nu)*\Delta T^2$, where ΔA is the absorbance change, $\Delta T = T_f - T_i$ is the T-jump amplitude; where T_f and T_i correspond to the final and initial temperatures, respectively; and $a(\nu)$ and $b(\nu)$ are constants that are determined by measuring the temperature dependence of the Fourier transform infrared (FTIR) spectra of the buffer solution.

3. Helix–Coil Transition Dynamics

Seminal works of Schellman et al., Lifson et al., and Poland et al. formed the basis of our understanding of the thermodynamics of monomeric helices *(50–53)*. The classical helix–coil transition theory is characterized by a nucleation event, wherein the first helical hydrogen bond formed between the amide carbonyl of residue i and the amide hydrogen of residue $i + 4$ is generated. A series of propagation steps then follow the nucleation step, in which the preformed helical structures become elongated. Because the nucleation step is entropically unfavorable, it has been suggested that it encounters the largest free energy barrier during the course of helix formation. Because the helix–coil transition takes place on the nanosecond time-scale, however, it was only recently that its kinetics were studied and the mechanism of this seemingly simple folding process was being understood *(9,13,16,22,33,34,54–56)*. Of course, many computational studies *(57–65)* have also significantly enhanced our understanding of the mechanism of the helix–coil transition, but an extensive review on this subject is beyond the scope of this chapter.

The laser-induced T-jump method, in conjunction with various spectroscopic techniques *(16,17,22,54–56)*, has been playing a major role in investigating the kinetic properties of the helix–coil transition. For example, Williams et al. *(16)* studied the folding kinetics of the Fs peptide (sequence: $[A]_5$-$[AAARA]_3$-A-NH$_2$) using a T-jump infrared technique and found that the relaxation process induced by a T-jump from 9.3 to 27.4°C could be described by a single exponential function with a time constant of approx 160 ns, indicating that the helix–coil transition is a submicrosecond event. However, by using a fluorescent probe (MABA) attached at the N-terminus of the Fs peptide, Thompson et al. *(17)* observed a much faster relaxation process (~20 ns at 303 K). They attributed this relaxation to rapid fraying of the helix ends in response to the T-jump. Using a "kinetic zipper" model, they have also calculated the decay of the average helix content and their results indicate that a slower rate should account for the transition between the helix-containing and nonhelix-containing structural ensembles, owing to the energy barrier associated with the nucleation process. Subsequently, these authors

(54) reported the observation of a slower relaxation process (220 ns at 300 K) in a 21-residue helical peptide, Ac-WAAAH$^+$(AAAR$^+$A)$_3$A-NH$_2$, with a tryptophan residue in position 1 to serve as the fluorescent probe. This relaxation is temperature dependent and has an apparent activation energy of approx 8 kcal/mol. Using ultraviolet resonance Raman as a probe, Lednev et al. *(55,56)* also observed a single exponential relaxation process that is weakly temperature dependent for the Fs peptide, following a 3-ns *T*-jump pulse *(66)*. Using hexafluoroisopropanol as a cosolvent, Andersen et al. have also been able to study the folding kinetics of a helical peptide from its cold denatured states *(67)*. Using a photoswitchable peptide, Hamm et al. *(33)* recently demonstrated that the helix–coil transition dynamics could be studied from the picosecond-to-microsecond time-scale, and their results highlighted the complexity of the kinetics of helix formation.

Using the *T*-jump IR method, we have studied the temperature dependence of the helix–coil transition rate of an alanine-based, 21-residue helical peptide and observed a relatively large activation energy for the *T*-jump-induced relaxation process, approx 15.5 kcal/mol *(18)*. Presumably, part of the observed energetic barrier is originated from the temperature dependence of the solvent viscosity, as suggested by the study of Hofrichter et al. *(68)*. However, this result also suggests that the free energy barrier encountered in the helix formation may not be dominated by the entropic cost associated with the nucleation events, as suggested by the classical helix–coil transition theory. Instead, an enthalpic penalty, probably associated with the process of desolvation or the breaking of nonnative interactions existed in the denatured state *(69)*, may also contribute significantly to the folding free energy barrier.

To further study the helix–coil transition kinetics in a site-specific manner, we have employed the isotope-editing strategy previously discussed *(49)*. Specifically, we studied the *T*-jump-induced relaxation kinetics of three alanine-based helical peptides that contain a block of ^{13}C-labeled alanines indicated by underlines:

NL peptide: Ac-YGSPE<u>AAA</u>KA$_4$KA$_4$r-CONH$_2$
ML peptide: Ac-YGSPEA$_3$K<u>AAAA</u>KA$_4$r-CONH$_2$
CL peptide: Ac-YGSPEA$_3$KA$_4$K<u>AAAA</u>r-CONH$_2$

where *r* represents D-arginine, serving as an efficient C-cap *(70)*. As shown (**Fig. 1**), the red-shifted amide I′ band of the ^{13}C-labeled carbonyls indeed provides site-specific structural information. For example, the ^{13}C = O signal of the CL peptide is much weaker than that of the ML and NL peptides, indicative of a frayed C-terminus *(71)*. In addition, we found that the *T*-jump-induced relaxation kinetics of these peptides show nonexponential behaviors and are sensitive to both initial and final temperatures (**Fig. 2**). Taken together, these results suggest that the nucleation process is fast and probably inseparable from the

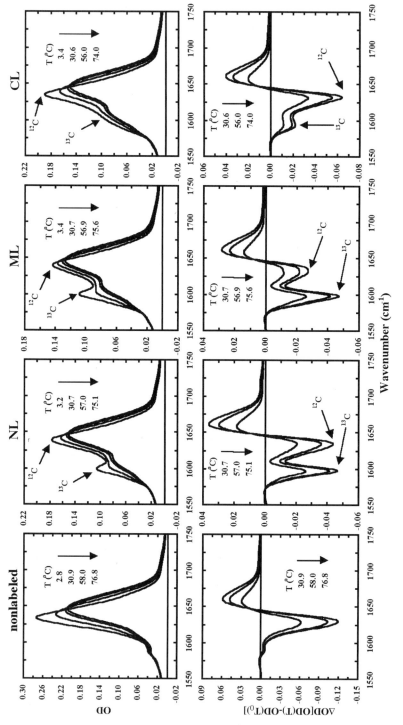

Fig. 1. Temperature-dependent equilibrium (top traces) and difference (bottom traces) FTIR spectra of the labeled and nonlabeled peptides, as indicated in the plot. Difference spectra were generated by subtracting the spectrum collected at the lowest temperature from the spectra collected at higher temperatures. The bands at approx 1600 and 1636 cm^{-1} are assigned to the amide I′ absorbance of the ^{13}C-labeled and nonlabeled residues, respectively. (Reproduced with permission from **ref. 71.**)

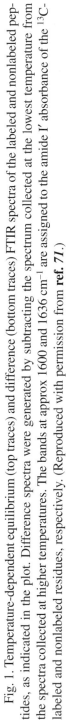

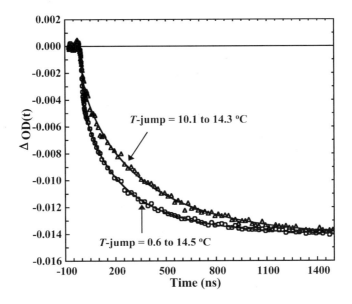

Fig. 2. *T*-jump induced relaxation kinetics of the ML peptide probed at 1600 cm^{-1}. The *T*-jump is from different initial temperatures to the same final temperature, as indicated. The smooth lines are fits to a stretched exponential function. The data obtained in response to a *T*-jump of 10.1–14.3°C have been scaled for comparison.

propagation event, and the helix formation may be described as a diffusion search process. Consistent with our observation *(71)*, several computer simulation studies *(62,65,72–74)* have also revealed the complex nature of the dynamics of helix formation.

To understand how peptide sequence affects the kinetics of the helix–coil transition, we further studied *(30)* a series of alanine-based peptides of different lengths, with either excellent or poor end-capping groups, as shown in the following:

AKA$_n$ peptides: YGAKAAAA(KAAAA)$_n$G, n = 2 – 6
SPE$_n$ peptides: YGSPEAAA(KAAAA)$_n$r, n = 2 – 5

Here, *r* also represents D-arginine and is one of the best C-caps *(70)*, and the Ser-Pro-Glu tripeptide (SPE) serves as a helix-stabilizing N-cap *(71)*.

Athough it has long been recognized that helix-capping sequences provide very significant thermodynamic stability to the α-helix *(75–80)*, their role in the kinetics of helix formation has not been previously studied. One possibility is that the role of end-caps is to lock in pre-existing helices, and the other is that they can effectively serve as helix initiators. These two opposing views lead to different experimental results. If the capping is not involved in initiation, the rate of helix initiation should not be much affected for peptides with different

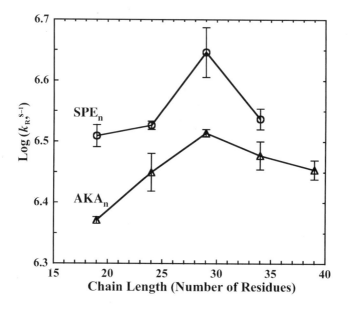

Fig. 3. Logarithm of the relaxation rates of AKA_n (Δ) and SPE_n (O) peptides follow-ing a *T*-jump from 1 to 11°C.

end-caps. On the other hand, if capping interactions are able to initiate helix formation then the rate of helix formation should be very profoundly affected. Our results show definitively that efficient end-capping sequences can not only stabilize pre-existing helices, but also promote helix formation through increasing the rate of helix folding (**Fig. 3**).

The peptide length is another factor that may also profoundly affect the dynamics of the helix–coil transition, as indicated by a number of theoretical studies *(58,81)*. If the kinetically and thermodynamically difficult step in the process of the coil-to-helix transition is initiation of a single turn of the helix, as one proceeds in peptide length from short to long the rate of helix formation should not be drastically affected. Only a small increase would be observed owing to a linear, length-dependent increase in the probability of initiation as the peptide is lengthened. However, our results show that the peptide length has a rather complicated effect on the kinetics of the helix–coil transition. For example, we found that the relaxation rates of these peptides following a *T*-jump of 1–11°C show rather complex behaviors as a function of the peptide length (**Fig. 3**), in disagreement with early theoretical predictions. These results are not readily explicable by theories in which alanine is taken to have a single helical propensity. However, if one were to assume that the propagation constant, *s*, depends on chain length *(82,83)*, the mean first-passage times of the coil-to-helix transition obtained, based on the model developed by Straub et al. *(81)*, show similar dependence on the peptide length as those observed experimentally *(30)*.

4. β-Hairpin Folding Dynamics

A β-hairpin structure is characterized by two antiparallel strands linked by a short turn or loop. Owing to its small size and simplicity, as well as the possible role of functioning as a folding nucleus *(84–88)*, the β-hairpin motif has been increasingly used to probe factors that govern the conformational stability as well as the folding mechanism of β-sheets *(89–91)*. The first experimental study of β-hairpin folding kinetics on the microsecond time-scale was carried out by Muñoz et al. *(23)*, who measured the folding time of a 16-residue β-hairpin derived from the C-terminal fragment (41–56) of protein G *(92)*. Subsequently, a large number of theoretical and computational studies have been performed *(93–104)*, aiming to understand the mechanism of β-hairpin folding. Many studies suggested that several factors, including the rigidity of the turn and the relative position of the hydrophobic cluster, may have rather complex effects on the folding free energy landscape and, therefore, the rate of β-hairpin formation.

The results of Muñoz et al. *(23)* indicated that the GB1 β-hairpin folds in a two-state manner with a folding time constant of 6 µs at 297 K. Subsequently, these authors developed a statistical zipper model *(93)* to explain the apparent two-state folding behavior of this β-hairpin and suggested that folding begins with the formation of the turn. Recently, Xu et al. *(24)* studied the folding kinetics of a 15-residue β-hairpin designed by Jiménez et al. *(105,106)*. This β-hairpin contains a type I + G1 β-bulge turn and shows a fairly broad thermal unfolding transition *(24)*. However, its folding rate, measured by the *T*-jump IR technique, is surprisingly fast. At 300 K, the folding time of this β-hairpin is approx 0.8 µs, only two to three times slower than that of α-helix formation. Therefore, the results of Muñoz et al. and Xu et al. suggest that the folding times of β-hairpins may vary greatly depending on the peptide sequence. Consistent with this picture, the recent work of Dyer and coworkers showed that the folding rate of a series of cyclic β-hairpins *(107)* is significantly faster than that of linear β-hairpins. Although for linear β-hairpins, the folding rate could be substantially accelerated by a shorter connecting loop *(108)*. In addition, it was found that the nature of the thermally denatured states can also have dramatic effects on the rate of β-hairpin folding *(109,110)*.

Although those studies previously described firmly demonstrated that β-hairpins fold on the submillisecond time-scale, the marked differences in the peptide sequence of those systems studied makes it difficult to explicitly determine the key factors that control the rate of β-hairpin folding. Thermodynamically, it is well known that a stable β-hairpin results from an intricate interplay among several factors, such as hydrogen bonding, electrostatic interaction, turn preference,

Table 1
Sequences of Trpzips and the GB1 Peptide[a]

Peptide	Sequence	T_m (°C)	Ref.
Trpzip1	SWTWEGNKWTWK	49.9	*(117)*
Trpzip2	SWTWENGKWTWK	71.9	*(117)*
Trpzip3	SWTWE^{D}PNKWTWK[b]	80.1	*(116)*
Trpzip3-I	SWTWDATKWTWK	16.8	*(116)*
Trpzip4	GEWTWDDATKTWTWTE	70.4	*(116)*
GB1	GEWTYDDATKTFTVTE	24.3	*(23)*

[a]Also shown are the thermal melting temperatures (T_m) of these β-hairpins (Adapted from **ref. *116***).
[b]^{D}P stands for D-Pro.

and hydrophobic packing of side chains *(111–115)*. For example, a stable β-hairpin conformation is often the result of a strong turn-promoting sequence and a stabilizing hydrophobic cluster. To provide a better understanding of the kinetic role of the turn and hydrophobic cluster in β-hairpin folding, we have recently examined the folding kinetics of a series of sequence-related β-hairpins *(116)*, i.e., tryptophan zippers *(117)*. As shown in **Table 1**, the 12-residue tryptophan zippers (trpzips) studied here differ only in the turn composition, whereas trpzip4 is a triple mutant of the GB1 β-hairpin in the hydrophobic cluster region. As expected, the thermal melting temperatures of these trpzips span a wide range, demonstrating the effect of sequence on stability (*see* **Table 1**) *(116,118)*.

Although trpzip4 and the GB1 β-hairpin differ from each other only in the composition of the hydrophobic cluster, comparing their folding kinetics will permit us to determine the kinetic role of the hydrophobic clusters found in these β-hairpins. Using the method of *T*-jump fluorescence, Eaton et al. *(23)* have shown that the GB1 β-hairpin folds in approx 6 μs at 297 K. Although trpzip4 is more stable, it folds with a similar rate. At 297 K, its folding time constant is extrapolated to be approx 15 μs (**Table 2**). On the contrary, their unfolding rates differ significantly. At 297 K, the GB1 β-hairpin unfolds in approx 6 μs *(23)*, whereas the unfolding time constant of trpzip4 is approx 234 μs. The similar folding rates exhibited by the GB1 β-hairpin and trpzip4 suggest that their folding free energy barriers are also quite similar. However, the slower unfolding rate of trpzip4 indicates that it has a larger unfolding free energy barrier, compared with the GB1 β-hairpin. Moreover, these results suggest that mutations to the hydrophobic side chains in the GB1 peptide do not significantly alter the free energy of the folding transition state. In other words, the native hydrophobic side chain–side chain contacts have not been formed in the transition state structural ensemble. Therefore, the role of a strong cross-strand hydrophobic cluster, such

Table 2
Folding/Unfolding Rate Constants and Equilibrium Entropic Changes of the 12-Residue Trpzips

	Trpzip1[a]	Trpzip2[a]	Trpzip3	Trpzip3-I	Trpzip4	GB1[b]
k_f^{-1} (μs)	6.3 ± 0.3	2.5 ± 0.05	1.7 ± 0.2	52.2 ± 4.5	14.9 ± 0.6	6.0
k_u^{-1} (μs)	18.3 ± 3.1	24.7 ± 3.0	12.2 ± 2.1	37.5 ± 3.4	234.0 ± 10.7	6.0
ΔS_f (cal K^{-1} mol^{-1})	-13.4	-5.7	-2.6	-31.9	-13.4	-39.0

At 23°C and those of trpzip4 and the GB1 peptide at 24°C. ΔS_f at temperature T was calculated using the following equation: $\Delta S_f = -[\Delta S_m + \Delta C_p \cdot \ln(T/T_m)]$. (Adapted from **ref. 116**.)
[a]Data from **ref. 104**.
[b]Data from **ref. 23**.

as the one found in trpzips, is to primarily prevent the unfolding of the folded β-hairpin conformation. Extensive φ-value analysis *(119)* on the folding kinetics of trpzip4 further confirmed this conclusion *(120)*.

The folding kinetics of the four 12-residue trpzips have been studied by either *T*-jump IR, or *T*-jump fluorescence, or both *(104,116)*. It has been shown that for trpzip2 and trpzip3 both methods yielded nearly identical results *(104,116)*. As shown in **Table 2**, the unfolding rate constants of these 12-residue trpzips are only different by a factor of three or less. However, their folding rate constants differ by as much as 30 times (e.g., comparing trpzip3 with trpzip3-I). Taken together, these results suggest that the turn sequence is a strong determinant of the β-hairpin folding rate but has a rather small effect on the unfolding rate, consistent with a β-hairpin folding model wherein the rate-limiting event in folding is the formation of the turn (or loop).

It is well established that the turn sequence is an important determinant of the turn conformation and thereby, other features of β-hairpin conformation, such as the pattern of interstrand hydrogen bonding and residue pairing *(121)*. Moreover, strong turn-promoting sequences, such as those containing a [D]P residue *(91,122)*, often correspond to a relatively low entropic cost upon turn formation. For example, Gellman et al. *(91)* have recently shown that the high turn-forming propensity of the [D]PG segment decreases the equilibrium entropic cost of β-hairpin formation relative to the more flexible NG segment. Interestingly, trpzip3, which contains a strong turn-promoting sequence, [D]PN, exhibits the fastest folding rate among the 12-residue trpzips studied. According to Gellman et al. *(91)*, the excellent turn preference of [D]PN may be attributed to the conformational rigidity of the [D]P residue, which effectively reduces the entropic penalty, ΔS_{turn}, upon turn formation. As indicated by results from this and other studies *(23,24,107)*, the free energy barrier of β-hairpin folding is dominated by the unfavorable entropic cost for the formation of the transition state. Therefore, within the framework of the folding transition state theory, it

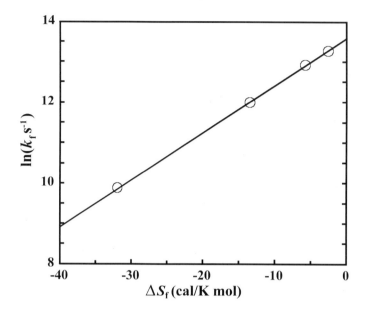

Fig. 4. ln(k_f) vs ΔS_f at 296 K for the 12-residue trpzips (**Table 2**). (Reproduced with permission from **ref. *116*.**)

is reasonable to assume that k_f is proportional to $\exp(\alpha \cdot \Delta S_{turn}/R)$, where α is a constant. Although it is difficult to explicitly determine ΔS_{turn} for a given β-hairpin, the following relationship is valid, $\Delta S_f = \Delta S_{turn} + \Delta S_{strand}$; where ΔS_f is the total folding entropic change obtained from thermal melting experiments, and ΔS_{strand} is the entropic change associated with the folding of the strands. Thus, k_f is proportional to $\exp[\alpha \cdot (\Delta S_f - \Delta S_{strand})/R]$. It is easy to show that the latter equation leads to the following relationship, $\ln(k_f) = \rho + \alpha \cdot (\Delta S_f - \Delta S_{strand})/R$, where ρ is a constant. For the four 12-residue trpzips previously discussed, an approximate but quite reasonable assumption is that they have the same ΔS_{strand}. Such an assumption would lead directly to a relationship wherein $\ln(k_f)$ is linearly proportional to ΔS_f. Indeed, such a linear correlation is observed for the four 12-residue trpzips (**Fig. 4**), where faster folding correlates with smaller entropic change. These results further strengthen our argument that the turn (or the turn region) is a key determinant of the folding rate of β-hairpins and clearly demonstrate that stronger turn-promoting sequences, such as those that contain a ^{D}P residue, increase the stability of β-hairpins by increasing the folding rate. Finally, it is worth pointing out that one should not simply interpret the term ΔS_{turn} invoked here as the differences in disordered loop entropy because it only corresponds to a fraction of the entropic change determined from the thermal unfolding studies. In addition, one should bear in mind that the thermally denatured states of β-hairpins may be quite compact and temperature dependent *(123,124)*. Therefore, the ΔS_f listed in **Table 2** may not totally measure the entropy difference between the native state and an extended (or random) conformational ensemble.

In summary, these results suggest a folding mechanism in which the turn (or loop) plays an important role in determining the rate of β-hairpin folding. Moreover, a good turn-promoting sequence or a strong interstrand hydrophobic cluster can help to stabilize the folded conformation of β-hairpins. However, the former increases the stability of a β-hairpin primarily by increasing its folding rate, whereas the latter affects the stability primarily by decreasing its unfolding rate.

5. Conclusion

The *T*-jump IR method provides a versatile means to study the conformational dynamics of peptides and proteins. Compared with the commonly used stopped-flow kinetic methods, the *T*-jump technique can initiate a conformational process on the nanosecond time-scale and, thus, offers a faster time resolution. The examples presented in this chapter illustrated how this technique can be applied to the study of the dynamics of some primary events in protein folding. In conjunction with the techniques of isotope-editing and site-directed mutagenesis, the *T*-jump IR method will allow the study of the early folding events in even greater detail.

Acknowledgments

We wish to thank our colleagues in the Gai research group for their contributions to this work, particularly Cheng-Yen Huang, Ting Wang, Yongjin Zhu, and Rolando Oyola. We also wish to thank our collaborators for their contributions, particularly Drs. W. F. DeGrado and Z. Getahun. Support for this work was provided by the National Institutes of Health (GM-065978, RR-001348) and the NSF (CHE-0094077).

References

1. Dyson, H. J. and Wright, P. E. (1996) Insights into protein folding from NMR. *Annu. Rev. Phys. Chem.* **47,** 369–395.
2. Eftink, M. R. and Shastry, M. C. R. (1997) Fluorescence methods for studying kinetics of protein-folding reactions. *Methods Enzymol.* **278,** 258–286.
3. Englander, S. W. (2000) Protein folding intermediates and pathways studied by hydrogen exchange. *Annu. Rev. Biophys. Biomol. Struct.* **29,** 213–238.
4. Eaton, W. A., Muñoz, V., Thompson, P. A., Henry, E. R., and Hofrichter, J. (1998) Kinetics and dynamics of loops, α-helixes, β-hairpins, and fast-folding proteins. *Acc. Chem. Res.* **31,** 745–753.
5. Dyer, R. B., Gai, F., Woodruff, W. H., Gilmanshin, R., and Callender, R. H. (1998) Infrared studies of fast events in protein folding. *Acc. Chem. Res.* **31,** 709–716.

6. Bieri, O. and Kiefhaber, T. (1999) Elementary steps in protein folding. *Biol. Chem.* **380,** 923–929.

7. Eaton, W. A., Muñoz, V., Hagen, S. J., Jas, G. S., Lapidus, L. J., and Henry, E. R. (2000) Fast kinetics and mechanisms in protein folding. *Annu. Rev. Biophys. Biomol. Struct.* **29,** 327–359.

8. Jones, C. M., Henry, E. R., Hu, Y., et al. (1993) Fast events in protein folding initiated by nanosecond laser photolysis. *Proc. Natl. Acad. Sci. USA* **90,** 11,860–11,864.

9. Volk, M., Kholodenko, Y., Lu, H. S. M., Gooding, E. A., DeGrado, W. F., and Hochstrasser, R. M. (1997) Peptide conformational dynamics and vibrational Stark effects following photoinitiated disulfide cleavage. *J. Phys. Chem. B.* **101,** 8607–8616.

10. Wittung-Stafshede, P., Lee, J. C., Winkler, J. R., and Gray, H. B. (1999) Submillisecond folding of monomeric λRepressor. *Proc. Natl. Acad. Sci. USA* **96,** 6587–6590.

11. Hansen, K. C., Rock, R. S., Larsen, R. W., and Chan, S. I. (2000) A method for photoinitating protein folding in a nondenaturing environment. *J. Am. Chem. Soc.* **122,** 11,567, 11,568.

12. Kumita, J. R., Smart, O. S., and Woolley, G. A. (2000) Photo-control of helix content in a short peptide. *Proc. Natl. Acad. Sci. USA* **97,** 3803–3808.

13. Huang, C.-Y., He, S., DeGrado, W. F., McCafferty, D. G., and Gai, F. (2002) Light-induced helix formation. *J. Am. Chem. Soc.* **124,** 12,674, 12,675.

14. Abbruzzetti, S., Crema, E., Masino, L., et al. (2000) Fast events in protein folding: structural volume changes accompanying the early events in the N→I transition of Apomyoglobin induced by ultrafast pH jump. *Biophys. J.* **78,** 405–415.

15. Barth, A. and Corrie, J. E. T. (2002) Characterization of a new caged proton capable of inducing large pH jumps. *Biophys. J.* **83,** 2864–2871.

16. Williams, S., Causgrove, T. P., Gilmanshin, R., et al. (1996) Fast events in protein folding: helix melting and formation in a small peptide. *Biochemistry* **35,** 691–697.

17. Thompson, P. A., Eaton, W. A., and Hofrichter, J. (1997) Laser temperature jump study of the helix ⇌ coil kinetics of an alanine peptide interpreted with a 'kinetic zipper' model. *Biochemistry* **36,** 9200–9210.

18. Huang, C.-Y., Klemke, J. W., Getahun, Z., DeGrado, W. F., and Gai, F. (2001) Temperature-dependent helix-coil transition of an alanine based peptide. *J. Am. Chem. Soc.* **123,** 9235–9238.

19. Hofrichter, J. (2001) Laser temperature-jump methods for studying folding dynamics. *Methods Mol. Biol.* **168,** 159–191.

20. Callender, R. and Dyer, R. B. (2002) Probing protein dynamics using temperature jump relaxation spectroscopy. *Curr. Opin. Struct. Biol.* **12,** 628–633.

21. Gruebele, M., Sabelko, J., Ballew, R., and Ervin, J. (1998) Laser temperature jump induced protein refolding. *Acc. Chem. Res.* **31,** 699–707.

22. Herberhold, H. and Winter, R. (2002) Temperature- and pressure-induced unfolding and refolding of Ubiquitin: a static and kinetic Fourier transform infrared spectroscopy study. *Biochemistry* **41,** 2396–2401.

23. Muñoz, V., Thompson, P. A., Hofrichter, J., and Eaton, W. A. (1997) Folding dynamics and mechanism of β-hairpin formation. *Nature* **390**, 196–199.
24. Xu, Y., Oyola, R., and Gai, F. (2003) Infrared study of the stability and folding kinetics of a 15-residue β-hairpin. *J. Am. Chem. Soc.* **125**, 15,388–15,394.
25. Sabelko, J., Ervin, J., and Gruebele, M. (1999) Observation of strange kinetics in protein folding. *Proc. Natl. Acad. Sci. USA* **96**, 6031–6036.
26. Hagen, S. J. and Eaton, W. A. (2000) Two-state expansion and collapse of a polypeptide. *J. Mol. Biol.* **301**, 1019–1027.
27. Sadqi, M., Lapidus, L. J., and Muñoz, V. (2003) How fast is protein hydrophobic collapse? *Proc. Natl. Acad. Sci. USA* **100**, 12,117–12,122.
28. Zhu, Y., Alonso, D. O. V., Maki, K., et al. (2003) Ultrafast folding of α-3D: a de novo designed three-helix bundle protein. *Proc. Natl. Acad. Sci. USA* **100**, 15,486–15,491.
29. Zhu, Y., Fu, X., Wang, T., et al. (2004) Guiding the search for a protein's maximum rate of folding. *Chem. Phys.* **307**, 99–109.
30. Wang, T., Zhu, Y., Getahun, Z., et al. (2004) Length dependent helix-coil transition kinetics of nine alanine-based peptides. *J. Phys. Chem. B.* **108**, 15,301–15,310.
31. Lu, H. S. M., Volk, M., Kholodenko, Y., Gooding, E., Hochrasser, R. M., and DeGrado, W. F. (1997) Aminothiotyrosine disulfide, an optical trigger for initiation of protein folding. *J. Am. Chem. Soc.* **119**, 7173–7180.
32. Chen, R. P., Huang, J. J., Chen, H.-L., Jan, H., et al. (2004) Measuring the refolding of β-sheets with different turn sequences on a nanosecond time scale. *Proc. Natl. Acad. Sci. USA* **101**, 7305–7310.
33. Bredenbeck, J., Helbing, J., Kumita, J. R., Woolley, G. A., and Hamm, P. (2005) α-Helix formation in a photoswitchable peptide tracked from picoseconds to microseconds by time-resolved IR spectroscopy. *Proc. Natl. Acad. Sci. USA* **102**, 2379–2384.
34. Bredenbeck, J., Helbing, J., Sieg, A., et al. (2003) Picosecond conformational transition and equilibration of a cyclic peptide. *Proc. Natl. Acad. Sci. USA* **100**, 6452–6457.
35. Zhu, Y., Wang, T., and Gai, F. (2005) Laser-induced *T*-jump method, a nonconventional photo-releasing approach to study protein folding. In: *Dynamic Studies in Biology: Phototriggers, Photoswitches and Caged Biomolecules*, (Goeldner, G. and Givens R. eds.), Wiley-VCH Verlag GmbH, Weinheim, pp. 461–478.
36. Palmer, K. F. and Williams, D. (1974) Optical properties of water in the near infrared. *J. Opt. Soc. Am.* **64**, 1107–1110.
37. Beechem, J. M. and Brand, L. (1985) Time-resolved fluorescence of proteins. *Annu. Rev. Biochem.* **54**, 43–71.
38. Adler, A. J., Greenfield, N. J., and Fasman, G. D. (1973) Circular dichroism and optical rotatory dispersion of proteins and polypeptides. *Methods Enzymol.* **27**, 675–735.
39. Zhang, C. F., Lewis, J. W., Cerpa, R., Kuntz, I. D., and Kliger, D. S. (1993) Nanosecond circular dichroism spectral measurements: extension to the far-ultraviolet region. *J. Phys. Chem.* **97**, 5499–5505.

40. Wen, Y. X., Chen, E., Lewis, J. W., and Kliger, D. S. (1996) Nanosecond time-resolved circular dichroism measurements using an upconverted Ti:sapphire laser. *Rev. Sci. Instrum.* **67,** 3010–3016.
41. Krimm, S. and Bandekar, J. (1986) Vibrational spectroscopy and conformation of peptides, polypeptides, and proteins. *Adv. Protein Chem.* **38,** 181–364.
42. Kitagawa, T. and Hirota, S. (2002) Raman spectroscopy of proteins. In: *Handbook of Vibrational Spectroscopy*, (Chalmers, J. M. and Griffiths, P. R., eds.), John Wiley and Sons, New York, pp. 3426–3446.
43. Spiro, T. G. and Czernuszewicz, R. S. (1995) Resonance Raman spectroscopy of metalloproteins. *Methods Enzymol.* **246,** 416–460.
44. Fabian, H. and Mantele, W. (2002) Infrared spectra of proteins. In: *Handbook of Vibrational Spectroscopy*, (Chalmers, J. M. and Griffiths, P. R., eds.), John Wiley and Sons, New York, pp. 3399–3425.
45. Barth, A. and Zscherp, C. (2002) What vibrations tell about proteins. *Q. Rev. Biophys.* **35,** 369–430.
46. Getahun, Z., Huang, C.-Y., Wang, T., De Leon, B., DeGrado, W. F., and Gai, F. (2003) Using nitrile-derivatized amino acids as infrared probes of local environment. *J. Am. Chem. Soc.* **125,** 405–411.
47. Susi, H. (1972) Infrared spectroscopy. Conformation. *Methods Enzymol.* **26,** 455–472.
48. Decatur, S. M. and Antonic, J. (1999) Isotope-edited infrared spectroscopy of helical peptides. *J. Am. Chem. Soc.* **121,** 11,914, 11,915.
49. Huang, C.-Y., Getahun, Z., Wang, T., DeGrado, W. F., and Gai, F. (2001) Time-resolved infrared study of the helix-coil transition using ^{13}C-labeled helical peptides. *J. Am. Chem. Soc.* **123,** 12,111–12,112.
50. Schellman, J. A. (1958) The factors affecting the stability of hydrogen-bonded polypeptide structures in solution. *J. Phys. Chem.* **62,** 1485–1494.
51. Zimm, B. H. and Bragg, J. K. (1959) Theory of the phase transition between helix and random coil in polypeptide chains. *J. Chem. Phys.* **31,** 526–535.
52. Lifson, S. and Roig, A. (1961) On the theory of helix-coil transition in polypeptides. *J. Chem. Phys.* **34,** 1963–1974.
53. Poland, D. and Scheraga, H. A. (1970) *Theory of Helix-Coil Transition in Biopolymers*, Academic Press, New York and London.
54. Thompson, P. A., Muñoz, V., Jas, G. S., Henry, E. R., Eaton, W. A., and Hofrichter, J. (2000) The helix-coil kinetics of a heteropeptide. *J. Phys. Chem. B.* **104,** 378–389.
55. Lednev, I. K., Karnoup, A. S., Sparrow, M. C., and Asher, S. A. (1999) Nanosecond UV resonance Raman examination of initial steps in α-helix secondary structure evolution. *J. Am. Chem. Soc.* **121,** 4076–4077.
56. Lednev, I. K., Karnoup, A. S., Sparrow, M. C., and Asher, S. A. (1999) α-Helix peptide folding and unfolding activation barriers: a nanosecond UV resonance Raman study. *J. Am. Chem. Soc.* **121,** 8074–8086.
57. Daggett, V. and Levitt, M. (1992) Molecular dynamics simulations of helix denaturation. *J. Mol. Biol.* **223,** 1121–1138.
58. Brooks, C. L. (1996) Helix-coil kinetics: folding time scales for helical peptides from a sequential kinetic model. *J. Phys. Chem.* **100,** 2546–2549.

59. Schaefer, M., Bartels, C., and Karplus, M. (1998) Solution conformations and thermodynamics of structured peptides: molecular dynamics simulation with an implicit solvation model. *J. Mol. Biol.* **284,** 835–848.
60. Duan, Y. and Kollman, P. A. (1998) Pathways to a protein folding intermediate observed in a 1-microsecond simulation in aqueous solution. *Science* **282,** 740–744.
61. Takada, S., Luthey-Schulten, Z., and Wolynes, P. G. (1999) Folding dynamics with nonadditive forces: a simulation study of a designed helical protein and a random heteropolymer. *J. Chem. Phys.* **110,** 11,616–11,629.
62. Hummer, G., Garcia, A. E., and Grade, S. (2001) Helix nucleation kinetics from molecular simulations in explicit solvent. *Proteins* **42,** 77–84.
63. Shimada, J., Kussell, E. L., and Shakhnovich, E. I. (2001) The folding thermodynamics and kinetics of crambin using an all-atom Monte Carlo simulation. *J. Mol. Biol.* **308,** 79–95.
64. Doshi, U. and Muñoz, V. (2004) Kinetics of α-helix formation as diffusion on a one-dimensional free energy surface. *Chem. Phys.* **307,** 129–136.
65. Sorin, E. J. and Pande, V. S. (2005) Exploring the helix-coil transition via all-atom equilibrium ensemble simulations. *Biophys. J.* **88,** 2472–2493.
66. Nishii, I., Kataoka, M., Tokunaga, F., and Goto, Y. (1994) Cold denaturation of the molten globule states of Apomyoglobin and a profile for protein folding. *Biochemistry* **33,** 4903–4909.
67. Werner, J. H., Dyer, R. B., Fesinmeyer, R. M., and Andersen, N. H. (2002) Dynamics of the primary processes of protein folding: helix nucleation. *J. Phys. Chem. B.* **106,** 487–494.
68. Jas, G. S., Eaton, W. A., and Hofrichter, J. (2001) Effect of viscosity on the kinetics of α-helix and β-hairpin formation. *J. Phys. Chem. B.* **105,** 261–272.
69. Chowdhury, S., Lei, H. X., and Duan, Y. (2005) Denatured-state ensemble and the early-stage folding of the G29A mutant of the B-domain of protein A. *J. Phys. Chem. B.* **109,** 9073–9081.
70. Schneider, J. P. and DeGrado, W. F. (1998) The design of efficient α-helical C-capping auxiliaries. *J. Am. Chem. Soc.* **120,** 2764–2767.
71. Huang, C.-Y., Getahun, Z., Zhu, Y., Klemke, J. W., DeGrado, W. F., and Gai, F. (2002) Helix formation via conformation diffusion search. *Proc. Natl. Acad. Sci. USA* **99,** 2788–2793.
72. Hummer, G., Garcia, A. E., and Grade, S. (2000) Conformational diffusion and helix formation kinetics. *Phys. Rev. Lett.* **85,** 2637–2640.
73. Elmer, S. and Pande, V. S. (2001) A new twist on the helix-coil transition: a non-biological helix with protein-like intermediates and traps. *J. Phys. Chem. B.* **105,** 482–485.
74. Bryngelson, J. D., Onuchic, J. N., and Wolynes, P. G. (1995) Funnels, pathways, and the energy landscape of protein folding: a synthesis. *Proteins* **21,** 167–195.
75. Presta, L. G. and Rose, G. D. (1988) Helix signals in proteins. *Science* **240,** 1632–1641.
76. Richardson, J. S. and Richardson, D. C. (1988) Amino acid preferences for specific locations at the ends of alpha helices. *Science* **240,** 1648–1652.

77. Regan, L. (1993) What determines where α-helices begin and end? *Proc. Natl. Acad. Sci. USA* **90,** 10,907–10,908.
78. Aurora, R., Creamer, T. P., Srinivasan, R., and Rose, G. D. (1997) Local interactions in protein folding: lessons from the α-helix. *J. Biol. Chem.* **272,** 1413–1416.
79. Aurora, R. and Rose, G. D. (1998) Helix capping. *Protein Sci.* **7,** 21–38.
80. Dasgupta, S. and Bell, J. A. (1993) Design of helix ends. Amino acid preferences, hydrogen bonding and electrostatic interactions. *Int. J. Pept. Prot. Res.* **41,** 499–511.
81. Buchete, N. -V. and Straub, J. E. (2001) Mean first-passage time calculations for the coil-to-helix transition: the active helix Ising model. *J. Phys. Chem. B.* **105,** 6684–6697.
82. Kennedy, R. J., Tsang, K.-T., and Kemp, D. S. (2002) Consistent helicities from CD and template t/c data for N-templated polyalanines: progress toward resolution of the alanine helicity problem. *J. Am. Chem. Soc.* **124,** 934–944.
83. Ohkubo, Y. Z. and Brooks, C. L. (2003) Exploring Flory's isolated-pair hypothesis: statistical mechanics of helix–coil transitions in polyalanine and the C-peptide from RNase A. *Proc. Natl. Acad. Sci. USA* **100,** 13,916–13,921.
84. Guo, Z. Y. and Thirumalai, D. (1995) Kinetics of protein folding: nucleation mechanism, time scales, and pathways. *Biopolymers* **36,** 83–102.
85. Gruebele, M. and Wolynes, P. G. (1998) Satisfying turns in folding transitions. *Nat. Struct. Biol.* **5,** 662–665.
86. Grantcharova, V. P., Riddle, D. S., Santiago, J. V., and Baker, D. (1998) Important role of hydrogen bonds in the structurally polarized transition state for folding of the src SH3 domain. *Nat. Struct. Biol.* **5,** 714–720.
87. Martinez, J. C., Pisabarro, M. T., and Serrano, L. (1998) Obligatory steps in protein folding and the conformational diversity of the transition state. *Nat. Struct. Biol.* **5,** 721–729.
88. Walkenhorst, W. F., Edwards, J. A., Markley, J. L., and Roder, H. (2002) Early formation of a beta hairpin during folding of staphylococcal nuclease H124L as detected by pulsed hydrogen exchange. *Protein Sci.* **11,** 82–91.
89. Serrano, L. (2000) The relationship between sequence and structure in elementary folding units. *Adv. Protein Chem.* **53,** 49–85.
90. Searle, M. S. (2001) Peptide models of protein β-sheets: design, folding and insights into stabilising weak interactions. *J. Chem. Soc., Perkin Trans.* **2,** 1011–1020.
91. Espinosa, J. F., Syud, F. A., and Gellman, S. H. (2002) Analysis of the factors that stabilize a designed two-stranded antiparallel β-sheet. *Protein Sci.* **11,** 1492–1505.
92. Blanco, F. J., Rivas, G., and Serrano, L. (1994) A short linear peptide that folds into a native stable β-hairpin in aqueous solution. *Nat. Struct. Biol.* **1,** 584–590.
93. Muñoz, V., Henry, E. R., Hofrichter, J., and Eaton, W. A. (1998) A statistical mechanical model for β-hairpin kinetics. *Proc. Natl. Acad. Sci. USA* **95,** 5872–5879.
94. Dinner, A. R., Lazaridis, T., and Karplus, M. (1999) Understanding β-hairpin formation. *Proc. Natl. Acad. Sci. USA* **96,** 9068–9073.
95. Pande, V. S. and Rokhsar, D. S. (1999) Molecular dynamics simulations of unfolding and refolding of a β-hairpin fragment of protein G. *Proc. Natl. Acad. Sci. USA* **96,** 9062–9067.

96. Klimov, D. K. and Thirumalai, D. (2000) Mechanisms and kinetics of β-hairpin formation. *Proc. Natl. Acad. Sci. USA* **97,** 2544–2549.
97. Bonvin, A. M. J. J. and van Gunsteren, W. F. (2000) β-Hairpin stability and folding: molecular dynamics studies of the first β-hairpin of tendamistat. *J. Mol. Biol.* **296,** 255–268.
98. Garcia, A. E. and Sanbonmatsu, K. Y. (2001) Exploring the energy landscape of a β-hairpin in explicit solvent. *Proteins* **42,** 345–354.
99. Zhou, R., Berne, B. J., and Germain, R. (2001) The free energy landscape for β-hairpin folding in explicit water. *Proc. Natl. Acad. Sci. USA* **98,** 14,931–14,936.
100. Klimov, D. K. and Thirumalai, D. (2002) Stiffness of the distal loop restricts the structural heterogeneity of the transition state ensemble in SH3 domains. *J. Mol. Biol.* **317,** 721–737.
101. Zhou, Y. and Linhananta, A. (2002) Role of hydrophilic and hydrophobic contacts in folding of the second β-hairpin fragment of protein G: molecular dynamics simulation studies of an all-atom model. *Proteins* **47,** 154–162.
102. Ma, B. and Nussinov, R. (2003) Energy landscape and dynamics of the β-hairpin G peptide and its isomers: topology and sequences. *Protein Sci.* **12,** 1882–1893.
103. Bolhuis, P. G. (2003) Transition-path sampling of β-hairpin folding. *Proc. Natl. Acad. Sci. USA* **100,** 12,129–12,134.
104. Snow, C. D., Qiu, L., Du, D., Gai, F., Hagen, S. J., and Pande, V. S. (2004) Trp zipper folding kinetics by molecular dynamics and temperature-jump spectroscopy. *Proc. Natl. Acad. Sci. USA* **101,** 4077–4082.
105. Santiveri, C. M., Santoro, J., Rico, M., and Jiménez, M. A. (2002) Thermodynamic analysis of β-hairpin-forming peptides from the thermal dependence of ^{1}H NMR chemical shifts. *J. Am. Chem. Soc.* **124,** 14,903–14,909.
106. de Alba, E., Jiménez, M. A., Rico, M., and Nieto, J. L. (1996) Conformational investigation of designed short linear peptides able to fold into β-hairpin structures in aqueous solution. *Fold. Des.* **1,** 133–144.
107. Maness, S. J., Franzen, S., Gibbs, A. C., Causgrove, T. P., and Dyer, R. B. (2003) Nanosecond temperature jump relaxation dynamics of cyclic β-hairpin peptides. *Biophys. J.* **84,** 3874–3882.
108. Dyer, R. B., Maness, S. J., Peterson, E. S., Franzen, S., Fesinmeyer, R. M., and Andersen, N. H. (2004) The mechanism of β-hairpin formation. *Biochemistry* **43,** 11,560–11,566.
109. Dyer, R. B., Maness, S. J., Franzen, S., Fesinmeyer, R. M., Olsen, K. A., and Andersen, N. H. (2005) Hairpin folding dynamics: the cold-denatured state is predisposed for rapid refolding. *Biochemistry* **44,** 10,406–10,415.
110. Xu, Y., Wang, T., and Gai, F. (2006) Strange temperature dependence of the folding rate of a 16-residue β-hairpin. *Chem. Phys.,* in press.
111. Blanco, F. J., Ramirez-Alvarado, M., and Serrano, L. (1998) Formation and stability of β-hairpin structures in polypeptides. *Curr. Opin. Struct. Biol.* **8,** 107–111.
112. Gellman, S. H. (1998) Minimal model systems for β-sheet secondary structure in proteins. *Curr. Opin. Chem. Biol.* **2,** 717–725.

113. Kobayashi, N., Honda, S., Yoshii, H., and Munekata, E. (2000) Role of side-chains in the cooperative β-hairpin folding of the short C-terminal fragment derived from streptococcal Protein G. *Biochemistry* **39**, 6564–6571.

114. Tsai, J. and Levitt, M. (2002) Evidence of turn and salt bridge contributions to β-hairpin stability: MD simulations of C-terminal fragment from the B1 domain of protein G. *Biophys. Chem.* **101-102**, 187–201.

115. Fesinmeyer, R. M., Hudson, F. M., and Andersen, N. H. (2004) Enhanced hairpin stability through loop design: the case of the protein GB1 domain hairpin. *J. Am. Chem. Soc.* **126**, 7238–7243.

116. Du, D., Zhu, Y., Huang, C.-Y., and Gai, F. (2004) Understanding the key factors that control the rate of β-hairpin folding. *Proc. Natl. Acad. Sci. USA* **101**, 15,915–15,920.

117. Cochran, A. G., Skelton, N. J., and Starovasnik, M. A. (2001) Tryptophan zippers: stable, monomeric β-hairpins. *Proc. Natl. Acad. Sci. USA* **98**, 5578–5583.

118. Owrutsky, J. C., Raftery, D., and Hochstrasser, R. M. (1994) Vibrational relaxation dynamics in solutions. *Ann. Rev. Phys. Chem.* **45**, 519–555.

119. Matouschek, A., Kellis, J. T., Serrano, L., and Fersht, A. R. (1989) Mapping the transition state and pathway of protein folding by protein engineering. *Nature* **340**, 122–126.

120. Du, D., Tucker, M. J., and Gai, F. (2006) Understanding the mechanism of β-hairpin folding via φ-value analysis. *Biochemistry* **45**, 2668–2678.

121. de Alba, E., Jiménez, M. A., and Rico, M. (1997) Turn residue sequence determines β-hairpin conformation in designed peptides. *J. Am. Chem. Soc.* **119**, 175–183.

122. Rose, G. D., Gierasch, L. M., and Smith, J. A. (1985) Turns in peptides and proteins. *Adv. Protein Chem.* **37**, 1–109.

123. Yang, W. Y., Pitera, J. W., Swope, W. C., and Gruebele, M. (2004) Heterogeneous folding of the trpzip hairpin: full atom simulation and experiment. *J. Mol. Biol.* **336**, 241–251.

124. Yang, W. Y. and Gruebele, M. (2004) Detection-dependent kinetics as a probe of folding landscape microstructure. *J. Am. Chem. Soc.* **126**, 7758–7759.

2

The Use of High-Pressure Nuclear Magnetic Resonance to Study Protein Folding

Michael W. Lassalle and Kazuyuki Akasaka

Summary

Recent development of high-pressure cells for a variety of spectroscopic methods has enabled the use of pressure as one of the commonly used perturbations along with temperature and chemical perturbations to study folding/unfolding reactions of proteins. Although various high-pressure spectroscopy techniques have their own significance, high-pressure nuclear magnetic resonance (NMR) is unique in that it allows one to gain residue-specific and atom-detailed information from proteins under pressure. Furthermore, because of a peculiar volume property of a protein, high-pressure NMR allows one to obtain structural information of a protein in a wide conformational space from the bottom to the upper region of the folding funnel, giving structural reality for the "open" state of a protein proposed from hydrogen exchange. The method allows a link between equilibrium folding intermediates and the kinetic intermediates, and manifests a new view of proteins as dynamic entities amply fluctuating among the folded, intermediate, and unfolded sub ensembles. This chapter briefly summarizes the technique, the principle, and the ways to use high-pressure NMR for studying protein folding.

Key Words: Pressure; protein folding; thermodynamic stability; volume change; structures of folding intermediates; high-pressure NMR; pressure-jump.

1. Introduction

Pressure perturbation is increasingly important in studies of protein folding, stability, conformational flexibility, and aggregation. A keyword search on HighWire and Medline with keywords: high pressure-protein-nuclear magnetic resonance (NMR) shows that there has been a dramatic increase in the number of pressure-related studies of proteins (**Fig. 1**). The dramatic increase is assisted by technical developments for high-pressure studies in a variety of spectroscopic measurements at high pressure including X-ray scattering (*1,2*), fluorescence (*3–6*), Fourier-transform infrared spectroscopy (*6–10*) and ultraviolet-visible derivative spectroscopy

From: *Methods in Molecular Biology, vol. 350: Protein Folding Protocols*
Edited by: Y. Bai and R. Nussinov © Humana Press Inc., Totowa, NJ

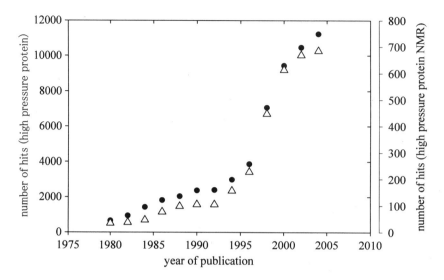

Fig. 1. Keyword search in HighWire and Medline as function of publication year. Circles: key words utilized: high pressure-protein. Triangle: key words utilized: high pressure-protein-nuclear magnetic resonance.

(11), kinetic processes under pressure *(12,13)*, and pressure-jump techniques *(14–16)*. In particular, the recent advancement in high-pressure NMR techniques has renewed interest in high-pressure studies of proteins, it allows to gain residue-specific or atom-detailed information from proteins under pressure.

2. Materials

Buffers to be used for pressure studies are preferred to have little variation of pKa with pressure. There is no complete buffer in this sense, but the following buffers are relatively inert to pressure variation. One must be aware, however, that those buffers inert to pressure variation tend to be more susceptible to temperature variation. Useful buffers are maleic acid ($\Delta V^o = -5.1$ cm^{3}mol^{-1}), MES ($\Delta V^o = 3.9$ cm^{3}mol^{-1}), Tris ($\Delta V^o = 4.3$ cm^{3}mol^{-1}), or imidazole buffers ($\Delta V^o = 1.8$ cm^{3}mol^{-1}) *(17)*. As chemical shift standard, the singlet signal of DSS is to be set at origin ($\delta = 0$). However, this often overlaps with some methyl proton signals in the ^{1}H NMR spectrum. The authors found that dioxane can be used as a convenient secondary standard, as it gives a sharp singlet at $\delta = 3.75$ ppm downfield from DSS nearly independently of pressure.

3. Methods

3.1. High-Pressure NMR Techniques

Recent contributions to high-pressure NMR study of proteins are mainly attributable to the groups of Jonas *(18)* and Akasaka *(19)*, in addition to those

Fig. 2. A pressure-resisting quartz cell and the protecting Teflon capsule used for the on-line cell high-pressure nuclear magnetic resonance system.

of Kalbitzer *(20)*, Wand *(21)*, and Markley *(22)*. The heart in designing high-pressure NMR techniques is how to design a pressure-resistant cell in a small signal detection space by maintaining the highest possible homogeneity in both static and radiofrequency magnetic fields without losing sensitivity of detection. Typically two cell designs have been utilized. The former group *(18)* preferred an autoclave design, which has the advantage of larger sample volume and high attainable pressure (~900 MPa). Here, receiver coils must be self-designed within the autoclave. The cell contains sufficient volume and applies very well to recording one-dimensional (1D)^{1}H NMR spectra of proteins. A major problem lies in the difficulty to obtain well-resolved two- or multidimensional spectra where multiple nuclei must be simultaneously excited within a highly homogeneous static and radiofrequency magnetic fields spanning the entire sample space. The latter group *(19)* developed pressure-resisting cells (**Fig. 2**) made up initially of glass, then of quartz (both hand-made) *(23)*, and used them on a commercial NMR probe. This is the on-line cell high-pressure NMR system described in some detail in the following section. Although the inner space of the pressure-resistive cell is so small (typically ~20 µL; **Fig. 2**) that the signal intensity is inherently low, the field homogeneity can be extremely good and allows all two- or multidimensional spectral measurements, giving excellent high resolution multidimensional spectra as those obtained with normal 5-mm tubes at 1 bar. Therefore, the on-line cell high-pressure method has a higher versatility than the autoclave method for protein studies. Some efforts have been made to increase the sample volume by utilizing sapphire cells with limited success *(20,21)*.

The maximum pressure range of the quartz cell method is normally limited from 200 to approx 400 MPa (because cells break at higher pressure), this relatively mild pressure range has been sufficient in many proteins to a variety of studies including folding intermediates *(24)*, conformational dynamics *(25)*, local disordering *(26)*, kinetic intermediates *(27)*, stability *(28)*, energy-landscapes *(29)*, and dissociation/association of proteins *(30)*. Although many proteins may not fully unfold below approx 400 MPa, 100% unfolding is actually not required in high-pressure NMR measurements because the onset of full unfolding is recognized if all the cross peaks begin to lose their intensities, say up to approx 20%, in a concerted matter in a typical well-resolved two-dimensional (2D) NMR spectra. The applicability of variable-pressure NMR to studies of folding intermediates is also assured by recognizing the general tendency for the decreasing order of tertiary structure as pressure is increased *(31)*. In this chapter, we will concentrate our effort solely to highlight the advantages of high-pressure NMR in the study of thermodynamic stability and folding, but the concept and the method of analysis are applicable to results of any high-pressure spectroscopic measurements.

3.1.1. The On-Line Cell High-Pressure NMR System

The heart of the success of the on-line cell high-pressure NMR method relies on the preparation of pressure-resistive sample cells *(23)* (**Fig. 2**). As mentioned, the starting material generally used is transparent quartz capsules or quartz capsules made by synthetic silica. First a perfect clean surface must be obtained; we achieved this goal by etching the glass inside and outside with a dilute hydrofluoric acid solution and by removing micro cracks with fire polishing. Then the quartz capsules are opened at scratched points, and the ends are heated to make a long tail (o.d. = 0.3–1 mm; length = 400–700 mm), a short sealed tail (o.d. = 0.1–0.5 mm; length = 1–2 mm), and a cell body. The short sealed tail enables a bottom design with a small tensile force in a small area. Finally, pyroxylin coating of the short tail, cell body, and long tail is performed. Our final cell has an inside diameter about 0.8–1.0 mm and an outside diameter of 3.0–3.7 mm.

The high-pressure cell assembly consists of a hand oil-pump, a pressure intensifier, and a Heise-Bourdon gage. A 6 m-long SUS-316 high-pressure tube (o.d./ i.d. = 3.0/0.6 mm) transmits the pressure to the top of the high-pressure cell-separator assembly. This assembly consists of a polytetrafluoroethylene (PTFE) safety jacket and the NMR tube described above with a long flexible tail that was bonded to an SUS-316 nozzle with a thermo hardening epoxy adhesive (Araldite AT1). The nozzle is connected through a PTFE connector tube to a PTFE inner tube, which is embedded in a BeCu separator cylinder. Within the PTFE inner tube, two PTFE pistons with a length of 3–4 mm and with diameter about 0.05 mm smaller than the i.d. of the PTFE inner tube are seated. The space between upper and lower pistons is filled with per-fluoro-polyether, as it is insoluble in both organic solvent and water and acts as separator between the pressure transmitting system and the NMR sample. An 80:20 mixture of motor oil and kerosene is used as pressure transmitting fluid. Both the NMR tube with flexible long tail and the PTFE inner tube up to the lower PTFE piston are filled with the sample solution, whereas the high-pressure tube up to the upper piston contains the pressure transmitting fluid.

3.2. Methods for Determining Thermodynamic Parameters From High-Pressure Experiment

3.2.1. Basic Equations for Two-State Transition

The thermodynamic stability of proteins has been much less explored in the pressure axis than in the temperature or chemical axis. However, if the solution conditions (chemical compositions) are fixed, the conformational stability of a protein $\Delta G = G_{unfolded} - G_{folded}$ is a simultaneous function of both temperature and pressure; thus defining a three-dimensional Gibbs free energy surface on the pressure–temperature plane. Several excellent papers *(29,32–34)* on multi-dimensional energy landscapes have been published.

The basic equations governing the stability of proteins on the p and/or T axes are given in the following:

1. By taking the pressure dependence of ΔG to the first order at constant temperature,

$$\Delta G = \Delta G^o_{1bar} + p\Delta V^0 \tag{1}$$

and to the second order,

$$\Delta G_p - \Delta G^o = \Delta V^0(p - p_o) - \frac{\Delta\beta}{2}(p - p_o)^2 \tag{1'}$$

2. By taking the "exact" expression for temperature dependence of ΔG and the pressure dependence of ΔG to the second order,

$$\Delta G = \Delta G^o - \Delta S^o(T - T^0) - \Delta C_p \left[T\left(\ln\frac{T}{T^o} - 1 \right) + T^o \right]$$

$$+ \Delta V^o(p - p^o) - \frac{\Delta\beta}{2}(p - p^o)^2 + \Delta\alpha(p - p^o)(T - T^o) \tag{2}$$

Performing a Taylor expansion and neglecting higher order terms with respect to temperature, this equation equals the following:

$$G = \Delta G^o - \Delta S^o (T - T^0) - \frac{\Delta C_p}{2T^o}(T - T^o)^2 + \Delta V^o (p - p^o) \tag{2'}$$

$$- \frac{\Delta \beta}{2}(p - p^o)^2 + \Delta \alpha (p - p^o)(T - T^o)$$

Utilizing **Eq. 1** implies that the isothermal compressibility change ($\Delta\beta$) can be assumed to be zero or negligibly small. If this is not the case, **Eq. 1′** must be used. Employing **Eq. 2** or **2′** leads to a multidimensional energy landscape under the assumption that in the temperature and pressure ranges under investigation, heat capacity (ΔC_p), thermal expansion coefficient ($\Delta\alpha$), and isothermal compressibility coefficient ($\Delta\beta$) are constant.

3.2.2. Experiment 1: Simultaneous Change of Pressure and Temperature: Elucidation of a Multidimensional Energy Landscape Using *Eq. 2*

Careful measurements of NMR spectra at varying pressure and temperature is particularly suited to determine the *p–T* energy landscape of proteins. This is because the folding and/or the unfolding fraction can be assessed accurately, as the NMR intensity for a perfect folded protein is intrinsically known for typical well-isolated NMR signals with no overlaps. As a target protein for our study to obtain an energy landscape on the *p–T* plane, we chose *Staphylococcal nuclease (29)*, which is known to closely obey the two-state transition between folded (F) and unfolded (U) state *(22,34,35)*, a requirement for the straightforward application of previously mentioned equations. 1D ^{1}H NMR spectra were measured in deuterated 20 m*M* MES buffer, pH 5.3, as a function of pressure between 3 and 330 MPa at various temperatures (**Fig. 3**). The well-isolated His8 ε proton signals were used to obtain conformational equilibrium K = [U]/[F], which, by using the relation $\Delta G = -RT\ln K$, allowed us to obtain experimental $\Delta G = G_{\text{unfolded}} - G_{\text{folded}}$ values as a simultaneous function of pressure and temperature. By least-squares-fitting the experimental ΔG values to **Eq. 2**, we determine all the thermodynamic parameters ΔG^0, ΔS^0, ΔV^o, ΔC^0, $\Delta\beta$, and $\Delta\alpha$, giving a three-dimensional energy landscape shown in **Fig. 4**.

The results demonstrate that combined temperature-pressure-dependent studies can help delineate the free energy landscape of proteins on *p–T* axes and, hence, help elucidate the features and thermodynamic parameters that are essential for the stability of the native conformational state of proteins.

3.2.3. Experiment 2: Pressure Dependence of 1D ^{1}H NMR: Elucidation of Thermodynamic Stability Using *Eq. 1*

A more general use of high-pressure NMR for determining thermodynamic stability of proteins ΔG^0 and the associated volume change ΔV^0 comes from the use of **Eq. 1**. Often small proteins exhibit reasonably good two-state unfolding.

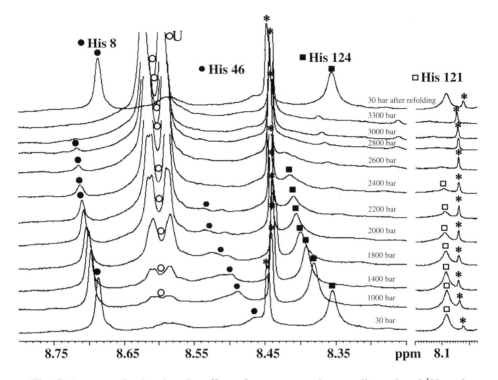

Fig. 3. An example showing the effect of pressure on the one-dimensional ^{1}H nuclear magnetic resonance spectrum of a globular protein *Staphylococcal* nuclease. * Indicates impurity signals. (Reprinted from **ref. *29***, with permission from Elsevier.)

In general, pressure effects on folded proteins appear in three major steps *(36)*. Step 1 is a nearly linear compression of the folded conformer. For example, in **Fig. 3**, the effect of compression on the folded conformer can be seen in the low-field shifts of three His ε proton signals. In step 2 a transition takes place from a folded conformer to an alternate conformer with an abrupt change in volume ΔV^o. Often, this step is manifested as a decrease of signals from the folded conformer with concomitant increase of new signals from the alternate conformer, in the case of *Staphylococcal* nuclease the unfolded state U. In **Fig. 3**, this effect is seen in the gradual loss of the native His ε proton signals with con-comitant appearance of the denatured His ε proton signals U as we increase the pressure. Step 3 is a compression of the unfolded conformer U. A linear plot of ΔG against pressure through **Eq. 1** gives simultaneously the volume change ΔV^o from the slope and the stability at 1 bar ΔG^0 by extrapolation of ΔG to 1 bar. The latter gives an alternative general method to the popular one—determining the stability of a protein structure with guanidium chloride or urea.

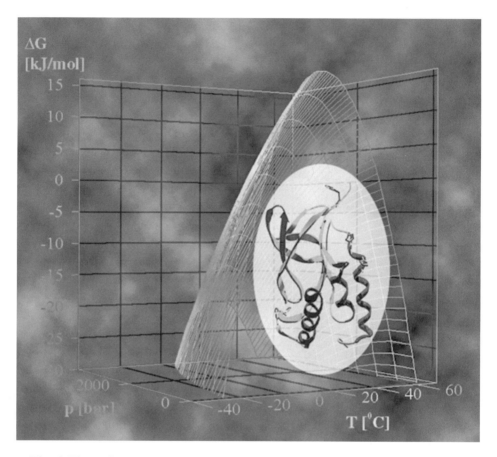

Fig. 4. Three-dimensional free energy landscape of *Staphyloccocal* nuclease with its folded structure.

The volume change ΔV^o is an extra parameter we obtain for protein unfolding uniquely from pressure experiment. The origin of the volume change ΔV^o has been discussed in various literatures *(37–39)* and is considered to originate from the loss of atom defects or cavities within the folded conformer, the hydration of hydrophobic residues upon unfolding and the hydration of charged residues upon unfolding. A consensus is attained such that, all effects combined, the resultant ΔV^o is negative for unfolding of globular proteins under closely physiological conditions *(40)*.

The effect of cavity on the stability ΔG^0 and the volume change ΔV^o has been experimentally examined on the wild-type c-Myb R2 sub-domain and its cavity-filling mutant (**Fig. 5**) using ^{1}H 1D NMR as a function of pressure *(28)*. The cavity of c-Myb R2 was filled by replacing a Val with a Leu and it was

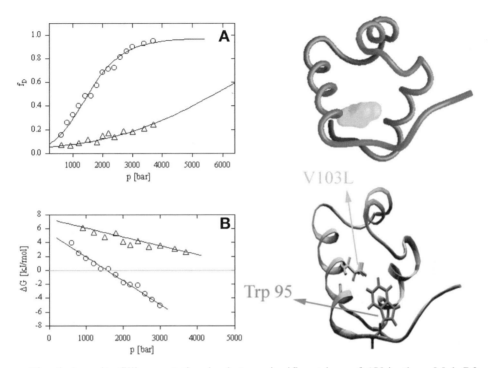

Fig. 5. A cavity-filling mutation leads to a significant loss of ΔV in the c-Myb R2 domain, showing the significant cavity contribution to ΔV. (**A**) Plot of the unfolded fractions against pressure for the wild-type (O) and the mutant (Δ). (**B**) Plot of the stability against pressure for the wild-type (O) and the mutant (Δ). Figures on the right shows ribbon models of the wild-type (upper) and the cavity-filling mutant (lower) of the c-Myb R2 domain. (Reproduced with permission from **ref. 28**.)

confirmed that the tertiary structure is unaltered *(41)*. We obtained ΔG^0 is increased from 5.36 to 7.34 kJ/mol by the cavity-filling mutation, whereas ΔV^o is changed from −33.9 to −12.6 mL/mol.

Among the three factors that contribute to ΔV^o, the difference in ΔV_{sol} (hydration term) and ΔV_c (van der Waals volumes of the constitutive atoms) between the two proteins are small enough to be neglected *(41)* and, therefore, high-pressure unfolding of the native and cavity-filling mutation reveals partial molar volume change (ΔV^o), whereas $\Delta\Delta V^o_{native\text{-}mutated}$ reflects the size of the cavity. Comparison of our calculated cavity size 35.3 Å with literature values 33.1 Å *(41)* shows a good agreement. The decisive effect of cavity on the volume change ΔV has been experimentally verified also in *Staphylococcal* nuclease *(38)*.

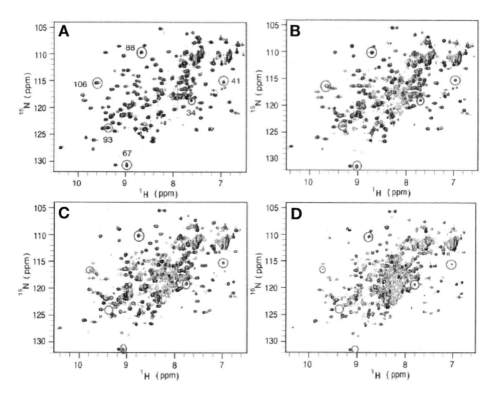

Fig. 6. Two-dimensional ^{15}N/^{1}H HSQC spectra of β-lactoglobulin at **(A)** 3 MPa; **(B)** 100 MPa; **(C)** 150 MPa; **(D)** 200 MPa at pH 2.0 and 36°C, measured at 750 MHz in 95% ^{1}H$_2$O/5% ^{2}H$_2$O. (Reprinted from **ref. 42**, with permission from Elsevier.)

3.2.4. Experiment 3: Pressure Dependence of 2D Heteronuclear NMR (e.g., ^{15}N/^{1}H HSQC): Characterization of Folding Intermediates

Detection of intermediates is often considered to be a key to understanding protein-folding process. These cannot be, in general, detected easily in ^{1}H NMR spectra even under variable pressure. We have to proceed to utilize NMR spectra for residue-specific structural information, which is possible by performing heteronuclear 2D NMR spectra such as ^{15}N-^{1}H HSQC as a function of pressure. By comparing molar volume changes (ΔV^o) and Gibbs free energy changes (ΔG^0) for individual amino acid sites based on **Eq. 1**, we can easily differentiate between an overall cooperative unfolding from local unfolding. By grouping residues having similar ΔG^0 and ΔV^o values together, structural and thermodynamic information can be obtained about locally unfolded or intermediate conformers of the protein *(27,42)*.

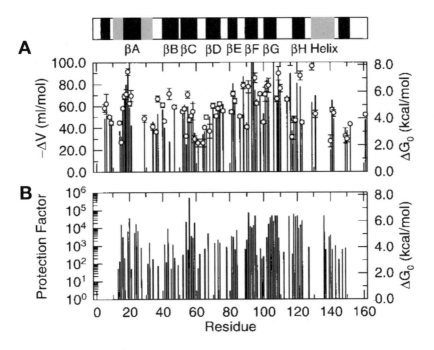

Fig. 7. Plots of (**A**) $\Delta G^0(-)$ and ΔV (O) determined from high-pressure nuclear magnetic resonance and (**B**) the protection factor from $^1H/^2H$ exchange, for individual residues of β-lactoglobulin. (Reprinted from **ref.** *42*, with permission from Elsevier.)

β-Lactoglobulin has a β-barrel structure, inside of which small hydrophobic molecules are bound for transportation. Kuwata et al. measured $^{15}N/^1H$ HSQC spectra of ^{15}N-labeled β-lactoglobulin as a function of pressure (**Fig. 6**). At 3 MPa, the well-dispersed cross peaks show the fully folded protein structure. Greater than 100 MPa, intensities of these cross peaks decrease and are replaced by a cluster of cross peaks in the central region of the spectra (the random-coil positions), showing an increasing degree of unfolding of the protein structure. The spectral changes were fully reversible with pressure. By examining cross peak intensities of $^{15}N/^1H$ HSQC spectra of ^{15}N-labeled β-lactoglobulin as a function of pressure, ΔG^0 and ΔV^o values were determined for individual amino acid sites according to **Eq. 1** (**Fig. 7**) *(42)*. The ΔG^0 values vary considerably from site to site, the noncore side of the barrel (βB, βC, βD, and βE) being comparatively less stable ($\Delta G^0 = 4.6 \pm 1.3$ kcal/mol) than the hydrophobic core side ($\Delta G^0 = 6.5 \pm 2.1$ kcal/mol).

The ΔG^0 values are compared with the free energy differences between the "closed" and "open" conformers independently obtained from hydrogen exchange experiments under the assumption of an EX_2 mechanism *(42)*. The

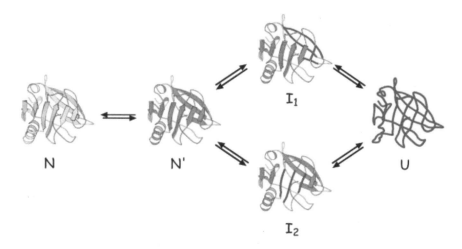

Fig. 8. The conformational equilibrium of β-lactoglobulin revealed by high-pressure NMR. (Reprinted from **ref. 42** with permission from Elsevier.)

good coincidence of ΔG values from high-pressure NMR (**Fig. 6A**) and hydrogen exchange (**Fig. 6B**) in **Fig. 6** indicates that high-pressure NMR is indeed useful to predict conformational fluctuations of a protein in a site-specific manner similarly to a hydrogen exchange experiment. However, the real merit of high-pressure NMR is that it gives not only the information for fluctuation, but also the information on individual structures of the "open" conformers.

The volume change also varies from site to site, the noncore side having a smaller volume change ($\Delta V^o = -57.4 +/- 14.4$ mL/mol) than the core side ($\Delta V^o = -90.0 +/- 35.2$ mL/mol). This gives an approximate picture of conformational fluctuation such that the noncore side of the barrel fluctuates to produce a group of intermediates I_1, whereas the hydrophobic core side of the barrel fluctuates to produce a group of intermediates I_2, which is schematically shown in **Fig. 8**. In the kinetic folding experiments at 1 bar using pulse-labeling $^1H/^2H$ NMR techniques *(43)*, a kinetic intermediate analogous to the pressure-stabilized equilibrium intermediate I_1 has been observed. Thus the equilibrium scheme shown in **Fig. 8** is likely to also represent the kinetic pathway of folding for β-lactoglobulin *(42)*.

Close identity of pressure-stabilized equilibrium intermediates with kinetic intermediates has also been observed for ubiquitin. Kitahara et al. *(27)* performed $^{15}N/^1H$ HSQC pressure unfolding measurements at 0°C on ^{15}N-uniformly labeled ubiquitin. The segment between residues 33 and 42 showed lower stability ($\Delta G^0 = 15.2 \pm 1.0$ kJ/mol, $\Delta V^o = -58 \pm 4$ mL/mol) than the rest of the protein ($\Delta G^0 = 31.3 \pm 4.7$ kJ/mol, $\Delta V^o = -85 \pm 7$ mL/mol), showing that the segment (residues 33–42) preferentially melts to produce an intermediate conformer I.

Kinetic intermediates for ubiquitin were detected by utilizing pulse-labeling $^1H/^2H$ NMR techniques combined with GdnHCl-induced refolding *(44)*. Parallel folding pathways were also suggested for this case, the major pathway occurring rapidly with a time constant of 10 ms, the other occurring slowly with a time constant of approx 10 s. In the latter case, the authors suggested an intermediate with Pro-37 and/or Pro 38 in a *cis* conformer is formed, which slowly (~10 s) converts into a *trans* conformer. The structure of conformer I is closely identical to that predicted for the slow folding intermediate (~10 s) in the kinetic folding experiment *(44)*. The close identity between the pressure-stabilized equilibrium intermediate with the kinetic intermediate is considered to be general rather than exceptional *(27)*.

3.2.5. Experiment 4: Pressure-Jump 1D and 2D NMR

Pressure-jump NMR measurements can be performed for slow folding proteins using the high-pressure NMR system primarily designed for equilibrium studies. This is because very often the folding or unfolding rate becomes extremely slow under pressure. In the case of P13^{MCTP1} *(14)*, just manually changing pressure within minutes is sufficient to observe pressure-jump kinetics intermediates in 1D or even in 2D NMR spectroscopy. For P13^{MCTP1}, it was shown from pressure-jump 2D $^{15}N/^1H$ HSQC measurements that the pressure-stabilized intermediate N_2 is closely related to the folding intermediate N_2 revealed by pressure jump to 300 MPa *(14)*. The authors concluded that prior to global unfolding the native structure of P13^{MCTP1} converts into a partially hydrated conformer N_2, which is still part of the folded ensemble.

Pressure-jump NMR measurements were recently performed on amyloid protofibrils, in which a slow dissociation of the fibrils into monomeric species and reassociation into fibrils were observed upon pressure-jump from 3 to 200 MPa *(45)*. The work symbolizes the utility of pressure in highly associating systems, which may be considered to be an extension of folding.

One of the merits of performing a pressure-jump experiment is that, when carried out as a function of pressure, it can give the activation volume for the reaction, crucial information concerning the structure of the transition state.

3.2.6. Experiment 5: Pressure-Dependent NOESY Experiment: Expressing Folding Intermediates in Atomic Coordinates

Recently, the technology of high-pressure NMR to gain structural information of an intermediate conformer has come to still a higher level of attainment. Kitahara et al. succeeded in obtaining a sufficient numbers of NOESY-based distance constraints along with torsion angle constraints to express the time-averaged structure of ubiquitin at 300 MPa, as well as that at 3 MPa in atomic

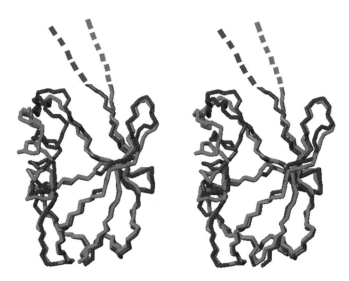

Fig. 9. Stereoview of the superposition of the two ubiquitin structures at 30 bar (black) and 3 kbar (gray) between which the protein fluctuates. (Reprinted from **ref. 46,** with permission from Elsevier.)

coordinates using the program CYANA *(46)*. The two structures of ubiquitin determined at 20°C at 3 and 300 MPa (**Fig. 9**) show globally similar folded structures, but large differences are found in average coordinates, particularly in the C-terminal segment 70–76 and in the helix. At least 3 Å differences are found in C_α atoms of the C-terminal side of the helix between the two structures, predicting large amplitude fluctuation of the molecule close to the C-terminal segment carrying the reactive site at its end. From the independent shift analysis, the two structures are considered to closely represent the two extreme structures (N_1 and N_2) between which the protein fluctuates at ambient pressure. The result of this fluctuation is a production of an "open" form of ubiquitin near its C-terminal end, whereas N_2 is considered suitable for interaction with enzymes covalently linking ubiquitin with a substrate protein. Conformer N_2 is too close to N_1 to be considered as a folding intermediate, careful analysis of ^{15}N longitudinal and transverse relaxation rates along with nuclear Overhauser effect gives the lifetime of N_2 to be on the order of 10 μs and only 1.2 kcal/mol energetically higher than N_1 at ambient pressure, thus, it is part of the native ensemble. So far, structure analysis by X-ray or by conventional NMR revealed only conformer N_1 but not the structure of N_2. But the method shown here can be extended to any intermediate conformers having partially folded segments for which distance constraints may be obtained from NOESY and J-couplings to determine time-averaged atomic coordinates. It is one of the highlights of high-pressure NMR that it can experimentally determine atomic coordinates of folding intermediates and reveal the biologic important structure.

4. Conclusion

The high-pressure NMR method has a great future for protein folding study. In applying it, one must keep in mind the following:

1. Among the two basic techniques of high-pressure NMR, the on-line cell technique is the method of choice for protein folding study. However, the technique is still in its infancy. More versatile cells must be developed before it is used widely.
2. There are basically two categories of use of high-pressure NMR to protein folding studies. One is the determination of thermodynamic stability (namely ΔG^0) of folded conformer over unfolding along with ΔV^o. For two-state transitions, this work can be best studied with 1D ^{1}H NMR. The method may be considered to be alternative to the guadinium chloride or urea denaturation techniques to determine the thermodynamic stability.
3. The second category of the use of high-pressure NMR is to obtain structural information about folding intermediates. This is quite difficult to perform with ^{1}H NMR. Usually, heteronuclear 2D or multidimensional NMR spectroscopy is to be used.
4. Information on the rate of conformational transition or folding can be obtained using NMR relaxation methods at different pressures or directly by using pressure-jump.

Acknowledgments

This chapter is primarily based on collaborative research. We thank the collaborators, especially Yamada, Sarai, Kuwata, Royer, Kitahara, and Yokoyama. Financial supports from the Ministry of Education, Sports, Culture, Science, and Technology (to K.A.) and from Japan Society for the Promotion of Science (to M. W. L. and K. A.) are gratefully acknowledged.

References

1. Fujisawa, T., Kato, M., and Inoko, Y. (1999) Structural characterization of lactate dehydrogenase dissociation under high pressure studied by synchrotron high pressure small-angle x-ray scattering. *Biochemistry* **38,** 6411–6418.
2. Winter, R. (2002) Synchrotron X-ray and neutron small-angle scattering of lyotropic lipid mesophases, model biomembranes and proteins in solution at high pressure. *Biochim. Biophys. Acta* **1595,** 160–184.
3. Ruan, K. and Balny, C. (2002) High pressure static fluorescence to study macromolecular structure-function. *Biochim. Biophys. Acta* **1595,** 94–102.
4. Ikeuchi, Y., Suzuki, A., Oota, T., et al. (2002) Fluorescence study of the high pressure-induced denaturation of skeletal muscle actin. *Eur. J. Biochem.* **269,** 364–371.

5. Di Venere, A., Salucci, M. L., van Zadelhoff, G., et al. (2003) Structure-to-function relationship of mini-lipoxygenase, a 60-kda fragment of soybean lipoxygenase-1 with lower stability but higher enzymatic activity. *J. Biol. Chem.* **278,** 18,281–18,288.

6. Herberhold, H., Marchal, S., Lange, R., Scheying, C., Vogel, R. F., and Winter, R. (2003) Characterization of the pressure-induced intermediated and unfolded state of red-shifted green fluorescent protein-a static and kinetic FTIR, UV-VIS and fluorescence spectroscopy study. *J. Mol. Biol.* **330,** 1153–1164.

7. Dzwolak, W., Kato, M., and Taniguchi, Y. (2002) Fourier-transform infrared spectroscopy in high pressure studies on proteins. *Biochim. Biophys. Acta* **1595,** 131–144.

8. Smeller, L., Meersmann, F., Fidy, J., and Heremans, K. (2003) High pressure FTIR study of the stability of horseradish peroxidase. Effect of heme substitution, ligand binding, Ca++ removal, and reduction of the disulfide bonds. *Biochemistry* **42,** 553–561.

9. Meersman, F., Smeller, L., and Heremans, K. (2002) Comparative fourier transform infrared spectroscopy study of cold-, pressure-, and heat-induced unfolding and aggregation of myoglobin. *Biophys. J.* **82,** 2635–2644.

10. Jung, C., Kozin, S. A., Canny, B., Chervin, J. C., and Hoa, G. H. (2003) Compressibility and uncoupling of cytochrome P450cam: high pressure FTIR and activity studies. *Biochem. Biophys. Res. Commun.* **312,** 197–203.

11. Lange, R. and Balny, C. (2002) UV-visible derivative spectroscopy under high pressure. *Biochim. Biophys. Acta* **1595,** 80–93.

12. Pappenberger, G., Saudan, C., Becker, M., Merbach, A. E., and Kiefhaber, T. (2000) Denaturant-induced movement of the transition state of protein folding revealed by high-pressure stopped-flow measurements. *Proc. Natl. Acad. Sci. USA* **97,** 17–22.

13. Jung, C., Bec, N., and Lange, R. (2002) Substrates modulate the rate-determining step for CO binding in cytochrome P450cam (CYP101). *Eur. J. Biochem.* **269,** 2989–2996.

14. Kitahara, R., Royer, C., Yamada, H., et al. (2002) Equilibrium and pressure-jump relaxation studies of the conformational transitions of P13MTCP1. *J. Mol. Biol.* **320,** 609–628.

15. Desai, G., Panick, G., Zein, M., Winter, R., and Royer, C. A. (1999) Pressure-jump studies of the folding/unfolding of trp repressor. *J. Mol. Biol.* **288,** 461–475.

16. Woenckhaus, J., Koehling, R., Thiyagarajan, P., et al. (2001) Pressure-jump small-angle x-ray scattering detected kinetics of staphylococcal nuclease folding. *Biophys. J.* **80,** 1518–1523.

17. Kitamura, Y. and Itoh, T. (1987) Reaction volume of protonic ionization for buffering agents. Prediction of pressure dependence of pH and pOH. *J. Solution Chemistry* **16,** 715–725.

18. Jonas, J. (2002) High-resolution nuclear magnetic resonance studies of proteins. *Biochim. Biophys. Acta* **1595,** 145–159.

19. Akasaka, K. and Yamada, H. (2001) On-line cell high-pressure nuclear magnetic resonance technique: application to protein studies. *Methods Enzymol.* **338,** 134–158.
20. Arnold, M. R., Kalbitzer, H. R., and Kremer, W. (2003) High-sensitivity sapphire cells for high pressure NMR spectroscopy on proteins. *J. Magnetic Resonance* **161,** 127–131.
21. Urbauer, J. L., Ehrhardt, M. R., Bieber, R. J., Flynn, P. F., and Wand, J. A. (1996) High-resolution triple-resonance NMR spectroscopy of a novel calmodulin peptide complex at kilobar pressures. *J. Am. Chem. Soc.* **118,** 11,329–11,330.
22. Royer, C. A., Hinck, A. P., Loh, S. N., et al. (1993) Effects of amino acid substitutions on the pressure denaturation of *staphylococcal* nuclease as monitored by fluorescence and nuclear magnetic resonance spectroscopy. *Biochemistry* **32,** 5222–5232.
23. Yamada, H., Nishikawa, M., Honda, M., Shimura, T., Akasaka, K., and Tabayashi, K. (2001) Pressure-resisting cell for high-pressure, high-resolution nuclear magnetic resonance measurements at very high magnetic fields. *Rev. Sci. Instrum.* **72,** 1463–1471.
24. Kamatari, Y. O., Kitahara, R., Yamada, H., Yokoyama, S., and Akasaka, K. (2004) High-pressure NMR spectroscopy for characterizing folding intermediates and denatured states of proteins. *Methods* **34,** 133–143.
25. Kuwata, K., Kamatari, Y. O., Akasaka, K., and James, T. L. (2004) Slow conformational dynamics in the hamster prion protein. *Biochemistry* **43,** 4439–4446.
26. Kuwata, K., Li, H., Yamada, H., and Legname, G. (2002) Locally disordered conformer of the hamster prion protein: a crucial intermediate to PrPSc? *Biochemistry* **41,** 12,277–12,283.
27. Kitahara, R. and Akasaka, K. (2003) Close identity of a pressure-stabilized intermediate with a kinetic intermediate in protein folding. *Proc. Natl. Acad. Sci. USA* **100,** 3167–3172.
28. Lassalle, M. W., Yamada, H., Morii, H., Ogata, K., Sarai, A., and Akasaka, K. (2001) Filling a cavity dramatically increases pressure stability of the c-myb R2 subdomain. *Proteins* **45,** 96–101.
29. Lassalle, M. W., Yamada, H., and Akasaka, K. (2000) The pressure-temperature free energy-landscape of *staphylococcal* nuclease monitored by ^{1}H NMR. *J. Mol. Biol.* **298,** 293–302.
30. Niraula, T. N., Konno, T., Li, H., Yamada, H., Akasaka, K., and Tachibana, H. (2004) Pressure-dissociable reversible assembly of intrinsically denatured lysozyme is a precursor for amyloid fibrils. *Proc. Natl. Acad. Sci. USA* **101,** 4089–4093.
31. Akasaka, K. (2003) Highly fluctuating protein structure revealed by variable-pressure nuclear magnetic resonance. *Biochemistry* **42,** 10,875–10,885.
32. Chan, H. S. and Dill, K. A. (1998) Protein folding in the landscape perspective: Chevron plots and non-Arrhenius kinetics. *Proteins* **30,** 2–33.
33. Smeller, L. and Heremans, K. (1997) Some thermodynamic and kinetic consequences of the phase diagram of protein denaturation. In: *High-Pressure Research in the Biosciences and Biotechnology* (Heremans, K., ed.), Leuven University Press, Leuven, Belgium, pp. 55–58.

34. Panick, G., Vidugiris, G. J. A., Malessa, R., Rapp, G., Winter, R., and Royer, C. A. (1999) Exploring the temperature-pressure phase diagram of *staphylococcal* nuclease. *Biochemistry* **38,** 4157–4164.

35. Panick, G., Malessa, R., Winter, R., Rapp, G., Frye, K. J., and Royer, C. A. (1998) Structural characterization of the pressure-denatured state and unfolding/refolding kinetics of *staphylococcal* nuclease by synchrotron small-angle x-ray scattering and fourier-transform infrared spectroscopy. *J. Mol. Biol.* **275,** 389–402.

36. Akasaka, K. (2003) Exploring the entire conformational space of proteins by high-pressure NMR. *Pure. Appl. Chem.* **75,** 927–936.

37. Chalikian, T. V. (2003) Volumetric properties of proteins. *Annu. Rev. Biophys. Biomol. Struct.* **32,** 207–235.

38. Frye, K. and Royer, C. A. (1998) Probing the contribution of internal cavities to the volume change of protein unfolding under pressure. *Protein Sci.* **7,** 2217–2222.

39. Imai, T., Harano, Y., Kovalenko, A., and Hirata, F. (2001) Theoretical study for volume changes associated with the helix-coil transition of peptides. *Biopolymers* **59,** 512–519.

40. Royer, C. (2002) Revisiting volume changes in pressure induced proteins unfolding. *Biochim. Biophys. Acta* **1595,** 201–209.

41. Ogata, K., Kanei-Ishii, C., Sasaki, M., et al. (1996) The cavity in the hydrophobic core of Myb-DNA-binding domain is reserved for DNA recognition and trans-activation. *Nat. Struct. Biol.* **3,** 178–187.

42. Kuwata, K., Li, H., Yamada, H., Batt, C.A., Goto, Y., and Akasaka, K. (2001) High pressure NMR reveals a variety of fluctuating conformers in β-lactoglobulin. *J. Mol. Biol.* **305,** 1073–1083.

43. Forge, V., Hoshino, M., Kuwata, K., et al. (2000) Is folding of β-lactoglobulin non-hierarchic? Intermediate with native-like β-sheet and non-native α-helix. *J. Mol. Biol.* **296,** 1039–1051.

44. Briggs, M. S. and Roder, H. (1992) Early hydrogen-bonding events in the folding reaction of ubiquitin. *Proc. Natl. Acad. Sci. USA* **89,** 2017–2021.

45. Kamatari, Y. O., Yokoyama, S., Tachibana, H., and Akasaka, K. (2005) Pressure-jump NMR study of dissociation and association of amyloid protofibrils, *J. Mol. Biol.* **349,** 916–921.

46. Kitahara, R., Yokoyama, S., and Akasaka, K. (2005) NMR snapshots of a fluctuating protein structure: Ubiquitin at 30 bar – 3kbar. *J. Mol. Biol.* **347,** 277–285.

3

Characterization of Denatured Proteins Using Residual Dipolar Couplings

Erika B. Gebel and David Shortle

Summary

The conventional view of denatured proteins as random coils is being challenged by nuclear magnetic resonance (NMR) experiments that can measure the relative orientation of backbone segments with respect to each other in a distance-independent manner. Known as the residual dipolar coupling (RDC), this NMR parameter is measurable on proteins that have been weakly aligned inside a NMR tube. The simple observation of dipolar couplings unequivocally establishes that a denatured protein is not a random coil. Structures of denatured states under differing experimental conditions can be compared by a simple scatter plot of the RDCs measured under each condition; direct structural interpretation, however, is much more problematic. Here some of the technical issues for achieving weak alignment of denatured proteins are addressed using three types of media: strained polyacrylamide gels, both neutral or highly charged, and neutral lipid liquid crystalline bicelles. All three alignment media are experimentally robust and compatible with 8 M urea and low pH, conditions frequently used to denature proteins.

Key Words: Denatured state; RDC; residual dipolar structure; alignment media; bicelles; strained polyacrylamide.

1. Introduction

The denatured states of proteins play a critical role in the energetics and kinetics of protein folding, yet remain poorly understood because of the paucity of techniques available to provide detailed information about their ensemble-average structure.

Measurements of the radius or volume of proteins denatured at high concentrations of urea or guanidine hydrochloride demonstrate that they exhibit the dimensions expected of a hypothetical random coil. Data from small angle X-ray scattering and gel filtration methods have consistently indicated that the radius of gyration as a function of chain length is in good agreement with the random

From: *Methods in Molecular Biology, vol. 350: Protein Folding Protocols*
Edited by: Y. Bai and R. Nussinov © Humana Press Inc., Totowa, NJ

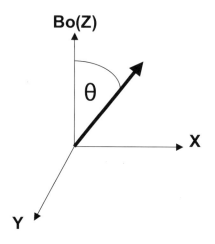

Fig. 1. The angle Θ between a bond vector (thick black arrow) of length r connecting nuclei i to j and the static magnetic field (Bo) of the spectrometer determines the magnitude of the residual dipolar couplings as defined in the following equation:

$$D_{ij} = \frac{-\mu_o \gamma_i \gamma_j h}{(2\pi r)^3} \left\langle \frac{3\cos^2\theta - 1}{2} \right\rangle$$

coil predictions (*1*). On the other hand, conventional nuclear magnetic resonance (NMR) experiments provide information about a denatured protein's local structure in turn regions and secondary structure. From this type of data, it appears that the denatured state of a protein must be a highly averaged ensemble of structures, but that randomness is not complete, with local residual helices and tight turns occasionally persisting at significant levels (*2*).

Recently, a new NMR parameter called the residual dipolar coupling (RDC) has shown promise for the characterization of long range or global structure in the denatured state (*3–8*). An RDC is observed as a small change in resonance frequency of one nucleus owing to the magnetic dipole of a nearby nucleus. As illustrated in **Fig. 1**, the RDC is dependent on the angle Θ defined by a vector between the two nuclei and the static magnetic field of the spectrometer. In free solution where a molecule tumbles randomly, this angle averages to zero and so does the RDC. Consequently, in order to measure RDCs, a small net alignment must be imparted to the molecule. For denatured proteins, the 1H–15N correlation spectrum is the easiest to resolve and assign, and as a result, RDCs derived from the amide proton–nitrogen bond vectors are the easiest to measure. A set of experimental RDCs provides partial information on the orientation of individual bond vectors with respect to a molecular frame of reference and with respect to each other, so that structural features of the ensemble average structure may be elucidated.

By itself, the observation of RDCs indicates that a denatured protein cannot be a random coil. In order to be aligned, the ensemble-averaged structure must either be asymmetric in shape or in its distribution of electrical charge. Furthermore, individual bond vectors must maintain on average a fixed relationship to the average global structure. If amino acid residues and hence the amide bond vectors were randomly oriented with respect to the rest of the polypeptide chain, they would be random with respect to the magnetic field of the NMR spectrometer and no RDC would be detectable. Unfortunately, direct structural interpretation from a single RDC data set is problematic. Techniques are being developed that could determine structural and dynamic properties of a denatured protein from RDC data alone *(9)*.

To measure RDCs, an asymmetric environment must be introduced to align the protein by restricting its orientational freedom. As described next, this can be accomplished through either steric or electrostatic interactions with the surrounding media. Measurement of the RDCs and data processing are specialized topics well covered in the NMR literature *(10–12)* and will not be treated here. Instead the focus will be on three sets of experimental conditions that seem most favorable for aligning denatured proteins: strained neutral polyacrylamide gels, strained and highly charged polyacrylamide gels, and electrically neutral lipid discs or bicelles formed by alkyl polyethylene glycol.

To characterize a denatured protein using RDCs, a series of preliminary experiments must be conducted to find conditions that confer sufficient alignment to generate RDCs distributed over a range of 10 to 15 Hz, yet do not interfere with the rapid tumbling required to maintain a narrow NMR lineshape (*see* **Note 1**). The following media have been successfully employed on a variety of proteins (*Staphylococcal* nuclease, Eglin C, ubiquitin, and others) and are compatible with high concentrations of urea and a pH of 2.0, common conditions used for denaturation.

2. Materials
2.1. Neutral Polyacrylamide Gel

1. 30% (w/v) Acrylamide stock solution for compressed gel with 1:20 bisacrylamide: 70 mg bisacrylamide, 1.43 g acrylamide, and 5 mL ddH$_2$O.
2. 30% (w/v) Acrylamide stock solution for stretched gel with 1:20 bisacrylamide: 35 mg bisacrylamide, 1.43 g acrylamide, and 5 mL ddH$_2$O.
3. 5% TEMED: 5 µL TEMED and 95 µL ddH$_2$O. Prepare fresh each day.
4. 5% Ammonium persulfate: 5 mg ammonium persulfate and 95 µL ddH$_2$O. Prepare fresh each day.

5. 1 *M* Sodium acetate buffer, pH 5.0: 82 g Na acetate and 1 L ddH$_2$O, pH to 5.0 with a saturated sodium hydroxide solution.
6. NMR buffer for charged polyacrylamide gels: 8 *M* urea, 50 m*M* Na acetate buffer pH 5.0, and10% D$_2$O. The urea should be added dry and each gram occupies a volume of 764 µL.
7. New Era Gel stretcher (New Era Enterprises Inc., Vineland, NJ). Select either the 6.0-, 5.4-, or 5.0-mm depending on protein system. From the New Era enterprises catalog, the 6.0- to 4.2-mm gel "works well with proteins like protein G," the 5.4- to 4.2-mm stretcher is "generally good for strong aligning proteins," and the 5.0- to 4.2-mm stretcher is "generally good for proteins that are both large and elongated." For this reason, the 5.0- to 4.2-mm gel stretcher is recommended for denatured proteins.
8. NMR tubes: open ended for gel stretching, Shigemi tubes (Shigemi, Inc., Allison Park, PA) for bicelles and compressed gels.

2.2. Highly Charged Polyacrylamide Gel

1. 40% (w/v) (Acrylamidopropyl)trimethylammonium chloride (APTMAC) stock solution for charged gels with 1:20 bisacrylamide: 1.9 g APTMAC, 0.1 g bisacrylamide, and 5 mL H$_2$O.
2. 40% (w/v) 2-Acrylamido-2-methyl propane sulfonate (AMPS) stock solution for charged gels with 1:20 bisacrylamide: 1.9 g AMPS, 0.1 g bisacrylamide, and 5 mL H$_2$O.
3. 40% (w/v) Acrylamide stock solution for charged gels with 1:20 bisacrylamide: 1.9 g acrylamide, 0.1 g bisacrylamide, and 5 mL H$_2$O.
4. 10X TBE buffer for charged gels: 21.8 g Tris, 11.1 g borate, 1.5 g EDTA, and 200 mL H$_2$O (check that pH is 8.2).
5. Teflon cylinders: made from 1-in. diameter Teflon rods purchased from McMaster-Carr (Chicago, IL) and bored with metric drill bits also purchased from McMaster-Carr.

2.3. Alkyl-PEG-Bicelles

1. 1-Octanol, Sigma-Aldrich (St. Louis, MO) 99+%, HPLC grade.
2. Pentaethylene glycol monooctyl ether (C8E5), Sigma-Aldrich, 99.4% pure.

3. Methods
3.1. Neutral Polyacrylamide Gels

The simplest and least problematic alignment media for unfolded proteins is the uncharged polyacrylamide gel *(13,14)*, a highly porous matrix long known to be chemically inert and minimally interactive with proteins in solution. Small polyacrylamide cylinders are easy to prepare and handle, and porous enough to permit free diffusion of proteins within the gel *(14)*. To render the gel matrix spatially anisotropic and, thus, capable of aligning a macromolecule by steric interactions *(13,14)*, the gel can either be radially compressed (i.e., stretched) using

Table 1
Components for Polyacrylamide Gels of Varying Concentrations

	30% AA stock	Water	5% TEMED	5% APS
6% AA gel	200 µL	740 µL	30 µL	30 µL
8% AA gel	267 µL	673 µL	30 µL	30 µL
10% AA gel	333 µL	607 µL	30 µL	30 µL
12% AA gel	400 µL	540 µL	30 µL	30 µL

a special apparatus or axially compressed using standard NMR equipment. In either case, the cavities in the gel are distorted from spherical to ellipsoidal symmetry, causing the long axis of the macromolecule to acquire a small preference for alignment along the long axis of the cavities through steric exclusion.

1. Combine components from **Table 1** using the 1:20 bisacrylamide:acrylamide solution for compressed gels and the 1:40 bisacrylamide:acrylamide solution for stretched gels (*see* **Note 2**).
2. The gel is cast in a Teflon cylinder (*see* **Note 3**) placed on square of Parafilm® and pressed down to create a seal. To cast a stretched gel, after initiating polymerization, rapidly transfer 255 µL acrylamide solution to the Teflon cylinder using a glass pipet. Try to avoid creating air bubbles. To cast a compressed gel, use 800 µL of solution added to the bottom of a 3.9- to 4.2-mm cast through a long stem glass pipet, raising the pipet upwards as solution is expelled. Use a 4.540-mm inside diameter NMR tube for a compressed gel and a 4.2-mm open NMR tube for the gel stretcher.
3. The gel should fully polymerize within 10 min. If it does not freely slide out of the Teflon cast, gently push the gel out with a blunt glass or plastic rod.
4. Transfer the gel to a 50-mL polypropylene Falcon tube filled with ddH_2O. Wash 12 h in ddH_2O using very gentle agitation, exchanging water after every 4 h or so. For compressed gels, place washed gel cylinder on a square of Parafilm and measure its length. Trim with a razor blade to a final length of exactly 21 mm.
5. Soak gel in NMR buffer for 8 h, exchanging buffer every 4 h. The NMR buffer will be determined by the requirements to denature the protein and collect good NMR spectra (i.e., low salt). 8 M urea and pH 2.0, alone or in combination, will denature most proteins.
6. In a 1.5-mL microfuge tube, soak gel overnight in 300 µL NMR buffer with 1.0 mM protein. This concentration will be reduced by approx 50% as the gel equilibrates with surrounding solution. The microfuge tube should be secured horizontally on a rotary mixer and slowly rotated overnight.
7. Carefully remove excess protein solution from the microfuge tube with a plastic pipet. Place the gel on a square of Parafilm and measure its length.
8. Gels:
   a. Compressed gel: add a small amount of buffer (~25 µL) to a Shigemi NMR tube, slide the gel into the tube, and then push it to the bottom with the

accompanying glass plunger. Compress gel by gently pressing down on the plunger with the index finger until the gel expands to meet the wall of the NMR tube, at which point resistance to further compression becomes very noticeable. Apply cellophane tape vertically to the NMR tube to secure the plunger in place. Measure the length of the compressed gel and calculate the degree of compression, which will in part determine the magnitude of the observed RDCs.

b. Stretched gel: slide the gel into the cylinder provided by the New Era Gel Stretcher kit and position the gel flush with the funnel end. Screw on the funnel. Push the plug into the other end of the gel cylinder. Attach piston and place an open-ended NMR tube into the open end of the funnel. Slowly turn the piston to propel gel into the NMR tube. Continue to turn the piston until the gel is fully in the NMR tube. If the gel should stop short, remove the piston and gel cylinder from the funnel and push the gel the rest of the way through the funnel with a blunt glass or plastic rod. Advance the gel until it is flush with the bottom of the NMR tube and then insert the NMR plug into the bottom. Insert a glass plunger into the top of the NMR tube until it touches the gel, and secure with tape.

3.2. Highly Charged Polyacrylamide Gels

Strained polyacrylamide gels formed by using acrylamide congeners with cationic or anionic groups provide a second mechanism to align a protein. In addition to the same steric exclusion phenomenon at work in neutral gels, the high-charge density creates electric field gradients that interact with the electric dipole and quadrupole moments *(15)* of the denatured protein. Thus, charged gels may give rise to significantly different molecular alignments when compared with neutral gels or bicelles. For a rapid estimate of the alteration of the alignment tensor, the RDCs measured in two different media can be graphed in a scatterplot and the correlation coefficient between the data sets compared (*see* **Fig. 2**). A correlation coefficient of greater than 0.9 indicates that the two media are conferring very similar alignments, whereas a correlation coefficient below 0.65 indicates a significant rotation of the axes of alignment has occurred.

Because of extensive swelling of charged gels and the need to dry them down to a small size before use, charged gels are considerably more difficult to work with than neutral gels. Thus, in most circumstances, the performance of neutral polyacrylamide gels should be evaluated first, with charged gels employed if more data is later required.

1. Polyacrylamide gels:

a. Negatively charged polyacrylamide gels: in a 1.5-mL microfuge tube, mix 12.3 µL 40% acrylamide stock solution, 12.3 µL 40% AMPS stock solution, 28.3 µL 10X TBE buffer, 82.6 µL ddH$_2$O, 4.2 µL 5% APS, and 1.4 µL TEMED. With a glass pipet, transfer to a 3.3-mm diameter cast that has been placed on Parafilm (*see* **Note 2**).

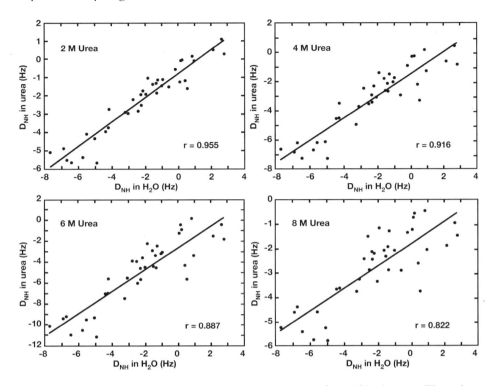

Fig. 2. Scatter plots of the N–H dipolar couplings D_{NH} for $\Delta131\Delta$ in urea. The values measured at the indicated urea concentration are plotted along the *y*-axis, whereas the values in H_2O are plotted along the *x*-axis. Gel cylinders were strained by compression.

    b. Positively charged polyacrylamide gels: in a 1.5-mL microfuge tube, mix 12.3 µL 40% acrylamide stock solution, 12.3 µL 40% APTMAC stock solution, 28.3 µL 10X TBE buffer, 82.6 µL ddH$_2$O, 4.2 µL 5% APS, and 1.4 µL TEMED. With a glass pipet, transfer to a 3.3-mm diameter cast that has been placed securely on Parafilm (*see* **Note 2**).

2. After 10 min, wash gels in a beaker filled with 500 mL ddH$_2$O. Replace water after 4 and 8 h, and continue overnight. Gels will undergo significant osmoelectric swelling (*see* **Note 4**).

3. Dry gels for 24 h at 30°C on Parafilm placed in the bottom of a plastic culture plate. APTMAC gels will develop a brownish tinge on drying (*see* **Note 5**).

4. Place gels in the bottom of a Shigemi NMR tube. Add 141 µL 1 m*M* protein NMR solution. Cover the top of the tube with Parafilm, leave at a slight angle, and equilibrate overnight.

5. Once gel has taken up the solution, gently compress gel with plunger until sides of gel are flush against the walls of the NMR tube with no bubbles. Addition of a small amount of buffer solution may facilitate the compression.

3.3. Alkyl-PEG Bicelles

A third medium demonstrated to align denatured proteins by the steric exclusion mechanism is a lipid liquid crystalline or bicelle phase formed by alkyl-polyethylene glycol (PEG) molecules *(16)*. This medium is easy to prepare, works well in 8 *M* urea, and is very efficient at aligning asymmetrically shaped molecules. However, because this medium contains hydrophobic lipids, binding between the denatured protein and the lipid discs may occur, manifest as line broadening if the interaction is weak and transient, or as phase separation if the interaction is strong and irreversible.

1. Combine 150 µL C8E5 with 550 µL ddH$_2$O and 300 µL D20 in a microfuge tube to make a 15% bicelle stock solution. Vortex vigorously for 1 min. Solution should be clear and slightly foamy.
2. Titrate in the 1-octanol by a series of microliter additions that decrease in volume. The final volume of octanol will be around 35 µL or a molar ratio between C8E5 and 1-octanol of 0.87 *(16)*. After each addition, vortex the solution vigorously for 1 min. The solution will become increasingly viscous until finally it forms a transparent and somewhat viscous phase. Watch carefully for this end point in the titration, as it is the indicator of the liquid crystalline phase (bicelles). The solution will have a faint blue opalescent tinge in bright, scattered light, indicating the formation of large particles.
3. Prepare a NMR sample by adding 100 µL bicelles directly to solid urea (amount depends on desired final molarity for denaturation). Add protein and buffer to attain a final concentration of 1 and 10 m*M*, respectively, in a volume of 300 µL to make a 5% bicelle sample (*see* **Note 6**).

4. Notes

1. The degree of anisotropy of the alignment media can sometimes be quantified by measuring the splitting of the HDO line *(7)*. For alkyl-PEG bicelles, a roughly linear relationship is observed, with 3, 5, and 7% bicelles giving values of 17.0, 30.8, and 45.3 Hz, respectively *(16)*. With strained neutral or charged gels, this splitting may or may not be observed, depending in part on the concentration of acrylamide used.
2. Decisions about gel percentage must be made on a case-by-case basis. Another issue to consider is the final concentration of urea. Urea will affect the degree to which a gel is capable of compression. For a given urea concentration, determine how much compression is necessary to obtain the desired RDC magnitude and what percentage gel provides this environment. The smaller the protein, the higher the gel percentage required to align the protein. For example, peptides may require an 18% gel, whereas a large denatured protein may only require an 8% gel to achieve the same degree of alignment.
3. Decisions about gel diameter should take into account (1) whether the gel is to be stretched or compressed, (2) the concentration of urea to be used, which will determine how much the gel will swell during equilibration, and (3) how a particular

denatured protein will align. Proteins that are more difficult to align will require a greater degree of compression or stretching to obtain a reasonable range of RDC values.

4. Gels are fragile, especially charged gels after they have been swollen in water. Do not touch them. Transfer them with the utmost concern for their structural integrity, using Teflon-coated spatulas and very gentle pushing. Gel manipulation should be practiced using the real conditions (but without protein) before a protein-containing sample is prepared.

5. Once dried, gels can be stored in 1.5-mL microfuge tubes for up to 2 mo at 4°C.

6. Bicelle solutions may become cloudy during NMR sample preparation, as octanol evaporates or is selectively adsorbed to glass and plastic surfaces. Addition of 1 μL of octanol to the NMR sample usually restores a clear solution of bicelles. However, the addition of too much octanol will also make the solution cloudy.

References

1. Kohn, J. E., Millet, I. S., Jacob, J., et al. (2004) Random-coil behavior and the dimensions of chemically unfolded proteins. *Proc. Natl. Acad. Sci. USA* **101,** 12,491–12,496.

2. Dyson, H. J. and Wright, P. E. (2004) Unfolded proteins and protein folding studied by NMR. *Chem. Rev.* **104,** 3607–3622.

3. Ackerman, M. S. and Shortle, D. (2001) Persistence of native-like topology in a denatured protein in 8 M urea. *Science* **293,** 487–489.

4. Ackerman, M. S. and Shortle, D. (2002) Robustness of the long-range structure in denatured Staphylococcal Nuclease to changes in amino acid sequence. *Biochemistry* **41,** 13,791–13,797.

5. Ohnishi, S. and Shortle, D. (2003) Effects of denaturants and substitutions of hydrophobic residues on backbone dynamics of denatured staphylococcal nuclease. *Protein Sci.* **12,** 1530–1537.

6. Ohnishi, S., Lee, A. L., Edgell, M. H., and Shortle, D. (2004) Direct demonstration of structural similarity between native and denatured Eglin C. *Biochemistry* **43,** 4064–4070.

7. Ding, K., Louis, J. M., and Gronenborn, A. M. (2004) Insights into conformation and dynamics of protein GB1 during folding and unfolding by NMR. *J. Mol. Biol.* **335,** 1299–1307.

8. Fieber, W., Kristjansdottir, S., and Poulsen, F. M. (2004) Short-range, long-range and transition state interactions in the denatured state of ACBP from residual dipolar couplings. *J. Mol. Biol.* **339,** 1191–1199.

9. Tolman, J. R. (2002) A novel approach to the retrieval of structural and dynamic information from residual dipolar couplings using several oriented media in biomolecular NMR spectroscopy. *J. Am. Chem. Soc.* **124,** 12,020–12,030.

10. de Alba, E. and Tjandra, N. (2004) Residual dipolar couplings in protein structure determination. In: *Protein NMR Techniques*, (Downing, A.K., ed.), Humana Press, Totowa, NJ, **278,** pp. 89–106.

11. Prestegard, J. H., Bougault, C. M., and Kishore, A. I. (2004) Residual dipolar couplings in structure determination of biomolecules. *Chem. Rev.* **104,** 3519–3540.

12. Lipsitz, R. S. and Tjandra, N. (2004) Residual dipolar couplings in NMR structure analysis. *Annu. Rev. Biophys. Biomol. Struct.* **33,** 387–413.
13. Tycko, F. J. and Blanco, Y. I. (2000) Alignment of biopolymers in strained gels: A new way to create detectable dipole-dipole couplings in high-resolution biomolecular NMR. *J. Am. Chem. Soc.* **122,** 9340.
14. Sass, H. J., Musco, G., Stahl, S. J., Wingfield, P. T., and Grzesiek, S. (2000) Solution NMR of proteins within polyacrylamide gels: diffusional properties and residual alignment by mechanical stress or embedding of oriented purple membranes. *J. Biomol. NMR* **18,** 303–309.
15. Ulmer, T. S., Ramirez, B. E., Delaglio, F., and Bax, A. (2003) Evaluation of backbone proton positions and dynamics in a small protein by liquid crystal NMR spectroscopy. *J. Am. Chem. Soc.* **125,** 9179–9191.
16. Ackerman, M. S. and Shortle, D. (2002) Molecular alignment of denatured states of Staphylococcal Nuclease with strained polyacrylamide gels and surfactant liquid crystalline phases. *Biochemistry* **41,** 3089–3095.

4

Characterizing Residual Structure in Disordered Protein States Using Nuclear Magnetic Resonance

David Eliezer

Summary

The importance of disordered protein states in biology is gaining recognition, and can be attributed in part to the participation of unfolded and partially folded states of globular proteins in normal and abnormal biological functions, such as protein translation, protein translocation, protein degradation, protein assembly, and protein aggregation *(1–5)*. There is also a growing awareness that a significant fraction of gene products from various genomes, including the human genome, fall into a category that includes low complexity, low globularity, or intrinsically unstructured proteins *(6–9)*. Unlike native states of globular proteins, disordered protein states, by definition, do not adopt a fixed structure that can be determined using classical high-resolution methods. Nevertheless, there has long been evidence that many disordered states contain detectable and significant residual or nascent structure *(10–16)*. This structure has been found to be important for nucleating local structure, as well as mediating long range contacts upon either intramolecular folding to the native state *(17–21)* or intermolecular folding with specific binding partners *(22–24)*, and is also predicted to influence intermolecular folding into structured aggregates *(25,26)*. The primary tool for the characterization of such structure is high-resolution solution state nuclear magnetic resonance (NMR) spectroscopy. Advances in NMR instrumentation and methods have greatly facilitated this task and in principle can now be accomplished by those without extensive prior experience in NMR spectroscopy. This chapter describes how this can be accomplished.

Key Words: Protein folding; residual structure; nascent structure; intrinsically unstructured; natively unfolded; unfolded state; denatured state; molten globule; folding intermediates; NMR; secondary shifts; random coil.

1. Introduction

Detailed structural information for the end point of typical protein folding reactions has been available for many decades. Only recently, however, has

From: *Methods in Molecular Biology, vol. 350: Protein Folding Protocols*
Edited by: Y. Bai and R. Nussinov © Humana Press Inc., Totowa, NJ

it become practical to address the structural properties of the ensemble of unfolded states from which folding initiates. Optical methods, such as circular dichroism, fluorescence, and Fourier-transform infrared spectroscopies of unfolded proteins frequently contain evidence for a small, but detectable amount of structure *(10)*, but further characterization of such structure was not often pursued. More recently, this has begun to change.

With the advent of multidimensional solution nuclear magnetic resonance (NMR) spectroscopy it became possible to resolve signals from individual sites in proteins, even in their denatured state *(12)*, and, therefore, to learn about the local environment and structure of these sites. There are two primary requirements for achieving this goal. First, the individual NMR resonances have to be mapped to the corresponding nuclei in the protein from which they originate. This is referred to as the resonance assignment process, and remained the primary hurdle for NMR studies of disordered proteins for some time. Recent advances in NMR methods combined with increased access to high-field and ultra-high-field NMR spectrometers, however, have provided the tools needed to make this a routine task, at least for backbone resonances, and the process will be described in detail in **Subheading 4**. Once backbone resonance assignments are available, measurable parameters that inform regarding local environment and structure are required. Many such parameters are useful in the high resolution structural characterization of well-folded proteins *(27)*, including chemical shifts, coupling constants, nuclear Overhauser effects (NOEs), relaxation rate constants *(28–31)*, and more recently residual dipolar couplings *(32,33)*. These same parameters can also be used to analyze the structural properties of disordered proteins, but the weak signatures of elements of residual structure require somewhat different considerations.

2. Production of Isotopically Labeled Protein

Like classical NMR structure determination, analyzing residual structure in proteins by NMR is greatly facilitated by the use of isotopically labeled protein, in which ^{14}N and ^{12}C are replaced with ^{15}N and ^{13}C. A requirement for this approach is the ability to produce the protein of interest recombinantly. Although isotopic labeling of proteins produced in yeast and in cell culture is slowly becoming feasible *(34,35)*, these systems are not described here as the great majority of proteins for NMR study are produced in *Escherichia coli*. Once recombinant protein can be expressed and purified from *E. coli*, producing isotopically labeled material is usually a simple exercise. To prevent incorporation of unlabeled isotopes into the protein, bacteria are grown on a minimal medium completely lacking in carbon or nitrogen, to which a single source of these elements is then added, typically in the form of ammonium-sulfate or -chloride for nitrogen and glucose for carbon. Usually, minimal media is made using

Table 1
Preparation of Isotopically Labeled Minimal Media

To make 1 L of 5X M9 salts
34 g of Na_2HPO_4 (or 36 g of $Na_2HPO_4[H_2O]$ or 64 g of $Na_2HPO_4[7H_2O]$)
15 g of KH_2PO_4
2.5 g NaCl
Bring up volume to 1 L with dH_2O and sterilize by autoclaving

To make 1 L of 15N- or 15N,13C-labeled minimal media
200 mL of 5X M9 salts
700 mL of sterile dH_2O
100 µL of sterile 1 M $CaCl_2$ (add only after other salts are completely dissolved!)
1 mL of sterile 1 M $MgSO_4$ (add only after other salts are completely dissolved!)
Dissolve and sterile filter into above:
1.0 g of labeled 15N-labeled ammonium chloride
4.0 g of unlabeled glucose (or 2.0–4.0 g of 13C-labeled glucose)
10 mL of 100X Basal Eagle vitamin mix
30 mL of dH_2O
Adjust to 1 L with dH_2O

---

an M9-recipe similar to the one shown in **Table 1** *(36)*. Protein yield is typically as much as 50% lower from cultures grown on minimal media, especially when limiting quantities of ^{13}C-labeled glucose are used.

Because disordered proteins are generally quite flexible, the relaxation rates that determine the efficiency of magnetization transfer during the triple-resonance experiments described below are quite favorable. This means that the benefits obtained from triple labeling with ^2H, ^{13}C, ^{15}N are not typically needed. If required, highly deuterated samples can be made by using D_2O instead of H_2O in preparing the M9 media, and uniformly deuterated samples can be produced by using ^2H, ^{13}C-labeled glucose as well as D_2O. Growing bacterial cultures in D_2O-based media leads to a significant (approximately twofold typically) reduction in growth rate, and often leads to a further decrease in protein yield. In order to reduce the amount of D_2O and isotopically labeled compounds needed for the production of labeled proteins, it is possible to grow cultures to midlog phase in rich media, spin down the cells, wash them, and then resuspend them in minimal media. The cells should then be given 30–60 min to recover before protein production is induced. Nearly uniform labeling can be achieved in this way with lower costs *(37)*.

3. NMR Sample Preparation

NMR experiments typically require approx 0.5 mL of $^{13}C,^{15}N$ double-labeled protein at a concentration of approx 0.3 mM. Higher concentrations can provide better signal to noise but should be used with caution, as disordered proteins often aggregate easily. NMR two-dimensional (2D) proton–nitrogen correlation (HSQC) spectra can be used as a diagnostic for aggregation in tandem with other techniques, such as gel filtration or light scattering. The use of NMR tubes with susceptibility-matched plugs (available from Shigemi, Inc. for example) allows sample volumes of 0.25–0.3 mL to be used if sample quantity is limiting. Because the backbone amide protons in disordered proteins are typically not protected from the solvent by either intramolecular hydrogen bonds or burial, samples should be prepared at lower pH if possible to limit the exchange of amide protons with solvent protons *(38,39)*. A pH of 4.0 is ideal for slowing exchange, but is often too low to maintain physiological relevance and can perturb the structure of interest. pH values of up to 7.5 or so can be used successfully by lowering the sample temperature to 10°C or even 5°C. Salt concentrations should be such that ionic strength remains below approx 0.2 mM. If cryogenic probe technology is to be used, sample concentrations can be reduced to approx 0.1 mM in favorable cases, although lower salt concentration and low-conductivity buffers may be necessary to achieve the necessary gain in sensitivity *(40)*.

4. Obtaining NMR Backbone Resonance Assignments for Disordered Proteins

In comparison to well-structured proteins, disordered proteins give rise to crowded, highly degenerate NMR spectra, increasing the challenge of assigning the individual resonances. Nevertheless, certain nuclei in polypeptide backbones retain significant dispersion that can be taken advantage of. In particular, the resonance frequencies or chemical shifts of both backbone ^{15}N nuclei *(41)* and backbone ^{13}CO nuclei *(42,43)* are highly sequence dependent, even in the absence of well-ordered structure. In contrast to the ^{15}N and ^{13}CO sites, the chemical shifts of the ^{13}Cα and ^{13}Cβ nuclei in the absence of well-ordered structure are quite degenerate. This, however, can also be made use of, because the combination of these two chemical shifts is highly effective at discriminating between different amino acid side chains *(44)*, and is, therefore, of great use in the assignment process *(45)*. With these considerations in mind, the following set of three-dimensional (3D) triple-resonance experiments can be considered as ideal for obtaining the data necessary for backbone resonance assignments of disordered proteins:

1. HN(CA)CO (correlates NH$_i$, N$_i$ and CO$_{i/i-1}$).
2. HNCO (correlates NH$_i$, N$_i$, and CO$_{i-1}$).
3. HNCACB (correlates NH$_i$, N$_i$, and Cα,Cβ$_{i/i-1}$).
4. CBCA(CO)NH or the equivalent HN(CO)CACB (correlates NH$_i$, N$_i$, Cα,Cβ$_{i-1}$).

These experiments are available in standard form in the pulse sequence libraries provided by the major manufacturers of NMR spectrometers, and can,

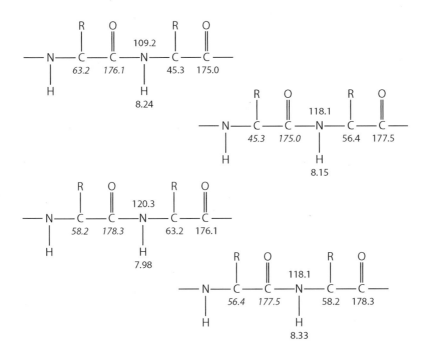

Fig. 1. Illustration of dipeptide spin systems constructed using backbone HN, N, Cα, and CO resonances for a hypothetical cyclical four-residue peptide. Inter-residue (i–1) chemical shift correlations are shown in italics.

therefore, be executed in a straightforward fashion. Although a minimal familiarity with the details of operating a NMR spectrometer is required, sufficient training can be obtained in 1 d, or even a few hours, from a knowledgeable facility manager or user. Processing of the resulting data to produce the 3D spectra is also straightforward, with a variety of software available for the task.

Each of these four experiments results in spectra that contains peaks, or correlations, that are each characterized by three resonance frequencies or chemical shifts: that of a backbone amide proton, of its covalently attached nitrogen, and of the CO, Cα, or Cβ nucleus that belongs to either the same residue (an "i" correlation) or the one immediately preceding it in the protein sequence (an "i–1" correlation). All together, the experiments provide the complete backbone and Cβ chemical shifts for each dipeptide spin system that gives rise to detectable signals. The data can be illustrated schematically for the HN, N, Cα, and CO shifts of a cyclical four-residue peptide as shown in **Fig. 1**.

The process of making sequence-specific resonance assignments involves linking individual dipeptide systems to each other by matching the overlapping chemical shifts, in this case the CO, Cα, and Cβ shifts. Because the Cα and Cβ

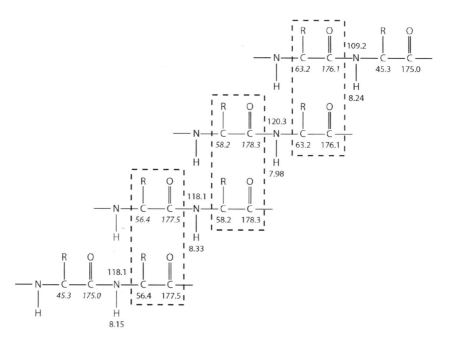

Fig. 2. Illustration of resonance assignment process relying on matching resonance frequencies between overlapping regions of different dipeptide spin systems.

shifts are highly degenerate, as previously discussed, they are less useful for linking dipeptides than the better dispersed CO shifts. However, the Cα and Cβ shifts can often be used to identify or constrain the specific side chains of each residue in the dipeptide spin system, and this information can be invaluable in placing the dipeptide, and any segments linked to it, into the known sequence of the protein. The process of linking the dipeptide spin systems is illustrated in **Fig. 2**. Once this basic concept is understood, various techniques can be used to accomplish the task at hand. One approach is to simply select an inter-residue (i–1) peak in one of the spectra and graphically search all other peaks in the appropriate spectrum for the corresponding intraresidue (i) peak. At the other end of the spectrum, one can generate a numerical table of all the correlations, identify the chemical shift of an inter-residue correlation, and search the numerical table for possible matches. Different people typically develop their own preferred approach. Note that although software packages for automated assignments of NMR backbone resonances of proteins (AutoAssign *[46]*, MARS *[47]*, and others) can be highly effective when applied to well-structured proteins, they are typically less successful with spectra from disordered proteins and significant manual effort is usually required.

5. Identifying Residual Structure
5.1. Chemical Shifts

Once backbone resonance assignments have been obtained, it is possible to begin to learn about the structural preferences of specific residues in the disordered protein of interest. The easiest and perhaps most sensitive parameters to analyze are chemical shifts, many of which are conveniently available immediately from the assignment process. Chemical shifts have proven to be sensitive and accurate predictors of secondary structure *(48–54)* and this property has been used to develop a robust algorithm to identify secondary structure in well-folded proteins *(55,56)*. Although this algorithm can be applied to several nuclei, its application to Cα shifts will be described here because they have proven to be the most sensitive and accurate predictors of secondary structure, being least affected by other factors *(57)*. The algorithm relies on calculating the differences between each observed chemical shift and the values that would be expected if a given residue were in an idealized random-coil state. The latter values have been estimated either from statistical analyses of chemical shift databases from well-structured proteins *(49,54,58)*, or from direct measurements of chemical shifts from residues in short peptides that contain little or no structural preferences *(53,59–62)*. The differences between the observed and tabulated random-coil shifts are referred to as secondary shifts. Once calculated, secondary shifts are used to predict the presence of helical or strand structure based on their sign, amplitude, and contiguity. For Cα secondary shifts, positive values correlate with helical structure and negative values with strand structure, and the Cα amplitude thresholds used to analyze structure in well-folded proteins are typically 0.5 PPM *(63)*. Uninterrupted stretches of positive or negative secondary shifts above the amplitude threshold indicate helical or strand structure, respectively. The contiguity criterion requires uninterrupted stretches of at least four nonstrand-indicating secondary shifts to establish a helix, and at least three nonhelix-indicating values to establish a strand, and in either case, the "density" of helix- or strand-indicating values should exceed 70%.

Although highly successful in identifying and delineating secondary structure elements in well-folded proteins, the previously mentioned algorithm is more difficult to apply to highly disordered proteins because the secondary structure elements are not well formed, being instead transient in nature, and the associated secondary shifts are of much lower amplitude. Nevertheless, Cα secondary shifts in particular still provide highly accurate information if care is taken. When dealing with residual or nascent structure, it is of crucial importance to use the "best" possible random-coil shifts. In contrast to examining well-formed structure, where secondary shifts are relatively large in magnitude (several PPM for Cα), residual structure leads to secondary shifts that are typically quite small (less than 1 PPM and often just a few tenths of a PPM for Cα), so errors in the random-coil shifts can have a profound effect on the results. In general, our experience has indicated that statistically derived random-coil shifts are not as accurate for analyzing poorly structured states as experimentally measured random-coil values. In contrast, the statistical values may be somewhat superior in the prediction of well-formed structure. Of the experimentally determined data sets, the most recent tabulation by Wishart et al. *(62)*, which was obtained at pH 5.0 and 25°C, is in our experience the most accurate for analyzing samples at near physiological pH values, while the data of Schwarzinger et al. *(64)* is most appropriate for proteins denatured in urea at low pH.

The latter data set was later accompanied by a detailed analysis of sequence-specific corrections to the residue-specific, random-coil shifts *(65)*. Although this analysis demonstrates clearly that these corrections are quite small for the Cα shifts, the corrections do improve the analysis of residual structure *(66)*. However, it should be noted that the corrections may be specific to the conditions used for the measurements, and it is not yet clear if they can be directly applied to residue-specific, random-coil values determined under other conditions. Sequence-specific corrections have also been carefully tabulated by Wang and Jardetzky using a statistical approach *(67)*, but these corrections too, in our experience, cannot be taken out of context and applied to experimentally determined random-coil values. Ultimately, the best possible results can be achieved only by directly measuring random-coil shifts from small peptides, including sequence-specific corrections, under conditions close to those of interest (but note that small peptides can adopt nonrandom structure *[66]* and such structure must be destabilized). In the absence of such pain-staking measurements, the more appropriate of the Wishart *(62)* or Schwarzinger *(64,65)* data sets should be used, but pH differences leading to the titration of side chain groups (Glu, Asp, or His) require corrections for their residues *(68)*. An additional note is that accurate determination of experimental carbon chemical shifts is facilitated by extensive zero filling of data in the carbon dimension. Although the intrinsic resolution of the data is not improved by this procedure, the effective increase in the quality of the interpolation is important for determining the positions of resonances.

Once secondary shifts are calculated using carefully determined Cα chemical shifts and the appropriate random-coil shifts, identifying residual secondary structure follows the same principles used for well-folded proteins. However, the smaller secondary shifts introduce a greater degree of ambiguity into the procedure. In particular, the amplitude threshold must be relaxed. However, the contiguity requirement should typically be increased to compensate. Longer stretches of secondary shifts of the same sign, but where several or even many fall below the amplitude threshold, can be reliable indicators of residual structure. Short (three to four residue) stretches of residual structure are more difficult to diagnose with certainty unless the secondary shifts are all above the threshold value. It should be noted that the availability of any complimentary functional, biochemical, or structural data for the protein of interest can greatly increase the confidence of the chemical shift analysis. This is based on the observation that residual structure is often most evident in regions that are well structured in other contexts (in the native state if one exists, or in a well-defined conformation adopted upon binding to an interaction partner), and that such regions therefore also tend to stand out in biochemical or functional studies employing site-specific or deletion mutagenesis *(24,69)*.

Although Cα secondary shifts are the most accurate indicators of secondary structure, other chemical shifts can also be helpful. Hα secondary shifts in particular are excellent indicators of secondary structure in well-folded proteins because of their great sensitivity to backbone φ/ψ angles. However, Hα shifts are also more sensitive than carbon shifts to other factors, such as ring currents, hydrogen bonding, and electrostatic effects *(57)*, which effectively make them "noisier" than Cα shifts when dealing with small amplitude deviations. Nevertheless, Hα shifts are a useful complement to Cα shifts. Similarly, CO and Cβ shifts are also sensitive to secondary structure. While CO shifts show excellent dispersion in well-folded states, much of this dispersion originates from sequence-dependence effects and is, therefore, retained in disordered states. These effects are difficult to correct for, making CO secondary shifts relatively noisy *(43)*. Cβ shifts are somewhat sensitive to strand structure, but are much less sensitive to helical structure *(57)*. They also display considerably less dispersion even in well-folded structures, providing less dynamic range for the detection of residual structure *(43)*. When residual structure is indicated by a consensus of secondary shifts from two or more nuclei, this again greatly increases the reliability of the analysis.

5.2. Nuclear Overhauser Effects

NOEs form the basis of classical structure determination by NMR by providing long-range constraints on global topology, and shorter-range constraints on local structure, including secondary structure *(27)*. Well-characterized patterns of NOEs have long been used to delineate secondary structure in proteins *(27)*. In disordered proteins, this type of analysis is not possible because of the fluctuating nature of any structural elements. Nevertheless, short-range NOE data can help to distinguish between helical and extended or strand structure, and can help to corroborate observations made based on chemical shifts as previously described. $H_N H_\alpha$(i,i-1) NOEs in particular are useful, because they are typically observable in highly unstructured proteins even at relatively low concentrations using standard ^{15}N-separated NOESY experiments. The distance reflected in this NOE is informative because it is considerably shorter in strand structure (~2.2 Å) than in helical structure (~3.5 Å) *(27)*. Fluctuations in the intensity of this NOE signal can, therefore, support chemical shift-based inferences regarding the existence and nature of residual or nascent structure. Sequential $H_N H_N$ NOEs can also be used in a similar fashion (this distance being longer in strand structure and shorter in helical structure), but these NOEs are typically weaker in poorly structured proteins and require much higher protein concentrations. $H_N H_N$ NOEs for disordered proteins are most easily measured using the HSQC-NOESY-HSQC experiment *(42,70)*.

A complication that arises when interpreting NOE data is that various relaxation processes exert a significant effect on NOE intensities. Therefore, it can be difficult to distinguish whether fluctuations in raw NOE intensities arise from differences in structure or dynamics. One method for partly compensating for this ambiguity is to normalize NOEs by the intensities of either diagonal peaks or other NOE peaks that originate nearby, with the rational being that the effects of local motions should, at least to a first approximation, be normalized out *(71)*. Specifically, $H_N H_\alpha(i,i\text{-}1)$ NOEs are sometimes normalized by the intraresidue $H_N H_\alpha(i,i)$ NOE to produce the so called $\sigma_{N\alpha}/\sigma_{\alpha N}$ ratio *(72)*, whereas sequential $H_N H_N$ NOEs are normalized by the diagonal (or self-NOE) signal to produce the σ_{NN} ratio *(73,74)*. This concern is most relevant when there is enough residual structure present to lead to significant variations in the backbone dynamics of the protein. For more highly unstructured proteins, the dynamics are often quite uniform, and the unnormalized NOE intensities can be informative *(69)*. It should also be noted that short-range NOEs can also be used to detect the presence of polyproline-II structure *(60)*, and that hydrophobic clusters can also be detected using short- and medium-range NOEs *(12)*.

5.3. Dynamics

Molecular motions profoundly influence the relaxation of NMR signals, and can, therefore, be characterized by measuring NMR relaxation rate constants *(75,76)*. Well-structured proteins typically exhibit rather uniform relaxation parameters that primarily reflect overall molecular tumbling. Intramolecular motions, however, can often be detected as departures from this uniform behavior, and can be highly informative regarding functionally important motions, such allosteric transitions *(77)* or enzyme active site rearrangements *(78,79)*. In disordered proteins, relaxation rates are strongly influenced by local motions, and can provide an additional mechanism for detecting regions of residual structure, especially the presence of hydrophobic clusters. In completely unstructured polypeptides, transverse relaxation rates (R2) data are relatively uniform, but when hydrophobic residues participate in either local- or long-range contacts, this tends to restrict the motions of the protein backbone in a manner that leads to exchange broadening of NMR signals *(80)*. This effect can often be detected as local regions, or clusters, of unusually high R2 values. The exchange contribution (Rex) to R2 can also be measured and analyzed directly. Rex is typically determined using CPMG-based relaxation dispersion methods at multiple field strengths *(81)*. Correlations between high R2 values or significant Rex values and a high degree of local hydrophobicity (which can be evaluated using Kyte-Doolittle *[82]* or similar algorithms) are good indicators of hydrophobic clusters *(19)*. If several hydrophobic clusters are detected, the possibility of intercluster contacts can be probed by studying the effects of perturbing mutations on the relaxation rate constants of residues belonging to distant clusters *(83)*. Such contacts can also be detected directly by measuring side chain NOEs in samples selectively protonated at methyl and aromatic sites *(84)*. Residual secondary structure, if present, may also be reflected in backbone relaxation parameters *(19,85)*.

5.4. Residual Dipolar Constants

Residual dipolar constants (RDCs) are a recent addition to the NMR toolbox used for structure determination *(32,33)*. RDCs provide a measure of the dipolar interaction between two (nuclear) magnetic moments, which is related both to the distance between them and to the orientation of the vector connecting them relative to the external magnetic field. In solution NMR, the dipolar couplings are averaged to zero by the isotropic tumbling of the solute molecules. If the symmetry of this molecular tumbling is broken, the dipolar couplings are reintroduced to an extent proportional to alignment of the solute along the preferred direction, and can be observed by NMR. The alignment must be weak enough that the tumbling time of the molecules is not unduly decreased (to prevent line broadening) but strong enough to produce measurable RDCs. A number of different aligning media have been described *(86)*, and many of them appear to work well for disordered proteins.

The geometrical dependence of RDCs is the basis for their use in the structure determination of well-ordered proteins. Measurements of RDCs, however, are also affected by molecular motions, and although this dependence can in principle be used to infer dynamics from RDCs, this is a challenging and ongoing effort *(87,88)*. Because of their highly dynamic nature, RDCs measured from poorly structured proteins are difficult to interpret. Nevertheless, recent work has shown that such measurements can be highly informative. In some cases, it appears that RDCs of non-native states can reflect long-range order and topology *(89)*, which can be similar to that present in the relevant native state *(90)*. When this is not the case, however, RDCs can be directly related to residual structure. In disordered systems, local elements of secondary structure appear to behave and align as semi-independent structural units *(91,92)*. Because NH bond vectors in helical and strand structure are approximately perpendicular to each other, the NH RDCs resulting from independently aligning helices and strands have opposite signs. Thus, NH RDCs can report on residual secondary structure *(92,93)*. In addition, regions of increased flexibility are expected to be less aligned on average and to exhibit below average RDC amplitudes. Such regions have been interpreted as molecular hinge regions *(92)*.

To date, RDC-based analyses of residual structure have relied primarily on NH RDCs. The most straightforward method for measuring these RDCs is the IPAP-HSQC experiment *(94)*. However, because 2D spectra of poorly structured proteins can be highly overlapped, the 3D HNCO-IPAP experiment *(95)* should also be considered.

5.5. Other Parameters

In addition to chemical shifts, NOEs, relaxation rate constants, and RDCs, additional NMR parameters can be useful in characterizing residual structure. Scalar coupling constants, for example, can be related to dihedral angles in both backbone and side chain groups *(27)*, and can be diagnostic of secondary structure and side chain conformations. Random-coil values of the ϕ-related vicinal coupling constant $^3JH_NH_\alpha$ and for the χ_1-related $^3JH_\alpha H_\beta$ have been reported *(96,97)*. Temperature coefficients of NH chemical shifts have been used to detect hydrogen-bonding interactions in disordered proteins *(98–100)*, although conformational transitions and other factors can complicate this approach for disordered proteins *(101)*. Measurements of the exchange rates of amide protons with solvent protons can be highly informative regarding hydrogen-bonded or solvent-excluding structure in partially folded proteins *(15,16)*, but are more difficult to obtain and interpret for more highly disordered states *(102,103)*. Although these additional observables can provide useful and unique information in certain cases and for specific sites, they typically offer somewhat more limited insights than the parameters previously described, and are, therefore, not described in further detail in this chapter.

6. Further Reading

NMR studies of disordered, non-native states of proteins have been reported for several decades, and a number of excellent reviews of the results and methodologies exist *(104–108)*. This chapter is intended both to provide an update, and to provide an avenue for nonspecialists to enter into this area of structural biology as it draws more interest and produces results with increasingly important biological implications. Although references to many of the primary sources relevant to this topic have been included, many important papers will unavoidably have been left out, and readers are encouraged to consult these reviews for further reading and additional perspectives.

References

1. Dobson, C. M. (2003) Protein folding and misfolding. *Nature* **426,** 884–890.
2. White, S. H. and von Heijne, G. (2004) The machinery of membrane protein assembly. *Curr. Opin. Struct. Biol.* **14,** 397–404.
3. Goldberg, A. L. (2003) Protein degradation and protection against misfolded or damaged proteins. *Nature* **426,** 895–899.
4. Frydman, J. (2001) Folding of newly translated proteins in vivo: the role of molecular chaperones. *Annu. Rev. Biochem.* **70,** 603–647.
5. Namba, K. (2001) Roles of partly unfolded conformations in macromolecular self-assembly. *Genes Cells* **6,** 1–12.
6. Dyson, H. J. and Wright, P. E. (2005) Intrinsically unstructured proteins and their functions. *Nat. Rev. Mol. Cell Biol.* **6,** 197–208.
7. Dunker, A. K., Obradovic, Z., Romero, P., Garner, E. C., and Brown, C. J. (2000) Intrinsic protein disorder in complete genomes. *Genome Inform.* **11,** 161–171.
8. Romero, P., Obradovic, Z., Li, X., Garner, E. C., Brown, C. J., and Dunker, A. K. (2001) Sequence complexity of disordered protein. *Proteins* **42,** 38–48.
9. Liu, J., Tan, H., and Rost, B. (2002) Loopy proteins appear conserved in evolution. *J. Mol. Biol.* **322,** 53–64.
10. Aune, K. C., Salahuddin, A., Zarlengo, M. H., and Tanford, C. (1967) Evidence for residual structure in acid- and heat-denatured proteins. *J. Biol. Chem.* **242,** 4486–4489.
11. Matthews, C. R. and Westmoreland, D. G. (1975) Nuclear magnetic resonance studies of residual structure in thermally unfolded ribonuclease A. *Biochemistry* **14,** 4532–4538.
12. Neri, D., Billeter, M., Wider, G., and Wuthrich, K. (1992) NMR determination of residual structure in a urea-denatured protein, the 434-repressor. *Science* **257,** 1559–1563.
13. Dobson, C. M., Evans, P. A., and Williamson, K. L. (1984) Proton NMR studies of denatured lysozyme. *FEBS Lett.* **168,** 331–334.
14. Shortle, D. and Meeker, A. K. (1989) Residual structure in large fragments of staphylococcal nuclease: effects of amino acid substitutions. *Biochemistry* **28,** 936–944.
15. Hughson, F. M., Wright, P. E., and Baldwin, R. L. (1990) Structural characterization of a partly folded apomyoglobin intermediate. *Science* **249,** 1544–1548.
16. Jeng, M. F., Englander, S. W., Elove, G. A., Wand, A. J., and Roder, H. (1990) Structural description of acid-denatured cytochrome c by hydrogen exchange and 2D NMR. *Biochemistry* **29,** 10,433–10,437.
17. Arcus, V. L., Vuilleumier, S., Freund, S. M., Bycroft, M., and Fersht, A. R. (1995) A comparison of the pH, urea, and temperature-denatured states of barnase by heteronuclear NMR: implications for the initiation of protein folding. *J. Mol. Biol.* **254,** 305–321.
18. Zhang, O. and Forman-Kay, J. D. (1995) Structural characterization of folded and unfolded states of an SH3 domain in equilibrium in aqueous buffer. *Biochemistry* **34,** 6784–6794.

19. Eliezer, D., Yao, J., Dyson, H. J., and Wright, P. E. (1998) Structural and dynamic characterization of partially folded states of apomyoglobin and implications for protein folding. *Nat. Struct. Biol.* **5,** 148–155.

20. Yi, Q., Scalley-Kim, M. L., Alm, E. J., and Baker, D. (2000) NMR characterization of residual structure in the denatured state of protein L. *J. Mol. Biol.* **299,** 1341–1351.

21. Bai, Y., Chung, J., Dyson, H. J., and Wright, P. E. (2001) Structural and dynamic characterization of an unfolded state of poplar apo-plastocyanin formed under nondenaturing conditions. *Protein Sci.* **10,** 1056–1066.

22. Radhakrishnan, I., Perez-Alvarado, G. C., Dyson, H. J., and Wright, P. E. (1998) Conformational preferences in the Ser133-phosphorylated and non-phosphorylated forms of the kinase inducible transactivation domain of CREB. *FEBS Lett.* **430,** 317–322.

23. Fuxreiter, M., Simon, I., Friedrich, P., and Tompa, P. (2004) Preformed structural elements feature in partner recognition by intrinsically unstructured proteins. *J. Mol. Biol.* **338,** 1015–1026.

24. Lacy, E. R., Filippov, I., Lewis, W. S., et al. (2004) p27 binds cyclin-CDK complexes through a sequential mechanism involving binding-induced protein folding. *Nat. Struct. Mol. Biol.* **11,** 358–364.

25. Bussell, R., Jr. and Eliezer, D. (2001) Residual structure and dynamics in Parkinson's disease-associated mutants of alpha-synuclein. *J. Biol. Chem.* **276,** 45,996–46,003.

26. Ahmad, A., Millett, I. S., Doniach, S., Uversky, V. N., and Fink, A. L. (2003) Partially folded intermediates in insulin fibrillation. *Biochemistry* **42,** 11,404–11,416.

27. Wuthrich, K. (1986) NMR of roteins and nucleic acids. J. Wiley and Sons, New York, pp. 292.

28. Nirmala, N. R. and Wagner, G. (1988) Measurement of 13C relaxation times in proteins by two-dimensional heteronuclear 1H-13C correlation spectroscopy. *J. Am. Chem. Soc.* **110,** 7557–7558.

29. Kay, L. E., Torchia, D. A., and Bax, A. (1989) Backbone dynamics of proteins as studied by 15N inverse detected heteronuclear NMR spectroscopy: application to staphylococcal nuclease. *Biochemistry* **28,** 8972–8979.

30. Clore, G. M., Driscoll, P. C., Wingfield, P. T., and Gronenborn, A. M. (1990) Analysis of the backbone dynamics of interleukin-1 beta using two-dimensional inverse detected heteronuclear 15N-1H NMR spectroscopy. *Biochemistry* **29,** 7387–7401.

31. Palmer, A. G., Rance, M., and Wright, P. E. (1991) Intramolecular motions of a zinc finger DNA-binding domain from Xfin characterized by proton-detected natural abundance 13C heteronuclear NMR spectroscopy. *J. Am. Chem. Soc.* **113,** 4371–4380.

32. Tolman, J. R., Flanagan, J. M., Kennedy, M. A., and Prestegard, J. H. (1995) Nuclear magnetic dipole interactions in field-oriented proteins: information for structure determination in solution. *Proc. Natl. Acad. Sci. USA* **92,** 9279–9283.

33. Tjandra, N. and Bax, A. (1997) Direct measurement of distances and angles in biomolecules by NMR in a dilute liquid crystalline medium. *Science* **278,** 1111–1114.

34. Pickford, A. R. and O'Leary, J. M. (2004) Isotopic labeling of recombinant proteins from the methylotrophic yeast Pichia pastoris. *Methods Mol. Biol.* **278,** 17–33.
35. Hansen, A. P., Petros, A. M., Mazar, A. P., Pederson, T. M., Rueter, A., and Fesik, S. W. (1992) A practical method for uniform isotopic labeling of recombinant proteins in mammalian cells. *Biochemistry* **31,** 12,713–12,718.
36. Sambroook, J., Fritsch, E. F., and Maniatis, T. (1989) *Molecular Cloning: A Laboratory Manual.* Cold Spring Harbor Laboratory Press, Plainview, NY.
37. Marley, J., Lu, M., and Bracken, C. (2001) A method for efficient isotopic labeling of recombinant proteins. *J. Biomol. NMR* **20,** 71–75.
38. Molday, R. S., Englander, S. W., and Kallen, R. G. (1972) Primary structure effects on peptide group hydrogen exchange. *Biochemistry* **11,** 150–158.
39. Bai, Y., Milne, J. S., Mayne, L., and Englander, S. W. (1993) Primary structure effects on peptide group hydrogen exchange. *Proteins* **17,** 75–86.
40. Kelly, A. E., Ou, H. D., Withers, R., and Dotsch, V. (2002) Low-conductivity buffers for high-sensitivity NMR measurements. *J. Am. Chem. Soc.* **124,** 12,013–12,019.
41. Braun, D., Wider, G., and Wuthrich, K. (1994) Sequence-corrected 15N "random coil" chemical shifts. *J. Am. Chem. Soc.* **116,** 8466–8469.
42. Zhang, O., Forman-Kay, J. D., Shortle, D., and Kay, L. E. (1997) Triple-resonance NOESY-based experiments with improved spectral resolution: applications to structural characterization of unfolded, partially folded and folded proteins. *J. Biomol. NMR* **9,** 181–200.
43. Yao, J., Dyson, H. J., and Wright, P. E. (1997) Chemical shift dispersion and secondary structure prediction in unfolded and partly folded proteins. *FEBS Lett.* **419,** 285–289.
44. Grzesiek, S. and Bax, A. (1993) Amino acid type determination in the sequential assignment procedure of uniformly 13C/15N-enriched proteins. *J. Biomol. NMR* **3,** 185–204.
45. Arcus, V. L., Vuilleumier, S., Freund, S. M., Bycroft, M., and Fersht, A. R. (1994) Toward solving the folding pathway of barnase: the complete backbone 13C, 15N, and 1H NMR assignments of its pH-denatured state. *Proc. Natl. Acad. Sci. USA* **91,** 9412–9416.
46. Zimmerman, D. E., Kulikowski, C. A., Huang, Y., et al. (1997) Automated analysis of protein NMR assignments using methods from artificial intelligence. *J. Mol. Biol.* **269,** 592–610.
47. Jung, Y. S. and Zweckstetter, M. (2004) Mars: robust automatic backbone assignment of proteins. *J. Biomol. NMR* **30,** 11–23.
48. Dalgarno, D. C., Levine, B. A., and Williams, R. J. (1983) Structural information from NMR secondary chemical shifts of peptide alpha C-H protons in proteins. *Biosci. Rep.* **3,** 443–452.
49. Gross, K.-H. and Kalbitzer, H. R. (1988) Distribution of chemical shifts in 1H nuclear magnetic resonance spectra of proteins. *J. Magn. Reson.* **76,** 87–99.
50. Szilagyi, L. and Jardetzky, O. (1989) [alpha]-Proton chemical shifts and secondary structure in proteins. *J. Magn. Reson.* **83,** 441–449.
51. Williamson, M. P. (1990) Secondary-structure dependent chemical shifts in proteins. *Biopolymers* **29,** 1423–1431.

52. Pastore, A. and Saudek, V. (1990) The relationship between chemical shift and secondary structure in proteins. *J. Magn. Reson.* **90,** 165–176.
53. Spera, S. and Bax, A. (1991) Empirical correlation between protein backbone conformation and C-alpha and C-beta C-13 nuclear magnetic resonance chemical shifts. *J. Am. Chem. Soc.* **113,** 5490–5492.
54. Wishart, D. S., Sykes, B. D., and Richards, F. M. (1991) Relationship between nuclear magnetic resonance chemical shift and protein secondary structure. *J. Mol. Biol.* **222,** 311–333.
55. Wishart, D. S., Sykes, B. D., and Richards, F. M. (1992) The chemical shift index: a fast and simple method for the assignment of protein secondary structure through NMR spectroscopy. *Biochemistry* **31,** 1647–1651.
56. Wishart, D. S. and Sykes, B. D. (1994) The 13C chemical-shift index: a simple method for the identification of protein secondary structure using 13C chemical-shift data. *J. Biomol. NMR* **4,** 171–180.
57. Wishart, D. S. and Case, D. A. (2002) Use of chemical shifts in macromolecular structure determination. *Methods Enzymol.* **338,** 3–34.
58. Wang, Y. and Jardetzky, O. (2002) Probability-based protein secondary structure identification using combined NMR chemical-shift data. *Protein Sci.* **11,** 852–861.
59. Richarz, R. and Wuthrich, K. (1978) Carbon-13 NMR chemical shifts of the common amino acid residues measured in aqueous solutions of the linear tetrapeptides H-Gly-Gly-X-L-Ala-OH. *Biopolymers* **17,** 2133–2141.
60. Bundi, A. and Wuthrich, K. (1979) 1H-NMR parameters of the common amino acid residues measured in aqueous solutions of the linear tetrapeptides H-Gly-Gly-X-L-Ala-OH. *Biopolymers* **18,** 285–297.
61. Merutka, G., Dyson, H. J., and Wright, P. E. (1995) 'Random coil' 1H chemical shifts obtained as a function of temperature and trifluoroethanol concentration for the peptide series GGXGG. *J. Biomol. NMR* **5,** 14–24.
62. Wishart, D. S., Bigam, C. G., Holm, A., Hodges, R. S., and Sykes, B. D. (1995) 1H, 13C and 15N random coil NMR chemical shifts of the common amino acids. I. Investigations of nearest-neighbor effects. *J. Biomol. NMR* **5,** 67–81.
63. Wishart, D. S. and Sykes, B. D. (1994) Chemical shifts as a tool for structure determination. *Methods Enzymol.* **239,** 363–392.
64. Schwarzinger, S., Kroon, G. J., Foss, T. R., Wright, P. E., and Dyson, H. J. (2000) Random coil chemical shifts in acidic 8 M urea: implementation of random coil shift data in NMRView. *J. Biomol. NMR* **18,** 43–48.
65. Schwarzinger, S., Kroon, G. J., Foss, T. R., Chung, J., Wright, P. E., and Dyson, H. J. (2001) Sequence-dependent correction of random coil NMR chemical shifts. *J. Am. Chem. Soc.* **123,** 2970–2978.
66. Schwarzinger, S., Wright, P. E., and Dyson, H. J. (2002) Molecular hinges in protein folding: the urea-denatured state of apomyoglobin. *Biochemistry* **41,** 12,681–12,686.
67. Wang, Y. and Jardetzky, O. (2002) Investigation of the neighboring residue effects on protein chemical shifts. *J. Am. Chem. Soc.* **124,** 14,075–14,084.
68. Howarth, O. W. and Lilley, D. M. (1978) Carbon-13-NMR of peptides and proteins. *Prog. NMR Spectroscopy* **12,** 1–40.

69. Eliezer, D., Barre, P., Kobaslija, M., Chan, D., Li, X., and Heend, L. (2005) Residual structure in the repeat domain of tau: echoes of microtubule binding and paired helical filament formation. *Biochemistry* **44**, 1026–1036.

70. Grzesiek, S., Wingfield, P., Stahl, S., Kaufman, J. D., and Bax, A. (1995) Four-dimensional 15N-separated NOESY of slowly tumbling perdeuterated 15N-enriched proteins. Application to HIV-1 nef. *J. Am. Chem. Soc.* **117**, 9594–9595.

71. Esposito, G. and Pastore, A. (1988) An alternative method for distance evaluation from NOESY spectra. *J. Magn. Reson.* **76**, 331–336.

72. Saulitis, J. and Liepins, E. (1990) Quantitative evaluation of interproton distances in peptides by two-dimensional overhauser effect spectroscopy. *J. Magn. Reson.* **87**, 80–91.

73. Wong, K. B., Freund, S. M., and Fersht, A. R. (1996) Cold denaturation of barstar: 1H, 15N and 13C NMR assignment and characterisation of residual structure. *J. Mol. Biol.* **259**, 805–818.

74. Freund, S. M., Wong, K. B., and Fersht, A. R. (1996) Initiation sites of protein folding by NMR analysis. *Proc. Natl. Acad. Sci. USA* **93**, 10,600–10,603.

75. Palmer, A. G., 3rd (1997) Probing molecular motion by NMR. *Curr. Opin. Struct. Biol.* **7**, 732–737.

76. Kay, L. E. (1998) Protein dynamics from NMR. *Nat. Struct. Biol.* **5** (**Suppl**), 513–517.

77. Kern, D. and Zuiderweg, E. R. (2003) The role of dynamics in allosteric regulation. *Curr. Opin. Struct. Biol.* **13**, 748–757.

78. Schnell, J. R., Dyson, H. J., and Wright, P. E. (2004) Structure, dynamics, and catalytic function of dihydrofolate reductase. *Annu. Rev. Biophys. Biomol. Struct.* **33**, 119–140.

79. Kern, D., Eisenmesser, E. Z., and Wolf-Watz, M. (2005) Enzyme dynamics during catalysis measured by NMR spectroscopy. *Methods Enzymol.* **394**, 507–524.

80. Schwalbe, H., Fiebig, K. M., Buck, M., et al. (1997) Structural and dynamical properties of a denatured protein. Heteronuclear 3D NMR experiments and theoretical simulations of lysozyme in 8 M urea. *Biochemistry* **36**, 8977–8991.

81. Palmer, A. G., 3rd, Kroenke, C. D., and Loria, J. P. (2001) Nuclear magnetic resonance methods for quantifying microsecond-to-millisecond motions in biological macromolecules. *Methods Enzymol.* **339**, 204–238.

82. Kyte, J. and Doolittle, R. F. (1982) A simple method for displaying the hydropathic character of a protein. *J. Mol. Biol.* **157**, 105–132.

83. Klein-Seetharaman, J., Oikawa, M., Grimshaw, S. B., et al. (2002) Long-range interactions within a nonnative protein. *Science* **295**, 1719–1722.

84. Crowhurst, K. A. and Forman-Kay, J. D. (2003) Aromatic and methyl NOEs highlight hydrophobic clustering in the unfolded state of an SH3 domain. *Biochemistry* **42**, 8687–8695.

85. Ochsenbein, F., Guerois, R., Neumann, J. M., Sanson, A., Guittet, E., and van Heijenoort, C. (2001) 15N NMR relaxation as a probe for helical intrinsic propensity: the case of the unfolded D2 domain of annexin I. *J. Biomol. NMR* **19**, 3–18.

86. Bax, A. (2003) Weak alignment offers new NMR opportunities to study protein structure and dynamics. *Protein Sci.* **12,** 1–16.
87. Tolman, J. R., Al-Hashimi, H. M., Kay, L. E., and Prestegard, J. H. (2001) Structural and dynamic analysis of residual dipolar coupling data for proteins. *J. Am. Chem. Soc.* **123,** 1416–1424.
88. Peti, W., Meiler, J., Bruschweiler, R., and Griesinger, C. (2002) Model-free analysis of protein backbone motion from residual dipolar couplings. *J. Am. Chem. Soc.* **124,** 5822–5833.
89. Shortle, D. and Ackerman, M. S. (2001) Persistence of native-like topology in a denatured protein in 8 M urea. *Science* **293,** 487–489.
90. Ohnishi, S., Lee, A. L., Edgell, M. H., and Shortle, D. (2004) Direct demonstration of structural similarity between native and denatured eglin C. *Biochemistry* **43,** 4064–4670.
91. Louhivuori, M., Paakkonen, K., Fredriksson, K., Permi, P., Lounila, J., and Annila, A. (2003) On the origin of residual dipolar couplings from denatured proteins. *J. Am. Chem. Soc.* **125,** 15,647–15,650.
92. Mohana-Borges, R., Goto, N. K., Kroon, G. J., Dyson, H. J., and Wright, P. E. (2004) Structural characterization of unfolded states of apomyoglobin using residual dipolar couplings. *J. Mol. Biol.* **340,** 1131–1142.
93. Fieber, W., Kristjansdottir, S., and Poulsen, F. M. (2004) Short-range, long-range and transition state interactions in the denatured state of ACBP from residual dipolar couplings. *J. Mol. Biol.* **339,** 1191–1199.
94. Ottiger, M., Delaglio, F., and Bax, A. (1998) Measurement of J and dipolar couplings from simplified two-dimensional NMR spectra. *J. Magn. Reson.* **131,** 373–378.
95. Goto, N. K., Skrynnikov, N. R., Dahlquist, F. W., and Kay, L. E. (2001) What is the average conformation of bacteriophage T4 lysozyme in solution? A domain orientation study using dipolar couplings measured by solution NMR. *J. Mol. Biol.* **308,** 745–764.
96. Plaxco, K. W., Morton, C. J., Grimshaw, S. B., et al. (1997) The effects of guanidine hydrochloride on the 'random coil' conformations and NMR chemical shifts of the peptide series GGXGG. *J. Biomol. NMR* **10,** 221–230.
97. West, N. J. and Smith, L. J. (1998) Side-chains in native and random coil protein conformations. Analysis of NMR coupling constants and chi1 torsion angle preferences. *J. Mol. Biol.* **280,** 867–877.
98. Eliezer, D., Chung, J., Dyson, H. J., and Wright, P. E. (2000) Native and non-native secondary structure and dynamics in the pH 4 intermediate of apomyoglobin. *Biochemistry* **39,** 2894–2901.
99. Yao, J., Chung, J., Eliezer, D., Wright, P. E., and Dyson, H. J. (2001) NMR structural and dynamic characterization of the acid-unfolded state of apomyoglobin provides insights into the early events in protein folding. *Biochemistry* **40,** 3561–3571.
100. Cao, W., Bracken, C., Kallenbach, N. R., and Lu, M. (2004) Helix formation and the unfolded state of a 52-residue helical protein. *Protein Sci.* **13,** 177–189.
101. Cierpicki, T. and Otlewski, J. (2001) Amide proton temperature coefficients as hydrogen bond indicators in proteins. *J. Biomol. NMR* **21,** 249–261.

102. Buck, M., Radford, S. E., and Dobson, C. M. (1994) Amide hydrogen exchange in a highly denatured state: Hen egg-white lysozyme in urea. *J. Mol. Biol.* **237,** 247–254.

103. Zhang, Y. Z., Paterson, Y., and Roder, H. (1995) Rapid amide proton exchange rates in peptides and proteins measured by solvent quenching and two-dimensional NMR. *Protein Sci.* **4,** 804–814.

104. Shortle, D. R. (1996) Structural analysis of non-native states of proteins by NMR methods. *Curr. Opin. Struct. Biol.* **6,** 24–30.

105. Smith, L. J., Fiebig, K. M., Schwalbe, H., and Dobson, C. M. (1996) The concept of a random coil. Residual structure in peptides and denatured proteins. *Fold Des.* **1,** R95–R106.

106. Dyson, H. J. and Wright, P. E. (2002) Insights into the structure and dynamics of unfolded proteins from nuclear magnetic resonance. *Adv. Protein Chem.* **62,** 311–340.

107. Redfield, C. (2004) Using nuclear magnetic resonance spectroscopy to study molten globule states of proteins. *Methods* **34,** 121–132.

108. Dyson, H. J. and Wright, P. E. (2005) Elucidation of the protein folding landscape by NMR. *Methods Enzymol.* **394,** 299–321.

5

Population and Structure Determination of Hidden Folding Intermediates by Native-State Hydrogen Exchange-Directed Protein Engineering and Nuclear Magnetic Resonance

Yawen Bai, Hanqiao Feng, and Zheng Zhou

Summary

Structural characterization of folding intermediates has been one of the important steps toward understanding the mechanism of protein folding. However, it has been very difficult to obtain high-resolution structures of folding intermediates. Such results have become available only very recently. Here, we review a procedure that uses the native-state amide hydrogen exchange-directed protein engineering method to populate partially unfolded intermediates and multidimensional NMR to solve the high-resolution stuctures of the intermediates.

Key Words: Protein folding; native-state hydrogen exchange; protein engineering; NMR structure.

1. Introduction

To understand how proteins fold, it is important to know the structure of folding intermediates in detail. For example, such information will be useful to answer the following important questions in protein folding:

1. Do proteins use partially unfolded intermediates to solve the large-scale conformational search problem?
2. Is the folding step continuous or discrete?
3. Are folding pathways degenerative?
4. Are there non-native interactions in the folding intermediates?

Although a great deal of experimental work has been done to identify and characterize partially unfolded intermediates that occur during the folding process, only very recently has it been possible to obtain detailed structural information at atomic resolution. The main hurdle has been that protein folding is a very fast

From: *Methods in Molecular Biology, vol. 350: Protein Folding Protocols*
Edited by: Y. Bai and R. Nussinov © Humana Press Inc., Totowa, NJ

process (<1 s for small single domain proteins with ~100 amino acids) and, therefore, intermediates only populate transiently during kinetic folding. The commonly used spectroscopic probes, such as fluorescence and circular dichroism, provide very limited information to define the structure of folding intermediates. Moreover partially unfolded intermediates do not populate significantly under equilibrium native conditions because the native state is the most stable structure.

Structural information on protein folding intermediates has often been obtained using the hydrogen/deuterium exchange pulse-labeling methods *(1,2)*, which can reveal the pattern of hydrogen bonding in transiently populated intermediates. Another way to obtain structural information on folding intermediates has been to populate partially unfolded states at equilibrium at low pH and characterize them using hydrogen exchange and nuclear magnetic resonance (NMR). The best example is apomyoglobin *(3,4)*. Detailed structural information on folding intermediates mainly has come from the native-state hydrogen exchange (NHX) method. This method detects partially unfolded intermediates that exist only in infinitesimal amounts under native conditions where the native state is the dominantly populated species *(5)*. By measuring the hydrogen exchange rates of amide protons and their dependence on denaturant concentration, the NHX method can characterize the hydrogen bonding pattern, stability, and surface exposure of partially unfolded intermediates. So far, multiple partially unfolded intermediates have been identified for a number of proteins, including cytochrome c (cyt c) *(5)*, ribonuclease H* *(6)*, apocytochrome b_{562} *(7)*, T4 lysozyme *(8)*, Rd-apocytochrome b_{562} (Rd-apocyt b_{562}) *(9)*, OspA *(10)*, barnase *(11)*, and a construct of the third domain of PDZ *(12)*. This method is particularly useful in detecting the intermediates that exist after the rate-limiting transition state, which can evade detection by the conventional kinetic and equilibrium methods.

2. Materials

1. D_2O (D 99%) can be purchased from Cambridge Isotope Laboratory Inc. (Andover, MA).
2. Deuterated guanidinium hydrochloride (GdmCl) is made by evaporating D_2O in 6 *M* GmdCl solution three times.
3. The Quick-change kit (Stratagene, San Diego, CA) is used for making mutations.
4. Isotopically enriched proteins for NMR structure determination are grown in *Escherichia coli* on M9 minimal media containing 1 g/L (U-^{15}N) $^{15}NH_4Cl$ and/or 4 g/L (U-^{13}C) glucose (Isotec) for double- (^{13}C, ^{15}N) and/or single- (^{15}N or ^{13}C) labeled proteins as the sole sources of nitrogen and carbon.
5. NMR samples including ^{13}C/^{15}N-, ^{15}N-, ^{13}C-labeled protein at a concentration of greater than 0.5 m*M* (95% H_2O and 5% D_2O) in certain buffers is required for the structure determination.

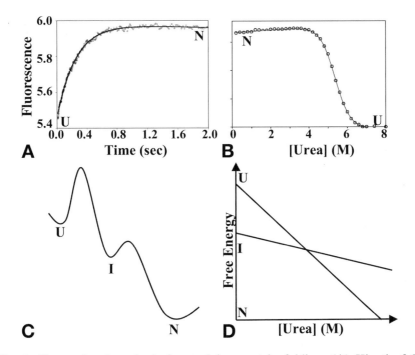

Fig. 1. Conventional methods for studying protein folding. (**A**) Kinetic folding monitored by fluorescence. (**B**) Equilibrium unfolding of proteins monitored by fluorescence. For small proteins (~100 amino acids or less), single exponential kinetics and cooperative transition are often observed in kinetic and equilibrium unfolding experiments. (**C**) Intermediates that exist after the rate-limiting transition state are not observable in the kinetic folding experiment. (**D**) Illustration of the switch of free energy levels between the unfolded state and the intermediate. At all denaturant concentrations, the intermediate is significantly less stable than either the native state or the unfolded state.

3. Methods

3.1. Conventional Methods for Studying Protein Folding

Figure 1 illustrates the conventional kinetic and equilibrium methods for studying protein folding. In kinetic folding experiments, unfolded protein molecules were prepared at high concentrations of denaturant, such as 6 *M* GdmCl or 8 *M* urea. Upon dilution of denaturant, proteins were returned to native conditions and folding was initiated. The process of folding is often monitored by fluorescence and circular dichoism. For small single domain proteins (<100 amino acids), single exponential kinetics is often observed (**Fig. 1A**) and the logarithm of the observed folding rates displays linear behavior as a function of denaturant concentrations (**Fig. 1A**, insert), suggesting the absence of detectable

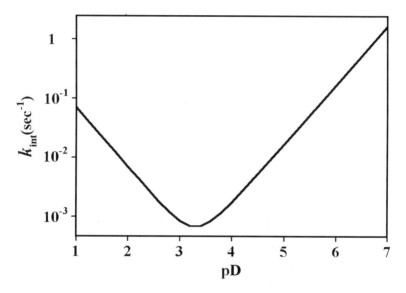

Fig. 2. Hydrogen exchange curve of poly-D,L-alanine.

kinetic folding intermediates. In equilibrium unfolding experiments, signals of the fluorescence or CD of protein samples are monitored at different denaturant concentrations. A single cooperative change of these signals is commonly observed for small proteins, suggesting the absence of a detectable equilibrium intermediate population at all denaturant concentrations.

The previously mentioned conventional methods, however, can miss folding intermediates. In kinetic folding studies, intermediates that exist after the rate-limiting transition state will not populate to a significant amount at any time of the folding process (**Fig. 1C**). Therefore, they cannot be detected in the kinetic folding experiment unless a highly sensitive approach is used, for example, a single-molecule experiment. The equilibrium method can miss folding intermediates as well. The free energy levels of the unfolded state and the intermediate can switch as the denaturant concentrations are increased. This switch results from the fact that global unfolding exposes more surface area than the partial unfolding does, which makes the unfolding free energy of the unfolded state more sensitive to the concentration of denaturant. At any denaturant concentrations, the intermediate may be less stable than either the native state or the fully unfolded state. Therefore, it is less populated than either the native state (at low concentrations of denaturant) or the unfolded state (at high denaturant concentrations).

3.2. Detection of Hidden Folding Intermediates by NHX

3.2.1. Amide Hydrogen Exchange Rates for Unfolded Polypeptides

Amide protons in polypeptides are chemically labile and exchangeable with hydrogen isotopes in solvent water such as:

$$>N - H + D_2O \rightarrow >N - H + DOH \tag{1}$$

Because of the extreme pKa values of main chain amides, the exchange of their hydrogens with solvent is relatively slow and is catalyzed only by the strongest of aqueous acids and bases (hydronium and hydroxide ion). Thus, the exchange rate is pH dependent. **Figure 2** illustrates the exchange rate constants as a function of pD (pD = pH_{read} + 0.4) *(13)* *(see* **Note 1**) for amide protons in unstructured poly-D/L-alanine. Here pH_{read} is the reading value from the pH meter. At pD 7.0 and 20°C, the exchange rate constant of an amide proton in an unfolded peptide, k_{int}, is affected mainly by the side chains of its two nearest amino acid residue neighbors. Both inductive *(14)* and steric blocking effects *(15)* are apparent. These effects have been characterized for all 20 amino acids using dipeptides as models *(15)*. k_{int} can now be predicted among a broad range of pH and temperature *(15,16)* *(see* **Note 2**). For unfolded polypeptides, the predicted k_{int} is likely to be within a factor of two of the measured k_{ex} *(17)*.

3.2.2. Amide Hydrogen Exchange in Folded Proteins

In folded proteins, many amide protons are protected from exchange because of hydrogen bonding and burial in the native structure. The protection factor (PF = k_{int}/k_{ex}) provides information on the native structure and stability. Here k_{ex} is the experimentally measured exchange rate constant. Linderstrøm-Lang and his colleagues pictured a two-state situation and assumed that amide hydrogens can exchange with solvent hydrogens only when they are transiently exposed to solvent in some kind of closed-to-open reaction, as indicated in **Eq. 2** *(18)*.

$$NH(closed) \overset{k_{op}}{\underset{k_{cl}}{\Leftrightarrow}} NH(open) \overset{k_{int}}{=>} \text{exchanged}; \quad K_{op} = k_{op}/k_{cl} \tag{2}$$

Here, k_{op} is the kinetic opening rate constant; k_{cl} is the kinetic closing rate constant. Under steady-state conditions, the exchange rate, k_{ex}, determined by the previously shown scheme is given by **Eq. 3**.

$$k_{ex} = k_{op}k_{int}/(k_{op} + k_{cl} + k_{int}) \tag{3}$$

There are two extreme cases for this reaction scheme with stable structures (k_{op} << k_{cl}). First, the closing reaction is much faster than the intrinsic exchange rate constants (k_{cl} >> k_{int}), termed as the EX2 condition. In this case, the exchange rate of any hydrogen, k_{ex}, is determined by its chemical exchange rate in the open form multiplied by the equilibrium opening constant, K_{op}.

$$k_{ex} = K_{op} \times k_{int} \tag{4}$$

This leads to free energy for the dominant opening reaction, as represented by the following equation:

$$\Delta G_{HX} = -RT \ln K_{op} = -RT \ln(k_{ex}/k_{int}) = -RT\ln(1/PF) \tag{5}$$

In this equation, R is the gas constant and T is the temperature. The free energy defined in **Eq. 5**, ΔG_{HX}, represents a combination of opening transitions from both structural unfolding and local fluctuations. Second, the closing reaction is much slower than the intrinsic exchange rate constant ($k_{cl} \ll k_{int}$), termed the EX1 condition. In this case, the exchange rate, k_{ex}, is equal to the opening rate constant k_{op}. For amide protons that can only exchange through global unfolding, the k_{op} will be the global unfolding rate constant. A more general presteady-state solution for a reaction scheme **Eq. 2** without any assumptions about the relative magnitudes of k_{op}, k_{cl}, and k_{int} was also solved *(19,20)*.

3.2.3. Native-State Hydrogen Exchange

Under native conditions, the native protein is the most highly populated form, but proteins must continually unfold and refold, and cycle through all possible higher energy forms. Partially unfolded intermediates whose stability lies between the native and unfolded states can be identified by HX measurements, even though they only exist at miniscule levels. This is so because the predominant native state makes no contribution to the HX rates that are measured. Hydrogens can exchange only when their protecting H-bond is broken in some higher energy state.

Figure 3A illustrates the hydrogen exchange processes that occur for a three-state system. Amide protons that are not strongly protected or deeply buried in the native protein can exchange through local structural fluctuations by breaking one or two hydrogen bonds without significantly exposing solvent-accessible surface area *(21–23)*. If the protein has a partially unfolded state that is more stable than the unfolded state, then the amide protons in the unfolded region of the intermediate can exchange in such partially unfolded states. All amide protons can also exchange from the fully unfolded states. The measured exchange rate constant is the sum of the exchange rate constants of all three processes, weighted by the unfolding equilibrium constants of the intermediate and the unfolded state:

$$k_{ex} = k_{loc} + K_{NI} \times k_{int} + K_{NU} \times k_{int} \tag{6}$$

i.e., the exchange terms from the intermediate and the unfolded state are controlled by the unfolding equilibrium constants. Here, k_{loc} represents the exchange process from the native structure. K_{NI} and K_{NU} are the unfolding equilibrium constants for the intermediate and the unfolded state.

In a NHX experiment, the hydrogen exchange rates are measured at different concentrations of denaturant. The denaturant is used to perturb the equilibrium constants and help to reveal the different exchange behavior for amide protons

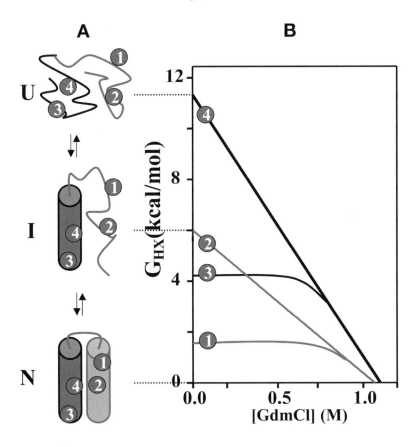

Fig. 3. Illustration of detection of folding intermediate by native-state hydrogen-exchange method. ΔG_{HX}s of amide protons vs the guanidine hydrochloride concentration [GdmCl] for a protein with one partially unfolded intermediate. Amide proton 2 dominantly exchanges in the partially unfolded intermediate. Amide proton 4 can only exchange through global unfolding. Because the above exchange processes involve the exposure of large solvent surface areas, the ΔG_{HX}s for these amide protons are sensitive to denaturant concentrations. The slopes of the ΔG_{HX}s vs denaturant concentration are proportional to the change of solvent accessible surface area of the partial unfolding (proton 2) and global unfolding (proton 4). Amide protons 1 and 3 can exchange dominantly through local structure fluctuations at low concentration of denaturant without exposing significant solvent accessible surface area. Therefore, their ΔG_{HX}s are independent of denaturant concentration until the exchange processes are taken over by the partial (for proton 1) and global (for proton 3) unfoldings, as represented by the ΔG_{HX}s of proton 2 and 4. The intercepts of ΔG_{HX}s for proton 2 and 4 represent the unfolding free energies of the partially unfolded and the fully unfolded states at zero denaturant *(5)*.

in different regions of the protein. When ΔG_{HX}s of different amide protons are plotted against the concentration of denaturant concentrations, different structural segments that unfold cooperatively as structural units will converge together to different ΔG_{HX} values before joining the global ΔG_{NU}. **Figure 3B** illustrates the hydrogen exchange pattern for a protein with a folding intermediate. Based on the exchange pattern, one can deduce the structure, stability, and exposed surface area of the intermediate *(5)*.

3.3. Population of Intermediates by NHX-Directed Protein Engineering

Intermediates identified by NHX have resolution at the level of residues. A structure at such resolution does not have the information on the topology and detailed interactions among side chains. To obtain the high-resolution structure of the intermediate identified by the native-state hydrogen exchange, **Fig. 4A** illustrates the principle of using protein engineering to destabilize the native state without affecting the partially unfolded states in order to populate the intermediate. It plotted the partially unfolded states detected by NHX for Rd-apocyt b_{562}, a four-helix bundle protein. Mutations that substitute larger stabilizing hydrophobic residues with glycines or charged residues in the unfolded region of the intermediates can be made to destabilize the native state while leaving the folded region of the intermediates unperturbed. Accordingly, the intermediate with the folded region unaffected by the mutation becomes the most stable state (**Fig. 4B**) and its structure can be determined using multidimensional NMR.

In the case of Rd-apocyt b_{562}, two partially unfolded intermediates have been identified by the NHX method. To selectively destabilize the native state and populate the lowest free energy intermediate (PUF2), hydrophobic core residues at five positions (W7D/L10G/L14G/V16G/I17G) in the unfolded regions of PUF2 were mutated to Asp and Gly. **Figure 4C** illustrates these mutation sites in the native structure.

3.4. Structure Determination of Folding Intermediates

3.4.1. Characterization of Folding Intermediates by Backbone ^{15}N Dynamics

An important step toward the structure determination of a partially unfolded intermediate is to test whether the intermediate is monomeric at the concentration where the structure of the intermediate is determined. A ^{15}N dynamics study may be used to do so. First, a series of three-dimensional experiments (CBCA[CO]NH, HNCACB, HCCH-TOCSY, and ^{15}N-edited TOCSY) can be collected for complete assignments of backbone resonance. Two sets of cross peaks with different intensities should show up for a partially unfolded state. The peaks in the unfolded region in the intermediate will

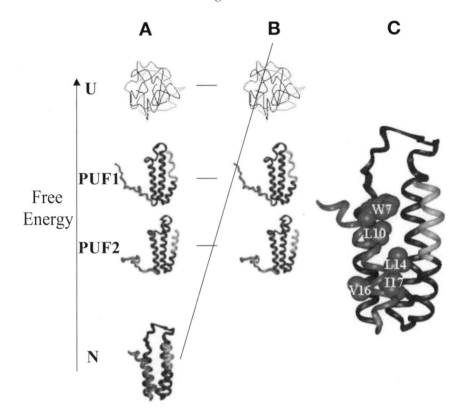

Fig. 4. Illustration of the use of native-state hydrogen exchange (NHX)-directed protein engineering to populate PUF2 of Rd-apocyt b_{562}. **(A)** Structures of the PUFs derived from the complete NHX results. **(B)** Glycine mutations that substitute hydrophobic core residues in the unfolded regions of PUF2 destabilize the native state without affecting the stability of the unfolded state, PUF1, and PUF2. Therefore, PUF2 becomes the most stable species. **(C)** Structural illustration of the mutated residues in the native structure of Rd-apocyt b_{562} for populating PUF2. The side chains of these residues are shown in CPK models. The Trp at position 7 was mutated to Asp. All other residues were changed to Gly.

have strong intensities, whereas the cross peaks with weaker intensities will occur to the folded region. Steady-state ^{1}H-^{15}N nuclear Overhauser effect (NOE) values can also be determined to define the folded and unfolded region of the intermediate by recording in the presence (NOE) and absence (NONOE) of ^{1}H saturation. NOE and NONOE experiments were normally interleaved in one experiment. NOE was calculated from the intensities of cross peaks by NOE = I_{NOE}/I_{NONOE}. NMR data were processed using NMRPipe *(24)* and analyzed using NMRview *(25)*.

The values of the transverse relaxation time (T_2) for the amide ^{15}N of the weak peaks can be used to determine whether the protein is monomeric. The possibility that a small fraction of the molecule may be forming dimers can be examined by diluting the sample 10- to 100-fold. If the chemical shift for each cross peak in the HSQC spectrum does not change, then a small fraction of dimers can be excluded.

In the case of the PUF2 mimic of Rd-apocyt b_{562}, the peaks with strong intensities in the HSQC spectrum have larger T_2 and smaller NOE values, indicating that the mutated regions have very flexible structures, presumably largely unfolded. The peaks with weaker intensities have smaller T_2 and larger NOE values, indicating folded regions. The T_2 values in the folded region of the molecule are approx 80 ms, consistent with a monomer protein of approx 12 kD and a significant part of the structure unfolded at 25°C. A dimer of approx 24 kD would have much shorter T_2 under the same condition. In addition, no chemical shift changes in the $^1H-^{15}N$ HSQC spectrum occur after dilution of a protein sample (~2 mM) by 100-fold, indicating that the mutant is a monomer at the concentration used for the ^{15}N dynamics studies.

3.4.2. Determination of Structure by Multidimensional NMR

Multidimensional NMR methods can be used to determine the structure of the intermediate populated by the NHX-directed protein engineering approach. Chemical shifts of ^1H, ^{15}N, and ^{13}C of a double-labeled ($^{15}N/^{13}C$) protein sample can be assigned using triple resonance methods. $^1H-^1H$ NOEs, backbone dihedral angles, and chemical shifts can be measured and used as constraints for the calculation of structures. Structural calculation can be done using NIH X-PLOR.

For example, in the case of PUF2 of Rd-apocyt b_{562}, an extended polypeptide chain of reasonable geometry was used as the initial template. ϕ- and ψ-angles were then randomized before each cycle of simulated annealing (SA) protocol using the constraints derived from NOE and backbone dihedral angles. Each SA structure was optimized by restrained refinement. Ten structures with lower energy, fewer NOE, and dihedral angle violations from 50 SA structures were selected for further refinement. For each starting structure, 10 refined structures (i.e., 100 structures) were calculated. The ones with lowest energy, no NOE, and dihedral angle violations from each set were used for a second round of refinement to calculate another 10 structures. This refinement procedure was repeated once more to obtain the final 10 structures with the lowest energy and no violation of restraints. These structures were further checked by PROCHECK-NMR (v3.5.4) *(26)*.

Figure 5 illustrates the structure of the PUF2 intermediate of Rd-apocyt b_{562} and the native state. The folded region of the structure is very well-defined with a RMSD of 1.11 Å for all heavy atoms. In contrast, the N-terminal helix with

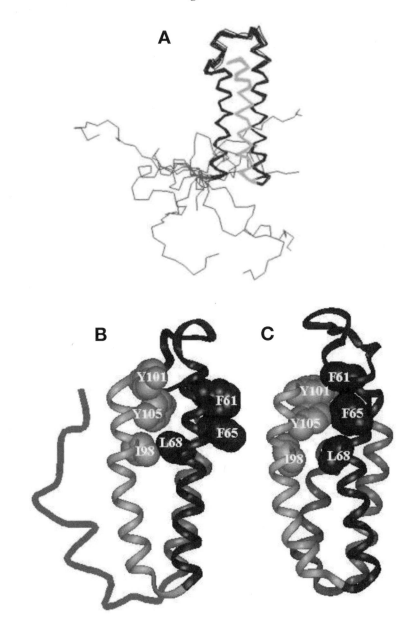

Fig. 5. The structures of the PUF2 mimic. (**A**) Overlay of 10 nuclear magnetic resonance C_α trace structures on the folded region of the PUF2 structure. (**B**) Typical residues with non-native interactions are shown with Corey–Pauling–Kolton models in the PUF2 mimic. (**C**) The corresponding structure of the native state.

Gly mutations has no identifiable long-range NOEs and shows significantly different conformations among the 10 NMR structures (*see* **Fig. 5A**). It has the native-like backbone topology. A comparison of the side chains in the PUF2 intermediate with those in the native state shows broad distributions of non-native side chain packing in the PUF2 intermediate (**Fig. 5B,C**).

4. Notes

1. The pH value measured in D_2O should be corrected by adding 0.4 U (pD = pH_{read} + 0.4).
2. An on-line program (http://www.fccc.edu/research/labs/roder/sphere/) is available for calculating intrinsic exchange rates for amide protons in unfolded polypeptide chains.

References

1. Udgaonkar, J. B. and Baldwin, R. L. (1988) NMR evidence for an early framework intermediate on the folding pathway of ribonuclease A. *Nature* **335,** 694–699.
2. Roder, H., Elove, G. A., and Englander, S. W. (1988) Structural characterization of folding intermediates in cytochrome c by H-exchange labelling and proton NMR. *Nature* **335,** 700–704.
3. Hughson, F. M., Wright, P. E., and Baldwin, R. L. (1990) Structural characterization of a partly folded apomyoglobin intermediate. *Science* **249,** 1544–1548.
4. Eliezer, D., Yao, J., Dyson, H. J., and Wright, P. E. (1998) Structural and dynamic characterization of partially folded states of apomyoglobin and implications for protein folding. *Nat. Struct. Biol.* **5,** 148–155.
5. Bai, Y., Sosnick, T. R., Mayne, L., and Englander, S. W. (1995) Protein folding intermediates: native-state hydrogen exchange. *Science* **269,** 192–197.
6. Chamberlain, A. K., Handel, T. M., and Marqusee, S. (1996) Detection of rare partially folded molecules in equilibrium with the native conformation of RNase H. *Nat. Struct. Biol.* **3,** 782–787.
7. Fuentes, E. J. and Wand, A. J. (1998) Local stability and dynamics of apocytochrome b_{562} examined by the dependence of hydrogen exchange on hydrostatic pressure. *Biochemistry* **37,** 9877–9883.
8. Llinas, M., Gillespie, B., Dahlquist, F. W., and Marqusee, S. (1999) The energetics of T4 lysozyme reveals a hierarchy of conformations. *Nat. Struct. Biol.* **6,** 1072–1078.
9. Chu, R. A., Pei, W. H., Takei, J., and Bai, Y. (2002) Relationship between the native-state hydrogen exchange and folding pathways of a four-helix bundle protein. *Biochemistry* **41,** 7998–8003.
10. Yan, S., Kennedy, S. D., and Koide, S. (2002) Thermodynamic and kinetic exploration of the energy landscape of Borrelia burgdorferi OspA by native-state hydrogen exchange. *J. Mol. Biol.* **323,** 363–375.
11. Vu, D., Feng, H., and Bai, Y. (2004) The folding pathway of barnase: the rate-limiting transition state and a hidden intermediate under native conditions. *Biochemistry* **42,** 3346–3356.

12. Feng, H., Vu, N. D., and Bai, Y. (2005) Detection of a hidden intermediate in the folding of the third domain of PDZ. *J. Mol. Biol.* **346,** 345–353.

13. Glasoe, P. F. and Long, F. A. (1960) Use of glass electrodes to measure acidities in deuterium oxide. *J. Phys. Chem.* **64,** 188–193.

14. Molday, R. S., Englander, S. W., and Kallen, R. G. (1972) Primary structure effects on peptide group hydrogen exchange. *Biochemistry* **11,** 150–158.

15. Bai, Y., Milne, J., Mayne, L., and Englander, S. W. (1993) Primary structure effects on peptide group hydrogen exchange. *Proteins* **17,** 75–86.

16. Connelly, G. P., Bai, Y., Jeng, M. F., and Englander, S. W. (1993) Isotope effects in peptide group hydrogen exchange. *Proteins* **17,** 87–92.

17. Huyghues-Despointes, B. M., Scholtz, J. M., and Pace, C. N. (1999) Protein conformational stabilities can be determined from hydrogen exchange rates. *Nat. Struct. Biol.* **6,** 910–912.

18. Hvidt, A. and Nielsen, S. O. (1966) Hydrogen exchange in proteins. *Adv. Protein Chem.* **21,** 287–386.

19. Qian, H. and Chan, S. I. (1999) Hydrogen exchange kinetics of proteins in denaturants: a generalized two-process model. *J. Mol. Biol.* **286,** 607–616.

20. Krishna, M. M. G., Hoang, L., Lin, Y., and Englander, S. W. (2004) Hydrogen exchange methods to study protein folding. *Methods* **34,** 51–65.

21. Mayo, S. L. and Baldwin, R. L. (1993) Guanidinium chloride induction of partial unfolding in amide proton exchange in RNase A. *Science* **262,** 873–876.

22. Qian, H., Mayo, S. L., and Morton, A. (1994) Protein hydrogen exchange in denaturant: quantitative analysis by a two-process model. *Biochemistry* **33,** 8167–8171.

23. Bai, Y., Milne, J., Mayne, L., and Englander, S. W. (1994) Protein stability parameters measured by hydrogen exchange. *Proteins* **20,** 4–14.

24. Delaglio, F., Grzesiek, S., Vuister, G., Zhu, G., Pfeifer, J., and Bax, A. (1995) NMRPipe: a multi-dimensional spectral processing system based on UNIX pipes. *J. Biomol. NMR* **6,** 277–293.

25. Johnson, B. A. and Blevins, R. A. (1994) NMRView: A computer program for the visualization and analysis of NMR data. *J. Biomol. NMR* **4,** 603–614.

26. Laskowski, R. A., MacArthur, M. W., Moss, D. S., and Thornton, J. M. (1993) PROCHECK: a program to check the stereochemical quality of protein structures. *J. Appl. Cryst.* **26,** 283–291.

6

Characterizing Protein Folding Transition States Using Ψ-Analysis

Adarsh D. Pandit, Bryan A. Krantz, Robin S. Dothager, and Tobin R. Sosnick

Summary

We discuss the implementation of Ψ-analysis for the structural characterization of protein folding transition states. In Ψ-analysis, engineered bi-histidine metal ion binding sites are introduced at surface positions to stabilize secondary and tertiary structures. The addition of metal ions stabilizes the interaction between the two known histidines in a continuous fashion. Measuring the ratio of transition state stabilization to that of the native state provides information about the presence of the metal binding site in the transition state. Ψ-Analysis uses noninvasive surface mutations and does not require specialized equipment, so it can be readily applied to characterize the folding of many proteins. As a result, this method can provide a wealth of high-resolution quantitative data for comparison with theoretical folding simulations. Additionally, investigations of other biological processes also may utilize metal binding sites and Ψ-analysis to detect conformational events during catalysis, assembly, and function.

Key Words: Protein folding; transition state; metal binding; biHis; psi-analysis; phi-analysis; heterogeneity.

1. Introduction

One of the main unanswered questions in molecular biology is how amino acid sequence codes for the three-dimensional structure of a protein. *De novo* attempts to determine structure from sequence have yet to succeed satisfactorily *(1)*. Elucidation of the energetic and structural steps on the protein folding pathway will likely play a major role in the improvement of structure prediction algorithms.

Most small globular proteins fold in a kinetically two-state manner (U↔N) where intermediates do not populate *(2–5)*. As a result, the transition state (TS)

From: *Methods in Molecular Biology, vol. 350: Protein Folding Protocols*
Edited by: Y. Bai and R. Nussinov © Humana Press Inc., Totowa, NJ

ensemble is the only intermediate point on the folding pathway readily amenable to characterization. Mutational φ-analysis has long been the accepted method for characterizing the TS structure of the folding pathway *(6,7)*. However, the interpretation of results, especially fractional φ-values, has become a subject of much debate *(8–15)*.

We have developed Ψ-analysis, a method complementary to φ-analysis, in order to better characterize TS ensembles *(8,14,16)*. Ψ-Analysis uses engineered bi-histidine (biHis) metal ion binding sites at known positions on the protein surface to stabilize secondary and tertiary structures. The addition of increasing concentration of metal ions stabilizes the interaction between the two histidine partners in a continuous fashion. As a result, this method quantitatively evaluates the ratio of metal-induced TS stabilization relative to that of the native state, represented by the Ψ-value. The translation of a measured Ψ-value to structure formation is straightforward, because the proximity of two specific histidine residues is probed. Hence, the method is particularly well suited for defining the topology and structure of TS ensembles. In the following sections, we will discuss the implementation of this method.

2. Materials

In addition to the standard materials used in folding experiments, metal-containing denaturant and buffer solutions are required. The Ψ-analysis method depends on metal binding to the deprotonated form of histidine (intrinsic pKa ~6.5), so experiments should be conducted at pH 7.5 or greater to maximize the binding stabilization energy, $\Delta\Delta G_{bind}$. Buffer selection depends primarily on proximity of its pKa to the target pH. However, testing of buffer–metal combinations before experimentation is recommended as some buffers, such as sodium phosphate, can lead to metal precipitation. We typically employ 50 mM Tris or HEPES at pH 7.5. Nevertheless, precipitation remains a persistent problem so solutions should be prepared fresh each day. We find it convenient to make a high concentration metal ion stock for diluting into solutions. All buffer solutions, especially those containing metal, should be checked for pH changes immediately prior to use.

1. Zinc chloride ($ZnCl_2$) stock solutions can be prepared at 0.25 M in 25 mM HCl and should be remade every month or two. Zinc chloride readily precipitates at high concentration and neutral pH, so buffers should be prepared with care. For example, when making a zinc-containing buffer, after adding the buffer stock solution, dilute to just under the final volume and add the zinc stock solution. The addition of acidic metal solution often lowers the pH, so it should be rechecked immediately before use.
2. Cobalt chloride ($CoCl_2$) stock solutions appear burgundy red and can be prepared at 1 M in water. Using buffers with cobalt at a guanidinium chloride (GdmCl) concentration of approx 5.5 M can cause metal precipitation and erratic kinetic results.

Excessive amount of chloride ions, as present in high concentration GdmCl buffers, will cause cobalt solutions to appear blue, which may affect results.

3. Cadmium chloride ($CdCl_2$) solutions can be prepared at 1 *M* in water. Cadmium is a heavy metal that can be toxic. Please refer to material safety data sheets for safe handling procedures, such as avoiding skin contact and inhalation. Solutions can be prepared in the same fashion as cobalt.

4. Nickel chloride ($NiCl_2$) solutions can be prepared at 0.25 *M* in 25 m*M* HCl. We have found nickel can be readily combined with many buffers as it is not as susceptible to precipitation.

3. Methods

In both φ-analysis and Ψ-analysis, the change in folding rate owing to an energetic perturbation identifies the degree to which this interaction is present in the TS ensemble. In mutational φ-analysis, the perturbation is a single side chain substitution. In Ψ-analysis, biHis sites are individually engineered onto the protein surface and its stability is manipulated using divalent metal ions. Generally, after insertion of a biHis site at a region of interest, the denaturant dependence of folding rates ("chevron analysis" *[6]*) is determined to characterize the overall folding behavior. A Brønsted *(17)* or Leffler *(18)* plot is then obtained to calculate the Ψ-value by determining the change in folding and unfolding rates as a function of metal concentration.

3.1. Engineering biHis Sites

The strategy of using engineered metal ion binding sites in biochemical studies has an extensive history *(19–28)*. Our past folding studies have used biHis sites located on the protein surface *(25)* rather than buried sites. For a surface site, the metal-induced stabilization is specific to a particular structural element, such as a helix or hairpin (*see* **Fig. 1A**), and the protein can be folded in the absence of the metal (*see* **Fig. 1B**). These properties generally are not applicable to proteins with buried sites, such as zinc finger proteins where metal ions are required for the cooperative folding of the entire protein *(22)*. These buried sites typically have four side chain ligands arranged in a precise geometry. The introduction of such sites often requires a substantial amount of protein design *(29–31)*, which is unnecessary for utilization of surface biHis sites.

The placement of surface biHis metal binding sites can be accomplished by inspecting the protein structure. Generally, sites require the imidazole nitrogens (N_ϵ) of the histidines to be located within 3–5 Å of each other, which can be accomplished using residues where the C_α–C_α distance is less than 13 Å *(25)*. The placement of histidine residues has a substantial effect on the binding affinity and the degree of energetic stabilization for each type of ion *(25)*. In helices, the histidines should be introduced four residues apart in *i* and *i+4*

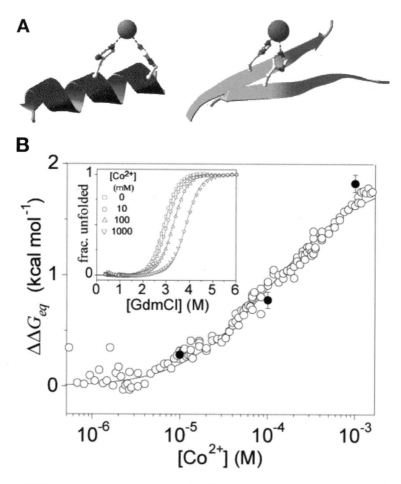

Fig. 1. BiHis sites and metal-induced stabilization. **(A)** Sample biHis sites located in an *i, i+4* arrangement on a helix and across two β-strands. **(B)** Increase in folding stability on the addition of divalent metal ions from folding kinetics (○) and from denaturant melts (●) for a helical site in ubiquitin variant *(14)*. **(Inset)** Standard GdmCl denaturation profiles at different [Co^{2+}] that are fit to a two-state equilibrium model. (Reprinted from **ref. *14***, with permission from Elsevier.)

positions (HXXXH, *see* **Note 1**). Metal sites can be introduced across a hairpin or β-strand and site selection does not appear to be very stringent. For example, a histidine on one strand of a β-sheet can form a biHis site with residues on either adjacent strand in ubiquitin (*see* **Fig. 2A**). However, β-sheets often are quite twisted and care must be taken to use two positions where the side chains are not angled away from each other. Previously, we have successfully replaced an interhelical salt-bridge with a biHis site, suggesting that

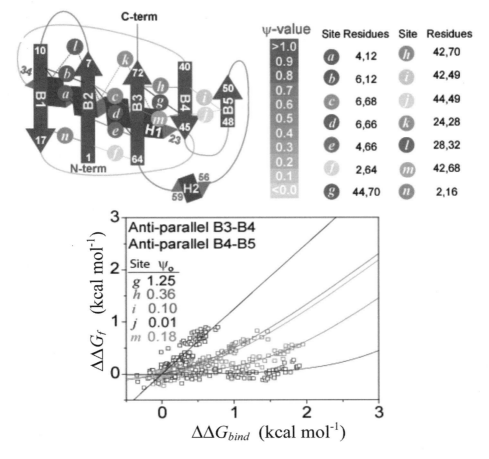

Fig. 2. Ψ-Analysis applied to Ub. **(A)** Schematic representation of biHis sites (circles with italic letters, each site was studied individually) and Ψ-values. The gray intensity represents the value of Ψ. Renderings were created in the Swiss-Prot Protein Viewer (Glaxo Wellcome). **(B)** Leffler plots for five sites illustrating the Ψ-values ranging from zero to unity. (Reprinted from **ref. *14*** with permission from Elsevier.)

pre-existing side chain interactions also may be good candidates for metal site replacement.

3.2. Equilibrium Studies

The suitability of each biHis site must be investigated to verify that introduction of the site does not perturb the structure of the protein either in the presence or absence of metal. Verifying the lack of structural perturbation with and without metal ions using NMR methods is possible, albeit very time consuming. In most cases, surface histidine substitutions have been found to have little effect on the structure or folding behavior of proteins, compared with the more common use of core residue substitutions in other folding studies. Nevertheless, the near-ultraviolet circular dichroism spectra can be measured with and without metal to confirm no gross structural changes occur as a result of metal binding.

Equilibrium denaturant titrations should be conducted at several metal concentrations, including near-saturating concentrations (e.g., ≥ 1 mM). These data provide a number of useful quantities:

- The energetic cost of inserting biHis mutation. Generally, biHis mutations are slightly destabilizing with reference to the wild-type, usually $\Delta\Delta G_{eq} < 2$ kcal/mol. This value is required later for correcting the Ψ-value to account for the biHis mutation (**Eq. 10**).
- Change in denaturant m^o-value. This parameter, which reflects the amount of surface burial in the folding transition, should be largely unchanged upon the addition of metal. A decrease in the m^o-value may indicate the formation of residual structure in the denatured state, or a perturbation of the native structure.
- Quantifying the maximal amount of metal-induced stabilization. The change in free energy of metal binding under saturating conditions identifies the experimental limits of metal-induced stabilization and the sensitivity to minor pathways. Values of $\Delta\Delta G_{bind}$ typically range from 0.5 to 3 kcal mol^{-1} and vary among divalent ions. We have found that cobalt stabilizes α-helices to a greater extent, whereas zinc and nickel tend to prefer β-sheet sites, although this correlation is empirical and several different metals should be tested at the outset (*see* **Note 2**). The use of the most stabilizing ion increases the accuracy in which fractional Ψ-values can be determined.
- The value of $\Delta\Delta G_{bind}$ is to be compared with that obtained from chevron analysis to confirm that metal binding is in fast equilibrium during the kinetic measurements, a requirement for implementation of the method.

3.3. Chevron Analysis

Although Ψ-analysis can be applied to more complicated reactions, for simplicity its application is illustrated here with a kinetically two-state system. First, a chevron (*6*) is measured on the biHis variant in order to compare the folding behavior of the mutant to that of the wild-type protein. The refolding and unfolding denaturant m_f- and m_u-values should be largely unchanged compared with the wild-type, signifying no major affect on surface area burial during the course of the reaction. Just as in mutational ϕ-analysis, data from systems with changing m-values should be interpreted cautiously as the biHis substitution may have altered the conformation of the denatured, native, or transition states. From a comparison of the chevrons for the wild-type and mutant proteins, a double-site $\phi^{wt\text{-}biHis}$ value can be calculated. However, translation of this ϕ-value to structure in the TS may be difficult, as the perturbation resulting from the introduction of the biHis site often is unclear.

Additional chevrons should be measured under near-saturating metal conditions to determine if metal binding grossly alters the folding pathway of the biHis variant (*see* **Fig. 3A**). This experiment is performed as usual except with buffers that contain divalent cations (*see* **Notes 3** and **4**). These kinetic studies can be used to calculate the maximal stabilization imparted by the binding of several different metal ions. Chevron data are acquired at a single high metal concentration for each cation, and the $\Delta\Delta G_{bind}$ is determined from the change in k_f and k_u (*see* **Note 5**). These values should match those obtained in the equilibrium measurements to confirm that cation binding is in fast equilibrium.

Once these chevrons have been acquired, a Ψ-value can be estimated by comparing the chevron measured without metal to one at saturating metal concentration. If the high-metal chevron shifts only the folding arm up (folding is faster) while leaving the unfolding arm unchanged, then metal binding stabilizes both the TS and the native state equally and the Ψ-value is unity. Conversely, if the presence of metal only shifts the unfolding rate down (unfolding is slower) then metal binding stabilizes only the native state, implying $\Psi \sim 0$. When both arms shift, the Ψ-value is fractional. A Ψ_0-value can be calculated using a two-point Leffler plot with data in the absence and presence of metal ion and fit to the single free parameter **Eq. 5**. More detailed metal-dependent folding measurements should be conducted to determine a more precise value of Ψ_0 (**Subheading 3.4.**).

3.4. Metal-Dependent Folding Kinetics

Additional chevrons can be measured to obtain more detailed information on the metal dependence of folding rates, each at a fixed metal ion concentration traversing the range of interest (e.g., 0.0, 0.1, 0.2, 0.4, 1 mM [Me^{2+}]) (*see* **Fig. 3A** and **Table 1**). Alternatively, the denaturant concentration can be fixed, while the folding and unfolding rates are measured at multiple, finely-spaced metal concentrations (*see* **Fig. 3B**). Specifically, the folding rates are measured at a single denaturant concentration under strongly folding conditions (*see* **Fig. 3B**, Line i), whereas the unfolding rates are measured at a single denaturant concentration under strongly unfolding conditions (*see* **Fig. 3A**, Line iii). When choosing the final denaturant concentration for the unfolding measurements, bear in mind that metal-induced stabilization shifts the chevron to the right. As a result, unfolding conditions in the absence of metal ions could become folding conditions with the addition of metal (*see* **Fig. 3A**, Line ii). For this reason, it is advisable to measure a chevron under saturating metal conditions first so that appropriate folding and unfolding conditions can be chosen at the outset.

This second strategy—varying metal concentration at a fixed denaturant value—effectively is a "vertical slice" through a multitude of denaturant chevrons in the presence of differing metal ion concentrations. The advantage of this

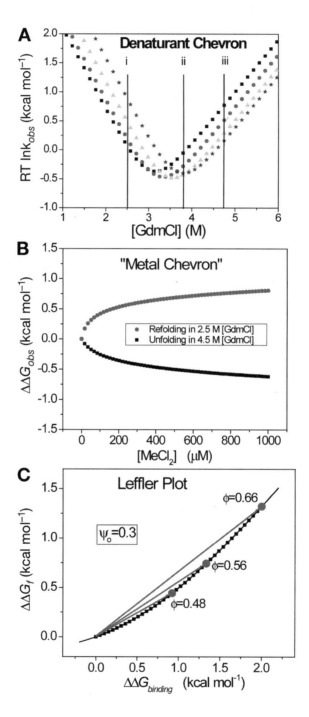

approach is that many more points are obtained for the Leffler plot, which is used to calculate the Ψ-value (*see* **Fig. 3C**). The advantage of the first method is that one can monitor changes in the *m*-values as a function of metal ion concentration, although this is usually not necessary once saturating metal chevrons are measured.

The implementation of the second strategy depends on the nature of the equipment used. Our studies have used four-syringe Biologic SFM-4 and SFM-400 stopped-flow apparatuses (www.bio-logic.fr). The metal-dependent folding measurements can be performed with this machine using a three-syringe mixing protocol where one syringe contains a protein/denaturant mixture. The two other syringes have identical denaturant concentrations, but one syringe contains metal ions (*see* **Table 2**). By varying the relative delivery volumes of these

---

Fig. 3. *(Opposite page)* **(A)** Kinetic analysis of biHis variant as a function of metal concentration. Sample chevron data representing the denaturant dependence of folding and unfolding rates. The left half of the curve represents a rapid jump from unfolding conditions to the denaturant concentration indicated on the *x*-axis (folding arm). The right half of the chevron represents the unfolding arm—kinetic relaxation rates under unfolding conditions. Chevron analysis of a biHis variant in the presence of 0 (square), 25 µ*M* (circle), 200 µ*M* (triangle), and 1 m*M* metal ion concentration (star). The folding and unfolding rates in water are determined by extrapolating both arms to the *y*-axis, which change as a function of metal. The parallel arms indicate no change in the denaturant dependence (*m*-values) as a function of metal. The vertical lines indicate (1) the final guanidinium chloride concentration [GdmCl] where refolding experiments should be performed, (2) where unfolding should be performed, and (3) denaturant concentration apparently ideal for unfolding experiments can become approach refolding conditions at higher metal ion concentrations. **(B)** Metal-dependent folding kinetics "Metal Chevron." Kinetic folding and unfolding data as a function of divalent metal ion concentration. The circles represent the kinetic response of rapid folding measurements with a final concentration of 2.5 *M* GdmCl as the concentration of metal is increased from 0 to 1 m*M*. Squares represent unfolding data taken at 4.5 *M* GdmCl final denaturant concentration. **(C)** Leffler plot. Squares indicate a change in the free energy of folding as a function of binding energy. Each metal chevron data at a different metal ion concentration (shown in panel **A**) yields one data point. The parameter Ψ_o is equal to the instantaneous slope at the origin (here shown with a value of 0.3) and is determined from fitting the data to the functional form given in **Eq. 5**. The slope approaches one with a sufficiently high amount of stabilization. This plot is analogous to traditional φ-analysis plots, although free energy relationships are generally assumed to be linear. The circles represent Leffler data generated from changes in relative free energy of folding and unfolding using denaturant chevrons at the three different metal concentrations (shown in panel **A**). These points can generate apparent φ-values using the two-point fit from the origin, although underlying curvature can lead to systematic errors in φ-value results.

Pandit et al.

Table 1
Sample Shot Protocol for a Denaturant Chevron at High Metal Ion Concentration[a]

| | | | Protein
4 M Gdm
6 mM Me^{2+} | |
| | | Buffer | 4 M Gdm | | |
[Metal] μM	[GdmCl] M	Syringe 1 (μL)	Syringe 2 (μL)	Syringe 3 (μL)	Total vol (μL)
1000	0.83	250	0	50	300
1000	1.50	200	50	50	300
1000	2.17	150	100	50	300
1000	2.83	100	150	50	300
1000	3.50	50	200	50	300
1000	4.17	0	250	50	300

[a]The metal concentration is designed to stay constant while the denaturant concentration changes. For a three-syringe protocol, syringe 1 contains buffer, syringe 2 contains 4 M GdmCl with buffer, and syringe 3 contains protein, 5 M GdmCl, 6 mM MeCl$_2$, and buffer.

Table 2
Sample Shot Protocol for Measuring Metal-Dependent Folding Rates[a]

| | | | 0 M Gdm + 1
mM Me^{2+} | Protein
5 M Gdm | |
| | | 0 M Gdm | | | |
[Metal] μM	[GdmCl] M	Syringe 1 (μL)	Syringe 2 (μL)	Syringe 3 (μL)	Total vol (μL)
0	0.83	250	0	50	300
166.7	0.83	200	50	50	300
333.3	0.83	150	100	50	300
500	0.83	100	150	50	300
666.7	0.83	50	200	50	300
833.3	0.83	0	250	50	300

[a]The shot protocol is designed to vary metal concentration while keeping denaturant concentration constant. Syringe 1 contains buffer, syringe 2 contains buffer with 1 mM MeCl$_2$, and syringe 3 contains protein in 5 M GdmCl.

otherwise identical buffers, a large amount of finely spaced data can be obtained over a range of ion concentrations using a single set of buffer solutions.

A standard two-syringe, fixed volume ratio stopped-flow apparatus can be used as well, although the solutions must be changed between each metal ion condition.

Often stopped-flow apparatus have limited dilution ratios resulting from inaccuracies arising from low-volume delivery. In our experimental setup, one set of buffers provides data over approximately one decade in metal ion concentration. To cover a wider range, only the metal-containing buffer needs be replaced with one of lower concentration. This change readily can be accomplished using dilutions of the metal buffer using the no-metal buffer, e.g., lower ranges are obtained by fivefold serial dilutions of the metal-containing buffer, e.g., 5 mM, 1 mM, 200 μM, and 40 μM (Me^{2+}).

Once refolding data has been collected over the range of metal ion concentration, the process must be repeated for the unfolding measurements. The data should be taken at the same metal concentrations as the folding data to allow calculation of $\Delta\Delta G_{bind}$ from the difference of $\Delta\Delta G_f$ and $\Delta\Delta G_u$ according to

$$\Delta\Delta G_{bind}([Me^{2+}]) = \Delta\Delta G_f - \Delta\Delta G_u = RT \ln (k_f^{M2+}/k_f) - RT \ln (k_u^{M2+}/k_u) \qquad (1)$$

where R is the gas constant and T is the absolute temperature, and k_f^{M2+} and k_u^{M2+} are the relaxation rates in the presence of the same concentration of cation. The folding rate at zero metal should also be well established, as this rate serves as the reference point in the Leffler plot.

3.4.1. Testing for Fast Equilibrium

If the folding rates are fast compared with metal binding or release, then binding may not be in fast equilibrium during the folding reaction. As a result, the folding and binding processes will be convoluted. For fast folding rates, metal ions may even no longer stabilize the TS as the biHis site is kinetically inaccessible for ion binding. Furthermore, if metal release rates are slower than unfolding rates, multiple populations will be observed in unfolding experiments; some molecules will unfold having metal bound, whereas other molecules will unfold as if there is no metal ion present in solution. When binding is no longer in fast equilibrium, the Ψ-analysis formalism will no longer be valid.

A required signature of fast metal binding equilibrium is the lateral translation of chevrons with increasing metal ion concentration. There should be no changes in the slope or kinks in the arms of plot, or additional kinetic phases. An additional test to check for fast equilibrium is confirmation that metal-induced stabilization determined from the equilibrium studies should match those from kinetic measurements (*see* **Eq. 1**). If metal binding is not in fast equilibrium, different metals can be tested as their binding properties may be more suitable. Also, folding rates can be manipulated by working at lower temperatures or higher denaturant concentrations.

3.5. Data Analysis

3.5.1. Equilibrium Denaturation Profiles

The folding transitions in the presence of metal ions are fit to the standard equation assuming a two-state equilibrium between the N(ative) and U(nfolded) states:

$$S([den]) = \frac{S_U + S_N \, e^{-(\Delta G + m\,[den])/RT}}{1 + e^{-(\Delta G + m\,[den])/RT}} \qquad (2)$$

where S_U and S_N are the signals of the U and N states, respectively. Parameters are fit using a nonlinear least-squares algorithm (e.g., Microcal Origin software package).

3.5.2. Metal Stabilization

The degree of stabilization owing to metal binding, $\Delta\Delta G_{bind}$, depends on a difference in metal dissociation constants between the biHis sites in the native state, K_N, and in the unfolded state, K_U (*see* **Fig. 1B**). The increase in protein stability on the addition of metal is fit to a linked equilibrium expression *(32)*

$$\Delta\Delta G_{bind}(M^{2+}) = RT \ln(1+[M^{2+}]/K_N) - RT \ln(1+[M^{2+}]/K_U) \tag{3}$$

3.5.3. Kinetic Chevron Analysis

The kinetic data are analyzed using chevron analysis of the denaturant dependence of folding rate constants *(6)*, where the standard free energy of folding, ΔG_{eq}, along with the standard activation free energy for folding, ΔG_f, and unfolding, ΔG_u, are linearly dependent on denaturant concentration.

$$\Delta G_{eq}([\text{Den}]) = \Delta G_{eq}^{\text{H2O}} + m^\circ[\text{Den}] = -RT \ln K_{eq} \tag{4A}$$

$$\Delta G_f([\text{Den}]) = -RT \ln k_f^{\text{H2O}} - m_f[\text{Den}] + constant \tag{4B}$$

$$\Delta G_u([\text{Den}]) = -RT \ln k_u^{\text{H2O}} - m_u[\text{Den}] + constant \tag{4C}$$

The denaturant concentration dependencies (*m*-values) report on the degree of surface area burial during the folding process *(33)*. When equilibrium and kinetic folding reactions are two-state and are limited by the same activation energy barrier, the equilibrium values for the standard free energy and surface burial can be calculated from kinetic measurements according to $\Delta G_{eq} = \Delta G_f - \Delta G_u$ and $m^\circ = m_u - m_f$.

3.5.4. Obtaining Ψ-Values From the Leffler Plot

The Leffler *(18)* plot is obtained from the change in the folding activation energy relative to the change in stability from metal ion binding (**Figs. 2B,3C**). $\Delta\Delta G_f$ is derived from the ratio of the folding rate in the presence (k_f^{M2+}) and absence of metal (k_f) using $\Delta\Delta G_f = RT \ln (k_f^{M2+}/k_f)$. $\Delta\Delta G_{bind}$ can be obtained from the equilibrium data (*see* **Eq. 3**) or from the change in the folding and unfolding rates taken at identical metal ion concentrations (*see* **Eq. 1**). When multiple chevrons are obtained, each at a different metal ion concentration, one Leffler data point is obtain for each chevron. When folding and unfolding data are obtained at fixed denaturant concentrations at varying metal ion concentrations, one Leffler data point is obtained for each metal concentration.

The plot will be either linear or curved depending on the degree of structure formation and heterogeneity in the TS. In either case, the data can be fit with the same single parameter equation:

$$\Delta\Delta G_f = RT \ln\left((1 - \Psi_o) + \Psi_o e^{\Delta\Delta G_{bind}/RT}\right) \tag{5}$$

where Ψ_0 is the instantaneous slope at the origin where data is obtained in the absence of metal ions. The instantaneous slope (the Ψ-value) at any point on the curve as a function of binding stability is given by:

$$\Psi = \frac{\partial\Delta\Delta G_f}{\partial\Delta\Delta G_{bind}} = \frac{\Psi_o}{(1 - \Psi_o)e^{-\Delta\Delta G_{bind}/RT} + \Psi_o} \tag{6}$$

3.5.5. Interpreting Ψ-Values

Once the value for Ψ_0 has been determined, the next task involves understanding its significance. The interpretation is clear in the two cases where the Leffler plot is linear, $\Psi_0 = 0$ or 1. For a value of unity, the biHis site is present (native-like) in the TS ensemble. For a value of zero, the site is absent (unfolded-like). In other cases, the Leffler plot will be curved as ligand binding continuously increases the stability of the TS ensemble (*34*). The curvature can be owing to TS heterogeneity, non-native binding affinity in a singular TS, or a combination thereof (*8,35*) (D. Goldenberg, private communication; *see also* Fersht [*36*] for a comparison of the Ψ- and ϕ-analysis methods using an alternative kinetic model which focuses on unfolded state population shifts while omitting any consideration of TS binding).

The proper interpretation of fractional Ψ-values involves an appreciation of the mathematical formalism behind **Eqs. 5** and **6**. Ψ-Analysis takes into account the shifts in the native, unfolded, and TS state populations resulting from the binding of the metal ion to each of these states (*see* **Fig. 4**). Folding rates are calculated assuming two classes of TSs depending on whether the biHis site is present ($k^{present}$) or absent (k^{absent}) (*see* **Fig. 5**). In the first class, TS^{present}, the biHis site is present in a native or near-native geometry with a dissociation constant $K_{TS}^{Present}$. In this case the associated backbone structure is folded, for example in a helical or β-sheet conformation. In the second class (TS_{absent}) the biHis site is essentially absent but has an effective dissociation constant K_{TS}^{Absent}, just as the unfolded state binds metal with a dissociation constant K_U (*see* **Eq. 3**). As a result, the model contains two TSs, each having distinct effective binding affinities $K_{TS}^{Present}$ and K_{TS}^{Absent}.

As per Eyring reaction rate theory (*37*), the overall reaction rate is taken to be proportional to the relative populations of the TS and U ensembles,

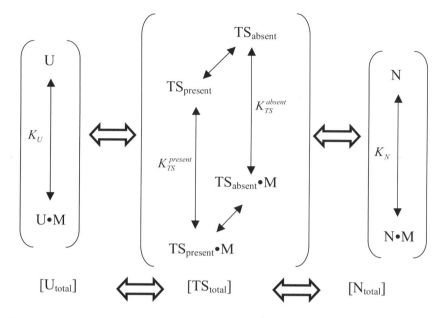

Fig. 4. Thermodynamic states considered in Ψ-analysis. The U, TS_{absent}, $TS_{present}$, and N states bind metal with affinities of, and, respectively. The degree of TS heterogeneity is defined as $\rho = [TS_{absent}]/[TS_{present}]$.

$k_f \propto [TS]/[U]$. The net folding rate is the sum of the rates going down each of the two routes, $K_f = k^{present} + k^{absent}$ or

$$k_f \equiv k_0^{present} \frac{1+[M]/K_{TS}^{present}}{1+[M]/K_U} + k_0^{present} \frac{1+[M]/K_{TS}^{absent}}{1+[M]/K_U} \qquad (7)$$

Fig. 5. *(Opposite page)* (**A**) Generalized Leffler plot for $\Psi_o = 0, 0.1$, and 1 (lower panel) and the derivative of each of the traces (upper panel). The $\Psi_o = 0.1$ trace is applicable to the scenario shown below with the gray lines illustrating the initial condition and after 2.86 kcal mol^{-1} of metal-induced stabilization. (**B**) Application of Ψ-analysis to a two-route scenario with a helical site with native binding affinity that is formed on 9% of the transition states prior to addition of metal. The absent route contains a TS that has the same binding affinity as the U state. The folding rate for the route with the biHis site present ($k^{present}$, lower pathway) increases from 1 to 100 upon the addition of 2.86 kcal mol^{-1} of metal ion binding energy at 20°C. This enhancement increases the flux down the metal ion stabilized route relative to all other routes (k^{absent}), from $\rho_o = k^{absent}/k_o^{present} = 10/1$ to metal-enhanced condition $\rho_M = 10/100$. The corresponding Ψ-values increase from $\Psi_o = 0.1$ to $\Psi_M = 0.9$. The binding energy required to stabilize a TS and switch a minor route to a major route identifies the barrier height for this route relative to that for all other routes. (Reprinted from **ref. 8** with permission from National Academy of Sciences, © 2004.)

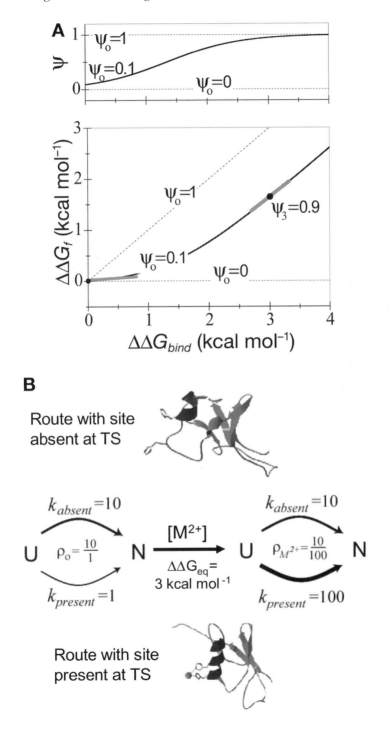

where $k_o^{present} \propto [TS_{present}]/[U]$ and $k_o^{absent} \propto [TS_{absent}]/[U]$ are the rates through each TS class prior to the addition of metal, with a ratio $\rho_0 \equiv k_o^{absent}/k_o^{present}$. By examining shifts in populations and assuming metal binding is in fast equilibrium, this treatment avoids any assumptions about possible pathways connecting the different bound and unbound states.

The Leffler plot will be linear or curved depending on the relative values of the binding constants and the degree of TS heterogeneity. The slope at the origin is

$$\Psi_0 = \frac{K_N}{K_N - K_U}\left(1 - \frac{K_U}{K_{TS}^{absent}}\right) - \frac{K_N K_U}{K_N - K_U}\left(\frac{1}{K_{TS}^{present}} - \frac{1}{K_{TS}^{absent}}\right)\frac{k_o^{present}}{k_o^{present} + k_o^{absent}} \tag{8}$$

As previously mentioned, there are two scenarios where the slope in the Leffler plot is linear. In the first scenario, the slope is zero across all metal concentrations ($\Psi = \Psi_o = 0$). This behavior occurs when metal ion binding does not increase the population of the biHis site in the TS ensemble relative to U. In this case, the entire TS ensemble lacks the binding site, or more rigorously, the site has the same binding affinity as the unfolded state. At the other limit, the slope is one ($\Psi = \Psi_o = 1$) indicating that the entire ensemble has the binding site formed with native-like affinity.

Otherwise, the Leffler plot will be curved as metal continuously increases the population configurations having the biHis site formed in the TS ensemble. In such cases the curvature can be due to TS heterogeneity, non-native binding affinity in a singular TS, or a combination thereof.

In the heterogeneous scenario, one can consider the simplified situation (*see* **Fig. 5**), where $TS_{present}$ has the biHis site present with native-like affinity ($K_{TS}^{present} = K_N$) while TS_{absent} has the site with the unfolded-like affinity ($K_{TS}^{absent} = K_U$). Here only the $TS_{present}$ state is stabilized upon the addition of metal ion. The height of the kinetic barrier associated with $TS_{present}$ decreases to the same degree as does the native state $k^{present} = k_o^{present} \, e^{\Delta\Delta G_{bind}/RT}$. The instantaneous slope simplifies to the fraction of the TS ensemble which has the biHis site formed at a given metal ion concentration:

$$\Psi = \frac{k^{present}}{k^{present} + k_o^{absent}} \tag{9}$$

In this simplified situation, the degree of pathway heterogeneity prior to the addition of metal ions is given by Ψ_o, the slope at zero stabilization. The Leffler plot exhibits upward curvature as the Ψ-value increases with added metal binding energy, which increases the fraction of the TS ensemble that has the biHis site present. Generally, Ψ-values continuously vary between 0 and 1 at the limits of for several different biHis variants can be combined to construct an accurate representation of the TS ensemble appropriate for the wild-type protein prior to mutation or metal binding.

3.5.7. Delineation Between the Heterogeneous and Homogeneous Scenarios

The question of whether fractional Ψ-values reflect TS heterogeneity or non-native binding affinity remains unresolved, and may be site dependent. We believe discrimination between these two models is possible through study of folding in the presence of two separate metal ions with different coordination geometries. The two ions are likely to manifest the same Ψ_o-value only in the case of TS heterogeneity, because the same fractional binding affinity is unlikely to be realized with both ions. However, Ψ_o-values should depend on the type of metal ion if the site is distorted.

Another test for heterogeneity involves altering the relative stability of the TS structure with the site present, e.g., via mutation far from the biHis site. If the Ψ-value responds accordingly, as we observed in the dimeric GCN4-coiled coil *(16)*, the heterogeneity model is the most parsimonious. For the coiled coil, the introduction of the destabilizing glycine (A24G) shifted the pathway flux away from this region so that most nucleation events occurred near the biHis site, which was located at the other end of the protein. As expected, the Ψ_o value increased to 0.5, indicating that half of the nucleation events occurred with the biHis site formed.

A quantitative comparison indicated the change in pathway heterogeneity recapitulated the destabilizing effect of the glycine substitution in GCN4 *(16)*. The A24G mutation increased the amount of flux going through the N-terminal biHis site. The ratio of the heterogeneity in these two molecules reflected the loss in stability for this mutation, $\Delta\Delta G_{eq} = RT \ln (\rho_{Ala}/\rho_{Gly}) = 2.5$ kcal/mol. This shift was consistent with the decrease in stability for the mutation backgrounds (1.7 – 2.4 kcal/mol) *(38)*. Hence, in this case, Ψ-analysis successfully quantified the level of TS heterogeneity. A homogeneous model with non-native binding affinity in the TS would require that the A24G mutation causes the biHis site to acquire native-like binding affinity. This is an unlikely scenario given the distance between the substitution and the biHis site. Potentially, binding sites introduced into well-defined helices will have native-like binding affinities in the TS, in which case, fractional Ψ-values will be due to TS heterogeneity.

4. Conclusion

The application of Ψ-analysis can provide detailed, site-resolved information on TS structures of protein folding pathways, as well as other conformational transitions. The use of this method only requires introduction of biHis sites on the surface of the protein and metal-dependent kinetic measurements, both of which are relatively undemanding. For the two limiting situations, $\Psi_o = 0$ or 1, infinite TS destabilization and stabilization, respectively. When the Ψ-value is 0.5, the site is formed half the time in the TS ensemble.

One can also consider a situation where the biHis site in $TS_{present}$ has non-native binding affinity ($K_{TS}^{present} \neq K_N$), whereas the site in TS_{absent} has unfolded-like affinity ($K_{TS}^{absent} = K_U$). Now, the initial slope is the degree of heterogeneity multiplied by an additional factor representing the differential binding affinity between $TS_{present}$ and N

$$\Psi_o = \frac{K_N}{K_{TS}^{present}} \frac{K_{TS}^{present} - K_U}{K_N - K_U} \frac{k_o^{present}}{k_o^{present} + k_o^{absent}} \tag{10}$$

Finally, curvature can also occur in a homogeneous scenario when the sole TS has non-native binding affinity *(8,35)*. Here the curvature reflects the stabilization of the TS relative to U with an initial slope of

$$\Psi_0 = \frac{K_N(K_{TS} - K_U)}{K_{TS}(K_N - K_U)} \tag{11}$$

Here, the curvature reflects the stabilization of the single TS relative to U. It is important to note that for this homogeneous scenario, the aforementioned interpretation of the two linear Leffler behaviors, $\Psi = 0$ or 1, remain unchanged.

The analysis of metal binding presented here is slightly different than that presented in our earlier papers *(14,16)* where curvature was associated only with the heterogeneous model. With the explicit inclusion of the binding affinities in the TS, $K_{TS}^{present}$ and K_{TS}^{absent}, the f-value ($f \equiv \Delta\Delta G_f/\Delta\Delta G_{eq}$) is no longer required. It is generally not constant as it depends on metal ion concentration, except in the two linear scenarios where $f = 0$ or 1.

3.5.6. Correcting for the Effects of the biHis Site

The introduction of the biHis substitution itself alters the stability of the native state by the amount $\Delta\Delta G_{eq}^{biHis}$. In the simplified, heterogeneous scenario where $K_{TS}^{present} = K_N$ and $K_{TS}^{absent} = K_U$, the Ψ_o value should be corrected in order to account for this change in stability

$$\Psi_o^{corr} = \frac{\Psi_o}{\Psi_o + e^{-\Delta\Delta G_{eq}^{biHis}/RT}(1 - \Psi_o)} \tag{12}$$

The resulting Ψ_o^{corr} is the instantaneous Leffler slope at which the metal ion binding energy is exactly offset by the change in stability from the biHis substitution. This correction is justified because both metal binding and the biHis substitution affect the same region of the protein. With this correction, the Ψ-values the region of the protein where the biHis site is introduced is either unfolded or folded, respectively. Fractional Ψ-values indicate the biHis site is either fractionally populated and/or distorted with non-native binding affinity in the TS. With the introduction of sufficient number of biHis sites, the topology of the entire TS structure can be identified. When combined with mutational studies, modeling, and other information, a complete picture of the TS ensemble can be determined.

5. Notes

1. BiHis sites can be mutated sequentially using a common method, the QuikChange protocol from Stratagene. However, engineering a biHis site into a helix is possible in one step using a single primer with both mutations encoded. This strategy places the mutagenic codons nine nucleotides apart with 10–15 complementary residues on either end, resulting in a very long mutagenic primer (>40 nt). Engineering in two-point mutations with one step does save time but bear in mind the polymerase chain reaction is less likely to be successful. When using this approach, it is best to lower the annealing temperature 5–10°C to improve the reaction efficiency.

2. To quickly test the amount of stabilization imparted by each type of metal ion, the protein can be placed in a denaturant solution where approx 20% of the molecules are folded ($K_{eq} = [N]/[U] = 1/4$). This level of denaturant can be obtained from a denaturation profile. The addition of high concentrations of metal will renature a fraction of the molecules according to the degree of metal-induced stabilization. From the increase in the equilibrium constant, K'_{eq}, the stabilization can be calculated $\Delta\Delta G_{bind} = -RT \ln (K'_{eq}/K_{eq})$.

3. Metal ions can be introduced into the folding reaction by including them in the syringe containing the protein solution, taking into account the dilution factor of the final mix (*see* **Table 1**).

4. The addition of metal can stabilize a protein to the point where higher than convenient levels of denaturant are required to unfold the protein. Rather than adding the metal only to the protein solution, the same experiment can be performed with metal ions in all buffers at the desired concentration. Alternatively, the nonprotein buffers both can contain metal at a concentration calculated to give the desired final value. The method suggested allows for the reuse of excess denaturant buffers in other experiments and reduces waste.

5. Several different cations can be readily tested using a configuration where metal ion is placed only in one syringe. The testing of different metal ions using this protocol requires only changing the contents of the single metal-containing syringe. However, metal concentrations in this syringe will be higher than if the ion was placed in all syringes and precipitation may become an issue.

Acknowledgments

We thank G. Bosco and A. Shandiz for comments, M. Baxa for assistance in deriving the equations, and all members of our group for extensive discussions and input. This work was supported by grants from the National Institutes of Health.

References

1. Venclovas, C., Zemla, A., Fidelis, K., and Moult, J. (2003) Assessment of progress over the CASP experiments. *Proteins* **53**, 585–595.
2. Jackson, S. E. (1998) How do small single-domain proteins fold? *Fold. Des.* **3**, R81–R91.
3. Krantz, B. A. and Sosnick, T. R. (2000) Distinguishing between two-state and three-state models for ubiquitin folding. *Biochemistry* **39**, 11,696–11,701.
4. Krantz, B. A., Mayne, L., Rumbley, J., Englander, S. W., and Sosnick, T. R. (2002) Fast and slow intermediate accumulation and the initial barrier mechanism in protein folding. *J. Mol. Biol.* **324**, 359–371.
5. Jacob, J., Krantz, B., Dothager, R. S., Thiyagarajan, P., and Sosnick, T. R. (2004) Early collapse is not an obligate step in protein folding. *J. Mol. Biol.* **338**, 369–382.
6. Matthews, C. R. (1987) Effects of point mutations on the folding of globular proteins. *Methods Enzymol.* **154**, 498–511.
7. Fersht, A. R., Matouschek, A., and Serrano, L. (1992) The folding of an enzyme. I. Theory of protein engineering analysis of stability and pathway of protein folding. *J. Mol. Biol.* **224**, 771–782.
8. Sosnick, T. R., Dothager, R. S., and Krantz, B. A. (2004) Differences in the folding transition state of ubiquitin indicated by phi and psi analyses. *Proc. Natl. Acad. Sci. USA* **101**, 17,377–17,382.
9. Feng, H., Vu, N. D., Zhou, Z., and Bai, Y. (2004) Structural examination of Phi-value analysis in protein folding. *Biochemistry* **43**, 14,325–14,331.
10. Sanchez, I. E. and Kiefhaber, T. (2003) Origin of unusual phi-values in protein folding: evidence against specific nucleation sites. *J. Mol. Biol.* **334**, 1077–1085.
11. Bulaj, G. and Goldenberg, D. P. (2001) Phi-values for BPTI folding intermediates and implications for transition state analysis. *Nature Struct. Biol.* **8**, 326–330.
12. Ozkan, S. B., Bahar, I., and Dill, K. A. (2001) Transition states and the meaning of Phi-values in protein folding kinetics. *Nature Struct. Biol.* **8**, 765–769.
13. Fersht, A. R. and Sato, S. (2004) Phi-Value analysis and the nature of protein-folding transition states. *Proc. Natl. Acad. Sci. USA* **101**, 7976–7981.
14. Krantz, B. A., Dothager, R. S., and Sosnick, T. R. (2004) Discerning the structure and energy of multiple transition states in protein folding using psi-analysis. *J. Mol. Biol.* **337**, 463–475.
15. Raleigh, D. P. and Plaxco, K. W. (2005) The protein folding transition state: what are phi-values really telling us? *Protein Pept. Lett.* **12**, 117–122.
16. Krantz, B. A. and Sosnick, T. R. (2001) Engineered metal binding sites map the heterogeneous folding landscape of a coiled coil. *Nature Struct. Biol.* **8**, 1042–1047.
17. Brønsted, J. N. and Pedersen, K. (1924) The catalytic decomposition of nitramide and its physico-chemical applications. *Z. Phys. Chem. A* **108**, 185–235.
18. Leffler, J. E. (1953) Parameters for the description of transition states. *Science* **107**, 340–341.

19. Dwyer, M. A., Looger, L. L., and Hellinga, H. W. (2003) Computational design of a Zn2+ receptor that controls bacterial gene expression. *Proc. Natl. Acad. Sci. USA* **100,** 11,255–11,260.
20. Liu, H., Schmidt, J. J., Bachand, G. D., et al. (2002) Control of a biomolecular motor-powered nanodevice with an engineered chemical switch. *Nat. Mater.* **1,** 173–177.
21. Goedken, E. R., Keck, J. L., Berger, J. M., and Marqusee, S. (2000) Divalent metal cofactor binding in the kinetic folding trajectory of Escherichia coli ribonuclease HI. *Protein Sci.* **9,** 1914–1921.
22. Kim, C. A. and Berg, J. M. (1993) Thermodynamic beta-sheet propensities measured using a zinc-finger host peptide. *Nature* **362,** 267–270.
23. Webster, S. M., Del Camino, D., Dekker, J. P., and Yellen, G. (2004) Intracellular gate opening in Shaker K+ channels defined by high-affinity metal bridges. *Nature* **428,** 864–868.
24. Lu, Y., Berry, S. M., and Pfister, T. D. (2001) Engineering novel metalloproteins: design of metal-binding sites into native protein scaffolds. *Chem. Rev.* **101,** 3047–3080.
25. Higaki, J. N., Fletterick, R. J., and Craik, C. S. (1992) Engineered metalloregulation in enzymes. *TIBS* **17,** 100–104.
26. Morgan, D. M., Lynn, D. G., Miller-Auer, H., and Meredith, S. C. (2001) A designed Zn2+-binding amphiphilic polypeptide: energetic consequences of pi-helicity. *Biochemistry* **40,** 14,020–14,029.
27. Jung, K., Voss, J., He, M., Hubbell, W. L., and Kaback, H. R. (1995) Engineering a metal binding site within a polytopic membrane protein, the lactose permease of Escherichia coli. *Biochemistry* **34,** 6272–3277.
28. Vazquez-Ibar, J. L., Weinglass, A. B., and Kaback, H. R. (2002) Engineering a terbium-binding site into an integral membrane protein for luminescence energy transfer. *Proc. Natl. Acad. Sci. USA* **99,** 3487-3492.
29. Benson, D. E., Wisz, M. S., and Hellinga, H. W. (1998) The development of new biotechnologies using metalloprotein design. *Curr. Opin. Biotechnol.* **9,** 370–376.
30. Dwyer, M. A., Looger, L. L., and Hellinga, H. W. (2003) Computational design of a Zn2+ receptor that controls bacterial gene expression. *Proc. Natl. Acad. Sci. USA* **100,** 11,255–11,260.
31. Regan, L. (1995) Protein design: novel metal-binding sites. *Trends Biochem. Sci.* **20,** 280–285.
32. Sharp, K. A. and Englander, S. W. (1994) How much is a stabilizing bond worth? *Trends Biochem. Sci.* **19,** 526–529.
33. Myers, J. K., Pace, C. N., and Scholtz, J. M. (1995) Denaturant m values and heat capacity changes: relation to changes in accessible surface areas of protein unfolding. *Protein Sci.* **4,** 2138–2148.
34. Sancho, J., Meiering, E. M., and Fersht, A. R. (1991) Mapping transition states of protein unfolding by protein engineering of ligand-binding sites. *J. Mol. Biol.* **221,** 1007–1014.

35. Krantz, B. A., Dothager, R. S., and Sosnick, T. R. (2004) Erratum to Discerning the structure and energy of multiple transition states in protein folding using psi-analysis. *J. Mol. Biol.* **347,** 889–1109.
36. Fersht, A. R. (2004) ϕ value versus Ψ analysis. *Proc. Natl. Acad. Sci. USA* **101,** 17,327, 17,328.
37. Eyring, H. (1935) The activated complex in chemical reactions. *J. Chem. Phys.* **3,** 107–115.
38. Moran, L. B., Schneider, J. P., Kentsis, A., Reddy, G. A., and Sosnick, T. R. (1999) Transition state heterogeneity in GCN4 coiled coil folding studied by using multisite mutations and crosslinking. *Proc. Natl. Acad. Sci. USA* **96,** 10,699–10,704.

7

Advances in the Analysis of Conformational Transitions in Peptides Using Differential Scanning Calorimetry

Werner W. Streicher and George I. Makhatadze

Summary

Differential scanning calorimetry can measure the heat capacity of a protein/peptide solution over a range of temperatures at constant pressure, which is used to determine the enthalpy function of the system. There are several experimental factors that can have a significant impact on the determined enthalpy and subsequent derived thermodynamic parameters. These factors are discussed in terms of sample and instrument preparation, as well as data collection and analysis.

Key Words: Conformational transitions; thermodynamics; heat capacity; differential scanning calorimetry; peptides; proteins.

1. Introduction

Heat affects accompany polypeptide folding/unfolding, as any other chemical reaction. These heat affects can be measured directly using differential scanning calorimetry (DSC). More specifically, the enthalpy function of the system can be determined by measuring the heat capacity at constant pressure, C_p, over a range of temperatures.

DSC operates in differential mode, which means that the heat capacity of the protein in aqueous solution is measured relative to the heat capacity of buffer. Ideally, when the heat capacities and volumes of both sample and reference cells are identical, a single peptide–buffer scan will be sufficient. However, in reality, the sample and reference cells are slightly different and this difference has to be taken into consideration by recording a buffer–buffer scan. Thus, prior to starting the peptide–buffer scan, it is very important to establish baseline reproducibility, i.e., its relative position and shape.

Figure 1 shows a typical DSC profile for a two-state unfolding process. The area under the heat capacity profile represents the enthalpy of unfolding, the

From: *Methods in Molecular Biology, vol. 350: Protein Folding Protocols*
Edited by: Y. Bai and R. Nussinov © Humana Press Inc., Totowa, NJ

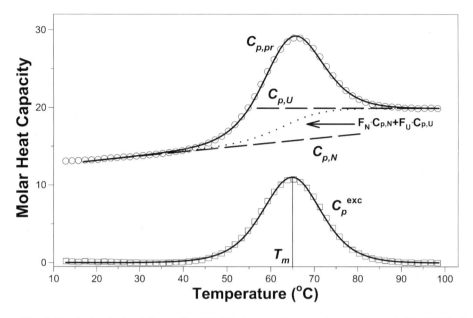

Fig. 1. Typical calorimetric profile (O) fitted according to a two-state model (solid-line). Also shown are the partial molar heat capacities of the native, $C_{p,N}$, and unfolded states, $C_{p,U}$, (dashed lines), the progress heat capacity, $F_N \cdot C_{pN} + F_U \cdot C_{pU}$ (dotted line), the excess heat capacity, C_p^{exc} experimental (□) and fitted (solid line). The thermodynamic parameters obtained as a result of the fit to a two-state model are $\Delta H = 200$ kJ/mol, $T_m = 65°C$, $\Delta C_p = 3.5$ kJ/(mol · K).

temperature at the maximum of the excess heat capacity profile is the transition temperature, and the difference in the heat capacities of the native and unfolded states defines the temperature dependence of the enthalpy and entropy functions, and, thus, the temperature dependence of peptide stability.

DSC also provides information about the modes of peptide unfolding, i.e., whether the unfolding process is two-state or multi-state. The effective enthalpy of transition, usually referred to as the van't Hoff enthalpy, is indicated by the sharpness of the heat capacity profile. The ratio of the experimental calorimetric enthalpy and the van't Hoff enthalpy provides information about the mode of the observed transition. A ratio equal to one indicates that the observed transition is a two-state process, proceeding from the native to the unfolded state without a significant population of intermediates (*see* **Fig. 1**). Deviation from unity indicates that the transition is more complicated *(1,2)*.

Recently, DSC was employed to analyze the thermodynamic properties of short monomeric peptides that form isolated α-helices (*see* **Fig. 2A**; *[3,4]*) or β-hairpins, in aqueous environments (*see* **Fig. 2B**; *[5]*). In the case of the α-helical peptide, the unfolded state is highly populated at temperatures higher than 90°C, whereas the helical state never becomes fully populated, even at low temperatures. In the case of the β-hairpin, the opposite is observed. The unfolded state is not fully populated even at 115°C, owing to its high transition temperature, whereas the folded state is well defined at low temperatures. The major challenge for these scenarios is to define the native and unfolded state baselines for the α-helical and β-hairpin peptides, respectively. The issue of the native-state baseline of the α-helical peptide can be solved by assuming that the heat capacity difference between the native (fully helical) and unfolded (coiled) states is very small (3,4). In this case, the heat capacities of the native and unfolded states are the same, which allows the temperature dependence of the unfolded state heat capacity to be calculated from the amino acid composition (6). The area between the experimental and calculated heat capacities represents the heat of helix unfolding. This heat can be normalized for the amount of helical structure at 0°C (estimated from circular dichroism or nuclear magnetic resonance experiments), to give the enthalpy of helix–coil transition. For the β-hairpin peptide, the unfolding transition appears to follow a two-state model, which allows the unfolded baseline to be obtained by fitting the data (5). The heat capacity for the unfolded state compares well with the heat capacity calculated using the amino acid composition of the peptide (4.6 ± 0.25 kJ/(mol · K) and 4.4 ± 0.2 kJ/(mol · K) at 75°C, respectively), further supporting the validity of the approach (*see* **Fig. 2**).

2. Materials

1. The DSC instruments designed to study biological systems are extremely sensitive and require small amounts of the material, 0.1–1.0 mg/mL of peptide/protein solution. Currently there are two commercial DSC instruments available, Nano-DSC from Calorimetric Science Corporation (Provo, UT) and VP-DSC from Microcal Inc. (Northhampton, MA). These instruments are fully automated for control, data collection, handling, and data analysis using personal computers. The VP-DSC is supplied with the ORIGIN graphics software.
2. A syringe with a precut needle is used to wash and load the cells (provided by the manufacturer). The needles are precut to a specific length so that the tip of the needle is barely above the bottom of the cell, when the overflow reservoir is in place.
3. Spectrophotometer and quartz cuvets for the cases in which the protein/peptide concentration is determined using ultraviolet spectroscopy. It is important to consider peptide quantitation at the peptide design stage and to include aromatic residues, if feasible, for this purpose (*see* **Note 1**).
4. Dialysis bags with the appropriate molecular weight cut-off (depending on the protein/peptide molecular weight).
5. Highly pure protein/peptide sample. Chemically synthesized peptides must be purified by reverse-phase high-performance liquid chromatography, which is typically performed in the presence of trifluoroacetic acid (TFA). To remove the TFA from the purified peptide, it is recommended that the lyophilized peptide be resuspended in deionized water and lyophilized, and the process repeated.

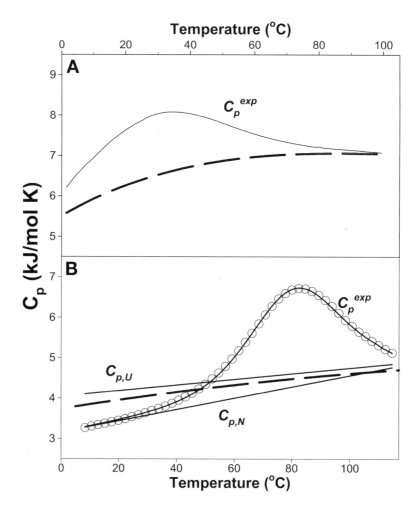

Fig. 2. Temperature dependence of the partial molar heat capacity. (**A**) α-Helical peptide AEARA6 (*4*). (**B**) β-Hairpin peptide trpzip4 (*5*). The thick dashed lines on both panels represent the heat capacity of the fully unfolded state calculated as described in **ref. 6**:

$$C_{p,U}^{calc}(T) = \sum_i n_i \cdot \hat{C}_{p,i}(T) + (N-1) \cdot C_{p,-CHCONH-}(T)$$ where N is the total number of

amino acid residues in the sequence, n_i is the number of i-th type amino acid residue, $\hat{C}_{p,i}(T)$ is the partial molar heat capacity for the side chain of the i-th type, and C_{p}–CHCONH–(T) is the partial molar heat capacity for the peptide unit. The values of $\hat{C}_{p,i}(T)$ and C_{p}–CHCONH–(T) have been derived for all twenty amino acid residues and for the peptide unit from the experimentally measured partial molar heat capacities of model compounds for a broad temperature range (*6*). (**B**) Also shows a fit to a two-state model using **Eqs. 2–9** with ΔH = 95 kJ/mol, T_m = 78°C, ΔC$_p$ = 0.8 kJ/(mol · K).

6. Buffer selection is of particular importance. Certain buffering components, with high enthalpies of ionization, could change the pH significantly as a function of temperature *(7)*. Buffers such as glycine (pH 2.0–3.5), sodium acetate (pH 3.5–5.0), sodium cacodylate (pH 5.5–7.0), and sodium phosphate (pH 6.0–7.5), are recommended (*see* **Note 2**).

3. Method

3.1. Instrument Preparation

1. The instrument should be turned on at least 12 h prior to the experiment and "thermal history" established for the particular experimental conditions, by running several baseline scans with the cells filled with appropriate buffer.
2. One of the most important user-defined parameters is the heating rate. Several considerations must to be taken into account. First, the increase in sensitivity is linear with respect to the heating rate, i.e., the sensitivity with a heating rate of 120°C per hour is twice that at 60°C per hour. One needs to keep in mind that the increase in sensitivity actually leads to the decrease in the signal-to-noise ratio. Second, if the expected transition is very sharp, e.g., occurs within a few degrees, a high heating rate will distort the shape of the heat absorption profile and lead to an error in the determination of all thermodynamic parameters for this transition and, in particular, the transition temperature. Third, the higher the heating rate, the less time the system has to achieve equilibrium. For slow unfolding/refolding processes it is preferred to use low heating rates. Usually small globular proteins and monomeric peptides exhibit fast folding/unfolding, and heating rates of 90–120°C per hour are acceptable. For larger proteins it is customary to use lower heating rates (30–60°C per hour). For fibrillar proteins, such as collagen or myosin, which exhibit very narrow transitions, a heating rate of 10–20°C per hour is more suitable.
3. The cells should be cleaned routinely (*see* **Note 3**).
4. Calibration of the instrument should be done periodically (once a year) using the procedure provided by the manufacturer.

3.2. Sample Preparation

1. Purified protein/peptide should be extensively dialyzed with several changes of buffer, every 6 h or more, against the appropriate buffer.
2. Prior to the experiment, insoluble particles should be removed by centrifugation at 13,000*g*. Filtration of the protein solution is not recommended.
3. Measure protein/peptide concentration (*see* **Note 1**).

3.3. Data Collection

1. Thoroughly wash both cells with buffer from the last dialysis change, and fill them with the same buffer without introducing air bubbles into the cells. To achieve this, all air bubbles should be removed from the syringe before loading. After inserting the needle into the calorimetric cell, allow the plunger to lower slowly until the solution appears in the overflow reservoir. At this point, start vigorously pumping

a small volume of the solution in and out of the cell. This vigorous pumping should dislodge trapped air bubbles from the cell.

2. Fill both cells, as in **step 1,** with buffer and run a buffer–buffer scan.
3. After a stable baseline has been achieved, refill the sample cell with protein/peptide solution and run a sample–buffer scan.
4. Rescan to check the reversibility of unfolding (*see* **Note 4**).

3.4. Data Analysis

1. Subtract the buffer–buffer scan from the peptide/protein–buffer scan to obtain $\Delta C_p^{app}(T)$, the heat capacity difference between sample and reference cells at temperature T.
2. Convert $\Delta C_p^{app}(T)$ into the partial heat capacity of the peptide/protein at temperature T, $\Delta C_{p,pr}^{exp}(T)$ as:

$$C_{p,pr}^{exp}(T) = \frac{C_{p,H_2O}}{\overline{V}_{H_2O}} \cdot \overline{V}_{pr} - \frac{\Delta C_p^{app}(T)}{m_{pr}} \tag{1}$$

where $\overline{V}_{pr}$ is the partial volume of the peptide/protein, m_{pr} is the mass of the petide/ protein in the calorimetric cell, $\overline{V}_{H_2O}(T)$ is partial molar volume of aqueous buffer, and C_{p,H_2O} is the heat capacity of aqueous buffer. The partial volume of the peptide/protein, $\overline{V}_{pr}$, can be calculated from the amino acid composition of the protein using an additivity scheme as describe (**8**). The parameter $C_{p,H_2O}/\overline{V}_{H_2O}$ can be considered independent of temperature and equal to 4.2 J/(K × cm^{-3}).

3. Depending on the peptide/protein, the partial specific heat capacity of the native state, $C_{p,N}$, at 25°C ranges from 1.25 to 1.80 J/K × g (**9**). The dependence of $C_{p,N}$ on temperature appears to be a linear function, with a slope from 0.005 to 0.008 J/K^{-2} g, which is also protein dependant (**9**). The partial specific heat capacity of the unfolded state, $C_{p,U}$, is always higher than the heat capacity of the native state. At 25°C, $C_{p,U}$ values for different proteins range from 1.85 to 2.2 J/K × g, whereas at 100°C, $C_{p,U}$ values are higher, from 2.1 to 2.4 J/K × g (**9**). Partial heat capacity of the unfolded state has a nonlinear dependence on temperature (e.g., **ref. 10**). It increases gradually (with the slope comparable to that for the native state) and approaches a constant value at 60–75°C. The heat capacity change upon peptide/protein unfolding, $\Delta C_p = C_{p,U} - C_{p,N}$, appears to be a temperature-dependent function. However, this dependence is weak in the temperature range of 0 to 70°C, so in a first approximation, ΔC_p can be considered constant.

4. Analysis of the DSC profiles, according to a certain model, can be done using the ORIGIN software from Microcal Inc. Alternatively, any nonlinear regression software (e.g., NONLIN, NLREG, SigmaPlot, KaleidaGraph) can be used to write user-defined scripts (**11**). An overview of the analysis of the complex non-two-state transitions is available (**1**). The following formalism is to be used for the simplest case when the unfolding is a monomolecular two-state process (*see* **Fig. 1**; *see* **refs. 2,12**). The heat capacity functions for the native and unfolded states are represented by the linear functions of temperature, T, expressed in Kelvin as:

$$C_{P.N}(T) = A_N \cdot (T - 273.15) + B_N \tag{2}$$

$$C_{P.U}(T) = A_U \cdot (T - 273.15) + B_U \tag{3}$$

The equilibrium constant of the unfolding reaction, K, is related to the Gibbs energy change upon unfolding as:

$$K = \exp\left(-\frac{\Delta G}{RT}\right) \tag{4}$$

The Gibbs energy of unfolding, ΔG, is defined as

$$\Delta G = \frac{T_t - T}{T_t} \cdot \Delta H_{fit}(T_t) + \Delta C_p \cdot (T - T_t) + T \cdot \Delta C_p \cdot \ln\left(\frac{T_t}{T}\right) \tag{5}$$

where ΔC_p is the heat capacity change upon unfolding taken to be independent of temperature, T_t is the transition temperature, and $\Delta H_{fit}(T_t)$ is the enthalpy of unfolding at T_t. The transition temperature is defined as the temperature at which the populations of the native, F_N, and unfolded, F_U, proteins are equal. The populations are defined by the equilibrium constant as:

$$F_N(T) = \frac{1}{1+K} \text{ and } F_U = \frac{K}{1+K} \tag{6}$$

The experimental partial molar heat capacity function, $\underline{C}_{p,pr}(T_t)$, is fitted to the following expression:

$$C_{p.pr}(T) = F_N(T) \cdot C_{P.N}(T) + C_p^{exc}(T) + F_U(T) \cdot C_{P.U}(T) \tag{7}$$

The excess heat capacity defined $C_p^{exc}(T)$ as:

$$C_p^{exc}(T) = \frac{\Delta H(T)^2}{R \cdot T^2} \cdot \frac{K}{(1+K)^2} \tag{8}$$

where the enthalpy function is defined as $\Delta \underline{H}(T)$

$$\Delta H(T) = \Delta H_{fit}(T_t) + \Delta C_p \cdot (T - T_t) \tag{9}$$

There are seven fitted parameters: T_t, $\Delta \underline{H}_{fit}$, ΔC_p, A_N, A_U, B_N, B_U.

In order to analyze the data according to the above equations, the reversibility of unfolding reaction should be established experimentally by reheating the sample. If more than 80% of the original signal is recovered, then the reaction can be considered to be reversible. For the analysis of the irreversible transitions, *see* **ref. 13**.

4. Notes

1. Protein/peptide concentration is a very important parameter as it is required for quantitative analysis according to **Eqs. 2–9**. The extinction coefficient can be calculated from the number of aromatic residues and disulfide bonds in a peptide/protein using the following empirical equation *(14)*:

$$\varepsilon_{280nm}^{0.1\%,1cm} = (5690 \cdot N_{Trp} + 1280 \cdot N_{Tyr} + 120 \cdot N_{SS})/Mw \qquad (10)$$

where *Mw* is the molecular mass of the peptide/protein in Daltons. A simple experimental procedure for estimating the extinction coefficient is described *(15)*. Alternatively, a method based on the absorption of light at 205 nm by the peptide bond can be used, as it is independent of amino acid composition *(16)*.

2. The change in pH, as a function of temperature, could lead to linked protonation effects between the buffer components and the protein/peptide. The main criterion for using the recommended buffers is that their ionization enthalpies are similar to those of the ionizable groups in the protein/peptide. Consequently, the deprotonation/protonation reactions should have little or no contribution to the overall enthalpy of unfolding. However, the change in ionization of the protein/peptide could lead to differences in the thermal unfolding process. This effect can be investigated by using buffer components that buffer in the same pH range, but, that have different enthalpies of ionization *(17)*.

3. The DSC cell should be cleaned regularly. This can be accomplished in most cases by filling the cells with 10% SDS and heating it up to 100°C followed by a thorough rinse with distilled water. Alternatively, the cells can be washed with 200 proof ethanol followed by washing with distilled water. Drying the cells is not recommended.

4. The reversibility of unfolding strongly depends on the upper temperature limit during the first scan. At high temperatures, irreversible modifications of proteins/peptides can occur (for example, hydrolysis, deamidation, and so on; *see* **ref. *18***).

Acknowledgments

This work was supported by a grant RO1-GM54537 from the National Institutes of Health.

References

1. Biltonen, R. L. and Freire, E. (1978) Thermodynamic characterization of conformational states of biological macromolecules using differential scanning calorimetry. *CRC Crit. Rev. Biochem.* **5,** 85–124.
2. Makhatadze, G. I. (2004) Thermal unfolding of proteins studied by calorimetry. In: *Protein Folding Handbook.* (Buchner, J. and Kiefhaber, T, eds.), Wiley-VCH, Weinheim, Germany, pp. 70–98.
3. Richardson, J. M., McMahon, K. W., MacDonald, C. C., and Makhatadze, G. I. (1999) MEARA sequence repeat of human CstF-64 polyadenylation factor is helical in solution. A spectroscopic and calorimetric study. *Biochemistry* **38,** 12,869–12,875.
4. Richardson, J. M. and Makhatadze, G. I. (2004) Temperature dependence of the thermodynamics of helix-coil transition. *J. Mol. Biol.* **335,** 1029–1037.
5. Streicher, W. W. and Makhatadze, G. I. (2006) Calorimetric evidence for a two-state unfolding of the beta-hairpin peptide trpzip4. *J. Am. Chem. Soc.* **128,** 30–31.
6. Makhatadze, G. I. and Privalov, P. L. (1990) Heat capacity of proteins. I. Partial molar heat capacity of individual amino acid residues in aqueous solution: hydration effect. *J. Mol. Biol.* **213,** 375–384.

7. Fukada, H. and Takahashi, K. (1998) Enthalpy and heat capacity changes for the proton dissociation of various buffer components in 0.1 M potassium chloride. *Proteins* **33,** 159–166.

8. Makhatadze, G. I., Medvedkin, V. N., and Privalov, P. L. (1990) Partial molar volumes of polypeptides and their constituent groups in aqueous solution over a broad temperature range. *Biopolymers* **30,** 1001–1010.

9. Makhatadze, G. I. (1998) Heat capacities of amino acids, peptides and proteins. *Biophys. Chem.* **71,** 133–156.

10. Makhatadze, G. I. and Privalov, P. L. (1995) Energetics of protein structure. *Adv. Protein Chem.* **47,** 307–425.

11. Ibarra-Molero, B., Loladze, V. V., Makhatadze, G. I., and Sanchez-Ruiz, J. M. (1999) Thermal versus guanidine-induced unfolding of ubiquitin. An analysis in terms of the contributions from charge-charge interactions to protein stability. *Biochemistry* **38,** 8138–8149.

12. Makhatadze, G. I. (1998) Measuring protein thermostability by differential scanning calorimetry. In: *Current Protocols in Protein Science, Vol. 2.* (Coligan, J. E., Dunn, B. M., Ploegh, H. L., Speicher, D. W., and Wingfield, P. T., editorial board) John Wiley and Sons, New York, NY, pp. 7.9.1.–7.9.14.

13. Sanchez-Ruiz, J. M. (1992) Theoretical analysis of Lumry-Eyiring models in differential scanning calorimetry. *Biophys. J.* **61,** 921–935.

14. Gill, S. C. and von Hippel, P. H. (1989) Calculation of protein extinction coefficients from amino acid sequence data. *Anal. Biochem.* **182,** 319–326.

15. Pace, C. N., Vajdos, F., Fee, L., Grimsley, G., and Gray, T. (1995) How to measure and predict the molar absorption coefficient of a protein. *Protein Sci.* **4,** 2411–2423.

16. Scopes, R. K. (1974) Measurement of protein by spectrophotometry at 205 nm. *Anal. Biochem.* **59,** 277–282.

17. Yu, Y., Makhatadze, G. I., Pace, C. N., and Privalov, P. L. (1994) Energetics of ribonuclease T1 structure. *Biochemistry* **33,** 3312–3319.

18. Volkin, D. B. and Klibanov, A. M. (1992) Alterations in the structure of proteins that cause their irreversible inactivation. *Dev. Biol Stand.* **74,** 73–80.

8

Application of Single Molecule Förster Resonance Energy Transfer to Protein Folding

Benjamin Schuler

Summary

Protein folding is a process characterized by a large degree of conformational heterogeneity. In such cases, classical experimental methods yield only mean values, averaged over large ensembles of molecules. The microscopic distributions of conformations, trajectories, or sequences of events often remain unknown, and with them the underlying molecular mechanisms. Signal averaging can be avoided by observing individual molecules. A particularly versatile method is highly sensitive fluorescence detection. In combination with Förster resonance energy transfer, distances and conformational dynamics can be investigated in single molecules. This chapter introduces the practical aspects of applying this method to protein folding.

Key Words: Protein folding; fluorescence spectroscopy; single molecule detection; Förster resonance energy transfer; FRET; diffusion; folding trajectories.

1. Introduction

The direct investigation of the folding of single protein molecules has only become feasible by means of new methods such as atomic force microscopy (AFM) *(1,2)* and optical single molecule spectroscopy *(3–9)*. These techniques offer a fundamental advantage beyond mere fascination for the direct depiction of molecular processes: they can resolve and quantify the properties of individual molecules or subpopulations inaccessible in classical ensemble experiments, where the signal is an average over many particles. Fluorescence spectroscopy is a particularly appealing technique, owing to its extreme sensitivity and versatility *(5,10,11)*. In combination with Förster resonance energy transfer (FRET) *(12–14)*, it enables us to investigate intramolecular distance distributions and conformational dynamics of single proteins. Time-resolved ensemble FRET can also be used to separate subpopulations and to obtain information on distance

From: *Methods in Molecular Biology, vol. 350: Protein Folding Protocols*
Edited by: Y. Bai and R. Nussinov © Humana Press Inc., Totowa, NJ

distributions *(15)*, but data interpretation is typically less model independent *(16)*. For kinetic ensemble studies, the reactions need to be synchronized, which is often difficult. An impressive example for the power of single molecule studies is single channel recording, which now dominates the study of ion channels and has revealed countless mechanisms that could not be obtained from ensemble experiments *(17)*.

Currently, the two most common single molecule methods to study protein folding are AFM and single pair FRET *(18)*. The first observations of the unfolding of single protein molecules were made using AFM *(1)* and laser tweezers *(19)*. Soon the first experiments using fluorescence followed *(20–22)*, which demonstrated the potential of single molecule spectroscopy for separating subpopulations and for obtaining dynamic information. Since then, single molecule spectroscopy has been used to identify the equilibrium collapse of unfolded protein under near-native conditions *(23)*, to study folding kinetics *(24)*, the dimensions of denatured proteins *(25)*, and to investigate previously inaccessible parameters of the free energy surface of protein folding *(23)*. Work on immobilized protein molecules has also allowed the study of reversible folding *(26)*, and has even enabled the direct observation of folding and unfolding trajectories *(27,28)*. For a recent review of the progress, concepts and theory of single molecule spectroscopy of protein folding, *see* **ref. 29**.

The basic idea of a protein folding experiment using FRET is very simple (**Fig. 1**): a donor dye and an acceptor dye are attached to specific residues of a protein. If a folded protein molecule resides in the volume illuminated by the focused laser beam, excitation of the donor dye results in rapid energy transfer to the acceptor dye because the dyes are in close proximity. Consequently, the majority of the fluorescence photons are emitted by the acceptor. Upon unfolding of the protein, the average distance between the donor and acceptor dyes will typically increase. As a result, the energy transfer rate is decreased, and the fraction of photons emitted by the acceptor is lower. The changes in fluorescence intensity from donor and acceptor can thus be used to distinguish between different conformational states of a protein.

1.1. Förster Resonance Energy Transfer

The quantitative relationship between the probability of transfer—the transfer efficiency—and the inter-dye distance is given by a theory developed by

---

Fig. 1. *(Opposite page)* Schematic structures of folded and unfolded protein labeled with donor (Alexa 488) and acceptor (Alexa 594) dyes. (**A**) Folded Csp*Tm*, a five-stranded, 66-residue β-barrel protein (PDB-code 1G6P) *(75)*, (**B**) unfolded Csp*Tm*. A blue laser excites the green-emitting donor dye, which can transfer excitation energy to the red-emitting acceptor dye.

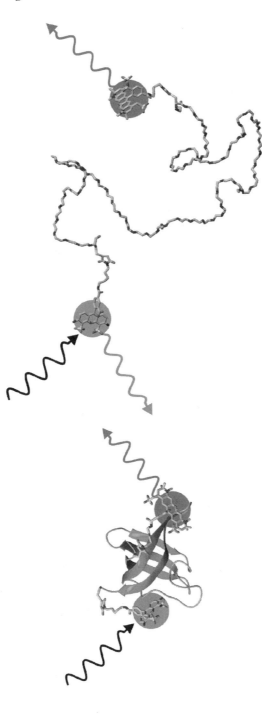

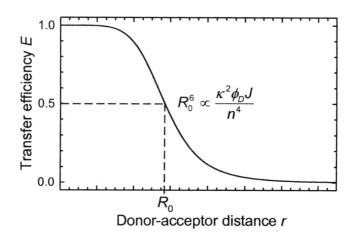

Fig. 2. Distance dependence of the transfer efficiency according to Förster theory (**Eq. 1**). The Förster radius R_0 is calculated according to **Eq. 2**. Owing to the characteristic $1/r^6$ dependence, the transfer efficiency is most sensitive for distance changes in the vicinity of R_0.

Theodor Förster in the 1940s *(12)*. According to Förster's theory, the transfer efficiency, E, for the dipole–dipole coupling between a donor and an acceptor chromophore depends on the inverse sixth power of the inter-dye distance, r:

$$E = \frac{R_0^6}{R_0^6 + r^6} \qquad (1)$$

where R_0 is the Förster radius, the characteristic distance that results in a transfer efficiency of 50% (**Fig. 2**). Because of the strong distance dependence of the efficiency, FRET can be used as a "spectroscopic ruler" on molecular length scales, typically between 2 and 10 nm. R_0 is calculated as (*see* **Note 1**)

$$R_0^6 = \frac{9000 \ln 10 \, \kappa^2 Q_D J}{128\pi^5 n^4 N_A} \qquad (2A)$$

where J is the overlap integral, Q_D is the donor's fluorescence quantum yield, n the refractive index of the medium between the dyes, and N_A is Avogadro's number *(12,30)*. The orientational factor is defined as $\kappa^2 = (\cos \Theta_T - 3 \cos \Theta_D \cos \Theta_A)^2$, where Θ_T is the angle between the donor emission transition dipole moment and the acceptor absorption transition dipole moment, Θ_D and Θ_A are the angles between the donor–acceptor connection line and the donor emission and the acceptor absorption transition moments, respectively. κ^2 varies between 0 and 4, but complete averaging of the relative orientation of the chromophores during the excited state lifetime of the donor results in a value of two-thirds the value most frequently used in practice (however, *see* **Subheading 3.3.**). Q_D and *n* need to be measured (*see* **Note 2**), and *J* is calculated from the normalized donor emission spectrum $f_D(\lambda)$ and the molar extinction coefficient of the acceptor $\varepsilon_A(\lambda)$ according to:

$$J = \int_0^\infty f_D(\lambda)\varepsilon_A(\lambda)\lambda^4 d\lambda \qquad (2B)$$

The accuracy of the calculation is ultimately limited by *n*, which is often nonuniform and difficult to estimate for a protein (but probably very close to *n* of the solvent for an unfolded protein), and ε_A, which cannot easily be determined independently and is provided by the manufacturer with an uncertainty of at least a few percent. Fortunately, the influence of such uncertainties is moderated by the fact that R_0 depends only on the sixth root of n^{-4} and *J* respectively, of these quantities (**Eq. 2A**).

Experimentally, transfer efficiencies can be determined in a variety of ways *(30)*, but for single molecule FRET, two approaches have proven particularly useful. One is the measurement of the number of photons *(31)* emitted from the donor and the acceptor chromophores, n_D and n_A, respectively, and the calculation of the transfer efficiency according to:

$$E = \frac{n_A}{n_A + n_D} \qquad (3)$$

where the numbers of photons are corrected for the quantum yields of the dyes, direct excitation of the acceptor, the detection efficiencies of the optical system in the corresponding wavelength ranges, and the crosstalk between the detection channels (*see* **Note 3**). A second approach to measure *E*, which can be combined with the first *(32)*, is the determination of the fluorescence lifetime of the donor in the presence (τ_{DA}) and absence (τ_D) of the acceptor, yielding the transfer efficiency as:

$$E = 1 - \frac{\tau_{DA}}{\tau_D} \qquad (4)$$

Frequently, we have to consider a distance distribution instead of a single distance, especially in unfolded proteins. If information about the distance distribution is available from simulations or independent experiments, it can be included in the analysis *(33)*. In general, it is important to be aware of the different averaging regimes, because the time-scales of both conformational dynamics of the protein and re-orientational dynamics of the dyes influence the way the resulting transfer efficiency has to be calculated *(30,33)*.

1.2. Outline of the Procedures

Performing a single molecule FRET experiment on protein folding requires several steps. First, protein samples have to be prepared for labeling, either by chemical synthesis or by recombinant expression in combination with site-directed mutagenesis. After identifying a suitable dye pair with the Förster radius in the desired range, the fluorophores need to be attached to the protein as specifically as possible to avoid chemical heterogeneity. The equilibrium and kinetic properties of the labeled protein should then be measured in ensemble FRET experiments and compared directly to unlabeled protein to ensure that the folding mechanism is not altered. For control experiments, it is helpful to prepare reference molecules, such as polyproline peptides or double-stranded DNA, with the same chromophores as the protein. After customizing the instrument for the sample, data can either be taken on freely diffusing molecules, or on immobilized molecules if observation times greater than a few milliseconds are desired. Finally, the data need to be processed to distinguish signal from background, to identify fluorescence bursts in diffusion experiments, and to calculate transfer efficiencies and the resulting distance changes.

2. Materials

2.1. Instrumentation

Considering that most molecular biologists or biochemists will not attempt to build their own single molecule instrument, it will be assumed in the following that a suitable system is already accessible, and instument design will not be described. The details of single molecule instrumentation can be found in several recent reviews *(5,10,34,35)*. An important development for the wide application of single molecule methods to the study of biomolecules is the recent availability of comprehensive commercial instrumentation *(36)*.

Experimental setups for single molecule FRET typically involve either confocal excitation and detection using a pulsed or continuous wave (cw) laser and avalanche photodiodes (APDs), or wide-field microscopy with two-dimensional (2D) detectors, such as intensified or electron-multiplying CCD cameras. Wide-field imaging allows the collection of data from many single molecules in parallel, albeit at lower time resolution and signal-to-noise ratio than in a confocal experiment using APDs. Imaging can be performed either via epi-illumination, where the exciting laser light is directed to the sample through the epi-illumination port of a conventional fluorescence microscope, or via evanescent field excitation, typically by total internal reflection of the excitation light at the water/glass interface, where the sample molecules are located.

Figure 3 shows a schematic with the main optical elements for confocal epifluorescence detection. A laser beam is focused with a high numerical aperture objective to a diffraction-limited focal spot that serves to excite the labeled

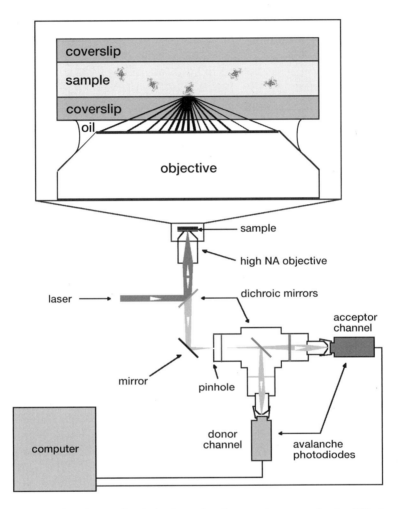

Fig. 3. Schematic of a confocal single molecule experiment on freely diffusing mole-cules (molecules not to scale). In this example, the signal is separated by wavelength into two detection channels corresponding to emission from donor and acceptor chromophores. With additional dichroic mirrors or polarizing beam splitters, the system can be extended to monitor polarization and/or emission from more dyes in parallel. In combination with a *xy*-stage, the system can be used for confocal imaging by sample scanning.

molecules. In the simplest experiment, the sample molecules are freely diffus-ing in solution at very low concentration, ensuring that the probability of two molecules residing in the confocal volume at the same time is negligible. When a molecule diffuses through the laser beam, the donor dye is excited and fluo-rescence from donor and acceptor is collected through the objective and gets focused onto the pinhole, a small aperture serving as a spatial filter. A dichroic mirror finally separates donor and acceptor emission into the corresponding detectors, from where the data are collected with multichannel scalers or suitable counting cards. The setup can be extended to sorting photons by additional colors, e.g., if more than two chromophores are used *(37,38)*, or by both wavelength and polarization *(32,39)*. The system can be coupled to a piezo xy flexure stage for sample scanning, which allows the acquisition of fluorescence images and the reproducible positioning of the laser beam on individual molecules immobilized on the surface. Even better spatial separation than with confocal one-photon excitation can be achieved by two-photon excitation *(40)*.

The advantage of observing freely diffusing molecules is that perturbations from surface interactions can largely be excluded, but the observation time is limited by the diffusion times of the molecules through the confocal volume. Typically, every molecule is observed for no more than a few milliseconds in the case of proteins. Alternatively, molecules can be immobilized on the surface and then observed for a more extended period of time, typically a few seconds, until one of the chromophores undergoes photodestruction. The complications in this case are interactions with the surface that can easily perturb the sensitive equilibrium of protein folding.

For sample design, especially for choosing the chromophores, it is important to be aware of the characteristics and limitations of the instrument, such as the laser lines available for excitation or the time resolution and signal-to-noise ratio achievable with the detectors. The lasers typically used range from simple cw systems with a single fixed wavelength to large, tunable, pulsed lasers that make a broad range of wavelengths accessible. With a pulsed source, fluorescence lifetimes become available in addition to intensities, which can provide additional information *(39,41)*.

2.2. Chemicals

Obviously, single molecule fluorescence experiments make great demands on buffer preparation and sample purity. Even though a single molecule is always pure, researchers rarely have the means to distinguish sample molecules unequivocally from contaminants. Some solutes, e.g., denaturants or osmolytes, may be present at concentrations of several molar. Highest purity buffer substances are therefore strictly required. Common buffers, such as phosphate salts and Tris, can be obtained in excellent purity from most major suppliers as spectrophotometric grade chemicals, other substances only from more specialized sources (e.g., GdmCl and Tween-20 [Pierce Biotechnology, Rockford, IL]). As a general rule, all solutions have to be tested in the single molecule instrument for fluorescent impurities prior to use. Quartz-bidistilled water is recommended; water from ion exchanger water purification systems ("MilliQ") is usually suitable, but needs to be monitored more regularly for contaminations.

Fluorophores for protein labeling can be obtained with a variety of reactive groups from several manufacturers, such as Molecular Probes/Invitrogen (Alexa Fluors), Amersham Biosciences (cyanine dyes), and others.

2.3. Chromatography

Purification of labeled proteins proceeds very much the same way as any other protein purification, but again, buffers should have very low fluorescence background. High-performance liquid chromatography (HPLC) or fast protein liquid chromatography (FPLC) systems with fluorescence and diode array absorption detectors can greatly simplify the identification of correctly labeled species.

3. Methods
3.1. Choosing the Fluorophores

Several criteria must be met by chromophores for single molecule FRET:

1. They must have suitable photophysical and photochemical properties, especially a large extinction coefficient ($\sim 10^5$ M^{-1}cm^{-1} or greater), a quantum yield close to 1, high photostability, a low triplet state yield, and small intensity fluctuations (which can result from intermittence, i.e., transitions between bright and dark states).
2. The absorption maximum of the donor chromophore must be close to a laser line available for excitation.
3. Good spectral separation of donor and acceptor emission is necessary to minimize direct excitation of the acceptor and to reduce crosstalk between the detection channels.
4. Acceptor absorption and donor emission spectra must give an overlap integral that results in a suitable Förster radius (calculated from **Eq. 2**). Keep in mind that the best sensitivity for distance changes can be obtained for distances close to R_0.
5. The dyes must be available with suitable functional groups for specific protein labeling (typically succinimidyl esters for amino groups or maleimides for sulfhydryl groups).
6. The dyes must be sufficiently soluble in aqueous buffers, otherwise they may induce protein aggregation, a problem that has been minimized by the introduction of charged groups in many of the popular dyes *(42,43)*.

Note that some of the fluorophores' properties may change on attachment to the protein. In many cases, it is thus advisable to screen a series of dye pairs. The most commonly used dyes are organic fluorophores developed specifically for sensitive fluorescence detection. Examples of common dye pairs are Cy3/Cy5 and Alexa 488/Alexa 594. Semiconductor quantum dots *(44,45)* are promising candidates because of their extreme photostability, but they are not yet available with single functional groups, so far they can only be used as donors because of their broad absorption spectra, and they are themselves of the size of a small protein, which increases the risk of interference with the folding process. As of today, tryptophan, the amino acid most commonly used for fluorescence detection in proteins, is not suitable for single molecule detection (unless the molecule contains a very large number of tryptophan residues *[46]*) owing to the low photostability of the indole ring.

3.2. Protein Labeling

To our misfortune, protein chemistry has not made it easy for us to investigate polypeptides in single molecule experiments (with the exception of the family of fluorescent proteins *[47,48]*). Specific placement of fluorophores on the protein ideally requires groups with orthogonal chemistry. For simple systems, such as short peptides, sequences can be designed to introduce only single copies of residues with suitable reactive side chains *(23,33)*. In chemical solid-phase peptide synthesis, protection groups and the incorporation of non-natural amino acids can be used to increase specificity, but for longer chains, chemical synthesis becomes inefficient and shorter chains have to be ligated *(49)* to obtain the desired product *(22)*.

The production of proteins of virtually any size and sequence by heterologous recombinant protein expression is the method of choice to obtain very pure material in sufficiently large amounts for preparative purposes. But the number of functional groups that can be used for specific labeling is then very limited. Sufficiently specific reactivity in natural amino acids is only provided by the sulfhydryl groups of cysteine residues, the ε-amino groups of lysine side chains, and the free α-amino group of the N-terminal amino acid. However, except for small peptides, the statistical and, therefore, often multiple occurrence of cysteine and especially lysine residues in one polypeptide prevents the specific attachment of labels. Increased specificity can be achieved by removing unwanted natural cysteines by site-directed mutagenesis or introducing cysteines with different reactivity owing to different molecular environments within the protein *(50)*. Labeling is usually combined with multiple chromatography steps to purify the desired adducts. Alternative methods *(51)* are native chemical ligation of recombinantly expressed and individually labeled protein fragments or intein-mediated protein splicing *(52)*, the specific reaction with thioester derivatives of dyes *(53)*, puromycin-based labeling using in vitro translation *(54)*, or introduction of non-natural amino acids *(55)*. Most of the latter methods are not yet used routinely, are not openly available, or must be considered under development.

Currently, the most common approach is to rely on cysteine derivatization. An outline for labeling a small protein with a FRET pair is given in the following:

1. Based on the three-dimensional structure of the protein, remove all solvent-accessible cysteine residues by site-directed mutagenesis and introduce two surface-exposed cysteines with a sequence separation resulting in a clear difference in FRET efficiencies for folded and unfolded protein, respectively.

2. Express the protein, purify it under reducing conditions, and concentrate it to at least 200 μ*M*.

3. Remove the reducing agent and adjust the pH by passing the protein over a desalting column equilibrated with 50 m*M* sodium phosphate buffer, pH 7.0. Ensure that the resulting protein concentration is at least 100 μ*M*.

4. React the protein with the first chromophore (*see* **Note 4**) by adding the maleimide derivative of the dye at a 1:1 *M* ratio, incubate 1 h at room temperature or at 4°C overnight.

5. Separate unlabeled, singly labeled, and doubly labeled protein by chromatography, e.g., by ion exchange chromatography, taking advantage of the negative charge on many common chromophores. In favorable cases, this method even allows the separation of labeling permutants *(23)*. Including low concentrations of detergents such as Tween-20 can reduce protein losses resulting from non-specific adsorption to the column material.

6. Concentrate singly labeled protein to at least 100 μ*M* and react with the second chromophore as in points 3 and 4. Make sure that the pH is adjusted properly.

7. Separate singly and doubly labeled protein as in point 5.

Interactions of the dye with the protein surface can interfere both with the photophysics of the chromophores and the stability of the protein. This needs to be taken into account both for the design of the labeled variants and the control experiments. Because of the substantial size of the fluorophores, they can usually only be positioned on the solvent-exposed surface of the protein if the folded structure is to be conserved. Even then, the use of hydrophobic dyes can lead to aggregation of the protein, and interactions with the protein surface can cause a serious reduction in fluorescence quantum yield. Important control experiments are equilibrium or time-resolved fluorescence anisotropy measurements *(22,23,33)*, which are sensitive to the rotational flexibility of the dyes and can therefore provide indications for undesirable interactions with the protein surface. It is also essential to ensure by direct comparison with unmodified protein that labeling has not substantially altered the protein's stability or folding mechanism *(22,23)*.

3.3. Controls

Several factors can complicate single molecule fluorescence experiments, for instance optical saturation and photobleaching, the influence of diffusion, possible interactions of the chromophore with the polypeptide (resulting in a reduction of quantum yields or lack of fast orientational averaging of the dyes), or a change of solvent conditions, which can affect the refractive index and the photophysics and photochemistry of the dyes. A suitably labeled control molecule that essentially provides a rigid spacer between the dyes, and whose conformation does not change under denaturing conditions can thus be valuable for avoiding misinterpretation of the results.

Two suitable types of molecules are double-stranded DNA *(13)* and polyproline peptides *(33)*. DNA dupleces are very stiff (persistence length of about 50 nm) and can simply be generated by annealing complementary oligonucleotide strands, which are commercially available with fluorophores already attached. For FRET experiments on proteins, however, it is desirable to use a polypeptide-based reference molecule because the type of attachment chemistry and the characteristics of the immediate molecular environment can influence the photophysical properties of the fluorophores *(56–58)*. Additionally, under some conditions used for protein denaturation, a DNA duplex will dissociate. Oligomers of proline in water form a type II helix with a pitch of 0.312 nm per residue and a persistence length of about 5 nm, providing a reasonably stiff spacer *(33,59–61)*. By including an amino terminal glycine residue and a carboxy terminal cysteine residue in the synthesis, the resulting α-amino group and the cysteine's sulfhydryl group can be labeled specifically with derivatives of suitable reactive dyes, such as succinimidyl esters and maleimides, respectively *(23,33)*. Polyproline peptides are therefore suitable reference molecules, but their chain dynamics have to be taken into account, especially for higher oligomers *(33)*.

Another issue that frequently complicates the quantitative analysis of FRET experiments is the orientational factor κ^2 (*see* **Subheading 1.1.**). Routinely, labeled protein samples should be analyzed with equilibrium or time-resolved fluorescence anisotropy measurements. Low anisotropy values are indicative of freely rotating chromophores; from time-resolved measurements, the rotational correlation time of the dyes can be determined directly *(62)*.

In general, it is always essential to compare the results from single molecule experiments quantitatively with ensemble data (*see* **Note 5**). Even though it may be tempting to analyze only the results from a few selected molecules, the overall result must agree with the ensemble measurement, and the criteria for singling out molecules for analysis have to be as objective and clearly defined as possible.

3.4. Other Technical Details

3.4.1. Filters

Customizing the single molecule instrument will involve the installation of suitable filters specific for the dye pair used. A compromise between maximum collection efficiency, minimal background from scattering, and cross-talk between the channels (especially donor emission leakage into the acceptor detector) has to be established. At least two dichroic mirrors are required, but the signal-to-noise ratio can usually be improved by additional filters, e.g., long-pass filters to reject scattered laser light, a laser line filter, and band-pass or long-pass filters for the individual detection channels. A broad range of filters and dichroic mirrors is available from companies such as Chroma or Omega Optical. Other variables involve the choice of excitation intensity and laser pulse frequency (which should be optimized carefully by systematic variation), the objective used (e.g., water immersion vs oil immersion [*see* **Note 6**]) and the size of the pinhole in a confocal setup.

3.4.2. Cover Glasses

Cells for single molecule measurements are usually assembled using glass cover slides with a thickness corresponding to the optical correction of the objective. Generally, fused silica results in lower background because of the high purity of the material. Impurities on the surface of cover slides can give rise to background, especially in experiments on immobilized molecules close to the surface of the glass. It can thus be crucial to clean them carefully. A wide variety of cleaning methods for cover slides are applied, e.g.,

1. Rubbing the glass carefully with acetone or isopropanol and rinsing.
2. Heating to 500°C or more with a flame or in a suitable oven.
3. Sonicating in a 1:1:5 mixture of 30% ammonium hydroxide, 30% hydrogen peroxide, and water.
4. Etching with a 10% solution of hydrofluoric acid in water for 5 min. (Hydrofluoric acid needs to be handled very carefully. The surface of the polished glass is degraded by etching, and the cover slide can usually not be reused, which is a disadvantage especially for expensive fused silica cover slides.)

3.4.3. Oxygen

The observation time of immobilized single molecules is ultimately limited by photobleaching, an intrinsic property of all organic dyes. Photobleaching is typically a result of excited state (probably triplet state) reactions with highly reactive molecules in solution, such as singlet oxygen. At the same time, oxygen is an efficient quencher of triplet states, whose population at excitation close to optical saturation can strongly decrease the overall fluorescence intensity. However, it may still be advantageous for some experiments to reduce the oxygen concentration. The popular combination of glucose oxidase and catalase in the presence of glucose *(63)* can obviously not be used under most denaturing conditions, but other methods for generating anaerobic conditions are available *(63)*.

3.5. Free Diffusion Experiments

Arguably the simplest single molecule experiment involves diluting a labeled protein sample to a concentration of about 10–100 pM and observing the signal from the confocal excitation and detection volume. In this concentration range, the probability of two protein molecules residing in the confocal volume at the same time is very small, and the signal bursts observed (**Fig. 4**) arise from individual molecules, provided aggregation can be excluded. As the molecules are only observed for about a millisecond each, bursts from hundreds to thousands of individual molecules are typically collected in several minutes to hours, depending on the protein concentration and the statistics required. The simplest way of analyzing the data is by binning them in intervals approximately equal to the average burst duration, typically about 1 ms. Photon bursts are then identified by a simple threshold criterion, and the counts from contiguous bins above the threshold are summed *(31)*. A slightly more sophisticated approach identifies the beginning and end of photon bursts with higher time resolution, using the corresponding increase and drop in the photon arrival frequency. The counts of a single burst are integrated, and a second threshold for the total number of photons is used to discriminate signal from noise and to select the largest bursts, which is particularly important if fluorescence lifetimes are to be computed from individual bursts *(32)*. Typically, burst sizes between 20 and 200 counts are reached.

Prior to further analysis of the identified bursts, several corrections need to be made. The background must be subtracted from the raw intensities, and the different quantum yields of donor and acceptor, the cross-talk between the channels, and direct excitation of the acceptor need to be taken into account. In the literature, it is common practice to address all of these corrections individually and simply add and subtract the corresponding contributions from each channel. Even though this is sufficient for most practical purposes, it neglects that some of these corrections are interdependent. A more general scheme for the correction of the raw signals is given in **Note 3**. If all parameters are calibrated correctly, the results from the number of emitted photons (**Eq. 3**) and fluorescence lifetimes (**Eq. 4**) should be in agreement.

The lifetimes and corrected photon counts obtained from the individual bursts can then be analyzed in histograms of transfer efficiencies, distances, polarization,

---

Fig. 4. *(Opposite page)* Example of data from an experiment on fluorescently labeled Csp*Tm* molecules *(23)* freely diffusing in a solution containing 1.5 *M* guanidinium chloride conditions close to the unfolding midpoint. One second of a fluorescence intensity measurement (total acquisition time 600 s) is shown, with large bursts of photons originating from individual molecules diffusing through the confocal volume. A histogram of transfer efficiencies calculated for the individual bursts from the entire measurement is shown in the inset. The histogram shows transfer efficiencies of $E \approx 0.9$ for folded and $E \approx 0.4$ for unfolded molecules. This allows changes in transfer efficiencies to be analyzed individually for the two subpopulations (*see* **Note 7** for the peak at $E \approx 0$ and **Note 8** for the width of the distributions). The measurement was done on a MicroTime 200 time-resolved fluorescence microscope (PicoQuant) with an excitation wavelength of 470 nm.

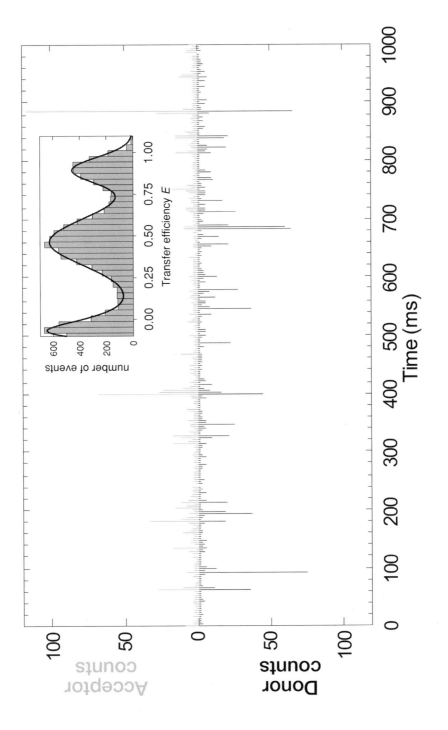

or burst size distributions, to name just a few examples (**Fig. 4**). In many cases, it is helpful to investigate the relation between several of these parameters, for instance in 2D histograms of fluorescence lifetime vs the transfer efficiency calculated from **Eq. 3**. As a type of 2D spectroscopy, this can lead to better separation of subpopulations or the identification of unique signatures for certain conformational states.

3.6. Experiments on Immobilized Proteins

An approach to observing individual proteins for an extended period of time is their immobilization on a surface. However, nonspecific interactions with the surface can easily disturb the folding reaction *(21)*. Strategies for minimizing such interactions include the optimization of surface functionalization *(26,64)* or the encapsulation of individual protein molecules in surface-tethered lipid vesicles *(27,28,65)*. Both methods have allowed the observation of single molecule protein folding reactions. The absence of binding to the surface can be tested by single molecule polarization measurements *(27,28,65)* or by quantitative comparison with experiments on freely diffusing molecules.

For immobilization experiments it can be helpful to prepare a small flow cell that allows rapid buffer exchange, the deposition of materials on the surface, and washing steps. Such flow cells can be assembled from two clean cover slides (*see* **Subheading 3.4.2.**) with double-sided tape, forming a channel several millimeters wide and about 100-μm deep. The following procedure outlines the deposition of labeled protein encapsulated in lipid vesicles *(65)*:

1. Sonicate a 500-μL suspension of 10 mg/mL egg-phosphatidylcholine (PC) and 0.2 mg/mL biotinylated phosphatidylethanolamine in the buffer to be used for the experiment to create multilamellar vesicles.
2. Extrude part of the suspension through a polycarbonate membrane with 100-nm pores *(66)* to create large unilamellar vesicles (LUVs), and repeat the procedure with the remaining sample, with about 1 μM of the protein to be encapsulated added to the suspension. This will result in statistical trapping of protein molecules, with the majority of vesicles being empty, and some vesicles containing one and only rarely two or more molecules.
3. Flow empty LUVs into the flow cell and incubate for a few minutes. The vesicles will form a supported bilayer on the surface. Wash the flow cell with buffer.
4. Introduce a solution of 1 mg/mL avidin into the flow cell, which will bind to the biotinylated lipid, and incubate for a few minutes. Wash the flow cell with buffer.
5. Flow in a dilute suspension of the LUVs with encapsulated protein. The vesicles will bind to the surface-adsorbed avidin, unbound vesicles and free protein can be removed by rinsing the flow cell.

Immobilized molecules prepared in this or other ways can be observed either with a confocal system in combination with sample scanning, or by wide field detection with evanescent field excitation. In the former case, the surface is first

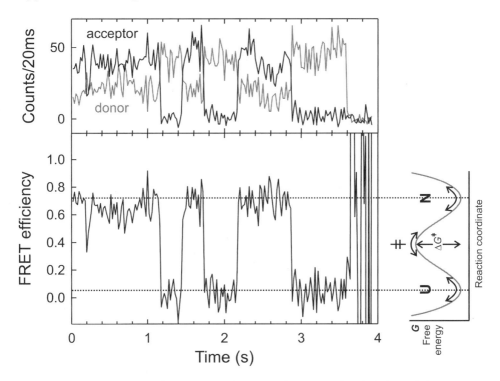

Fig. 5. Example of data from an experiment on fluorescently labeled, vesicle-encapsulated Csp*Tm* molecules immobilized on a surface *(28)*. The experiment was performed at 2.0 *M* guanidinium chloride, the denaturation midpoint of Csp*Tm*. Under these conditions, the protein would be expected to remain in the folded and unfolded states for extended periods of time, with rapid, intermittent jumps across the barrier between the two corresponding free energy minima. The top panel shows the fluorescence intensity trajectories recorded from the donor and acceptor chromophores of an individual protein. The anticorrelated changes in their emission intensities result in clear jumps of the transfer efficiency (bottom panel), reflecting the expected behavior of a two-state protein.

scanned to identify individual molecules, which are then targeted by the laser individually and observed sequentially. In the latter case, the information of all molecules in the field of view is obtained simultaneously, albeit with lower time resolution, and usually with an inferior signal-to-noise ratio. In both cases, trajectories of fluorescence intensities are obtained, terminated by photobleaching of a chromophore, which typically occurs after several milliseconds to seconds, depending on the exciting laser intensity. Transfer efficiencies or other parameters are calculated from the trajectories and corrected in a similar way as for the bursts from freely diffusing molecules (*see* **Subheading 3.5.**). An important criterion for identifying transitions between states is the anticorrelated signal change between donor and acceptor channels (**Fig. 5**), which is expected for a change in distance between the chromophores. The resulting trajectories of transfer efficiencies can potentially be analyzed by a wide range of methods pioneered in the field of single channel recording *(17)*.

3.7. Limitations of Single Molecule FRET

The number of photons that can be detected from an individual fluorophore is limited by photobleaching to some 10^5 (under ideal conditions). Roughly 100 photons are needed for a signal-to-noise ratio of 10, which means that on the order of up to 1000 observations can be made, in principle sufficient to observe many conformational transitions of a protein. The observation time can be extended by periodically interrupting the laser excitation.

In the other extreme, time resolution is ultimately limited by the photon emission rate, which cannot be greater than the decay rate of the electronically exited state of the chromophores, typically on the order of 10^9 per second. Together with other complications, such as the population of triplet states or low photon collection efficiencies, resolving processes on time-scales of a few microseconds and faster is currently hardly feasible from single events. A possible solution is to make full use of the photon emission statistics from many events, as in fluorescence correlation spectroscopy (FCS) *(67)* and similar experiments, which have the potential to resolve dynamics on microsecond time-scales *(68)* and below *(69)*. FCS is a powerful complementary method for studying protein folding dynamics, and as the setup is essentially identical to that of a confocal single molecule instrument, correlations can often be obtained from the same type of measurements as described in **Subheading 3.5**.

How accurately can distances be measured by single molecule FRET? According to **Eq. 1**, the transfer efficiency is most sensitive for distance changes close to R_0. Theoretically, the precision of a transfer efficiency value determined from a single molecule is only limited by shot noise, i.e., the variation of count rates owing to the quantization of the signal *(31,70)*. For instance, from 100 photons, the standard deviation of E at a distance $r = R_0 = 5$ nm would then be approx 5%, corresponding to an uncertainty in distance of approx 0.2 nm. If more photons are collected or if the signal from many molecules is averaged, a correspondingly higher precision would be achievable. In practice, the presence of background will limit the signal-to-noise ratio and impair the precision—even more so, if the transfer efficiencies are not close to R_0. In view of the possible systematic errors, especially the uncertainties in the calculation of R_0 (*see* **Subheading 1.1.**), FRET is currently more powerful for detecting distance changes than for accurately measuring absolute distances. However, important advances of the technique are still being made *(71,72)*, and together with a wider variety of reference molecules *(33)*, the accuracy of single molecule FRET measurements will probably be improved further. The study of protein folding will certainly benefit.

4. Notes

1. Note that the equation for R_0 given in the popular textbook of Cantor and Schimmel *(73)* is not correct.

2. The quantum yield of a chromophore can change upon attachment and must therefore be determined from a protein sample labeled with only one dye.

3. The relation between the raw photon counts $n_{A,0}$ and $n_{D,0}$, as measured in the two detection channels for acceptor and donor emission, respectively, and the corrected values n'_A and n_D can be expressed by the matrix equation

$$\begin{pmatrix} n_{A,0} \\ n_{D,0} \end{pmatrix} = \begin{pmatrix} a_{11} & a_{12} \\ a_{21} & a_{22} \end{pmatrix} \begin{pmatrix} n'_A \\ n_D \end{pmatrix} + \begin{pmatrix} b_A \\ b_D \end{pmatrix},$$

where the matrix a_{ij} describes the cumulative effect of the differences in quantum yields, the different collection efficiencies of the detection channels, and crosstalk ("bleed-through"), i.e., acceptor emission detected in the donor channel and donor emission detected in the acceptor channel. b_A and b_D are the background count rates in the acceptor and the donor channel, which can be estimated from a measurement on blank buffer solutions.

   The elements of matrix a_{ij} can be determined for a specific single molecule instrument (except for a scaling factor α) from a measurement of two samples containing donor and acceptor dye, respectively, with a concentration ratio equal to the ratio of their extinction coefficients at the excitation wavelength (ensuring that, at identical laser power, the same mean number of excitation events take place per unit time in both samples). By inverting the resulting matrix, the correction matrix $c_{ij} = a_{ij}^{-1}$ is obtained, which transforms the background-corrected raw counts $n_{A,0} - b_A$ and $n_{D,0} - b_D$ into the corrected values n'_A and n_D. Note that the factor α remains unknown, but cancels if intensity ratios are computed, as in the case of the transfer efficiency. Also note that this correction procedure can easily be extended to more than two channels by using a matrix of higher rank. Finally, n'_A has to be corrected for direct excitation of the acceptor according to, $n_A = n'_A - (n'_A + n_D)/(1 + \varepsilon_D/\varepsilon_A)$, where ε_D and ε_A are the extinction coefficients of donor and acceptor, respectively, at the excitation wavelength. Ideally, these corrections should already be taken into account for burst identification.

4. Adhere to the labeling instructions given by the manufacturer, especially the notes provided by Molecular Probes that are very detailed and helpful.

5. These experiments will take up the great majority of the labeled protein samples, which should be taken into account for the preparation scale.

6. With oil immersion objectives, the focal volume must be positioned very close to the cover slide surface to minimize chromatic aberration. In this case it is particularly important to use fused silica cover slides to reduce background from glass luminescence.

7. The signal from molecules with a transfer efficiency of zero is a notorious phenomenon in free diffusion experiments, which can of course be from incomplete labeling

or impurities, but may also be caused by light-induced inactivation of the acceptor, as suggested by experiments where the sample is flowed *(24)*, but the exact origin of this component of the signal is still a matter of debate. There is no problem if the transfer efficiency of the intact molecules under study is sufficiently different from zero (as in the example of **Fig. 3**), or if the intact molecules can be separated from the zero transfer events in combination with other observables, such as lifetime or polarization. An elegant general solution is the alternating excitation of donor and acceptor *(72,74)*, which independently probes the acceptor chromophore and allows all molecules with an inactive acceptor to be excluded from the analysis.

8. The width of the transfer efficiency histogram has often been found to be greater than expected from shot noise. This observation has been made on a variety of molecular systems *(22,23,33)*, indicating that additional, unidentified instrumental or physical factors may contribute to the width. Note that a distribution of transfer efficiencies does not necessarily imply a distribution of distances. In this type of measurement, subpopulations with different distances can only be separated if their dynamics are slow on the observation time-scale (in this example milliseconds).

Acknowledgments

This work has been supported by the Deutsche Forschungsgemeinschaft, the VolkswagenStiftung, and the Schweizerischen Nationalfonds. Part of **Fig. 3** was provided by Everett Lipman. The author would like to thank Daniel Nettels for discussion, especially concerning data analysis.

References

1. Rief, M., Gautel, M., Oesterhelt, F., Fernandez, J. M., and Gaub, H. E. (1997) Reversible unfolding of individual titin immunoglobulin domains by AFM. *Science* **276,** 1109–1112.
2. Engel, A., Gaub, H. E., and Müller, D. J. (1999) Atomic force microscopy: a forceful way with single molecules. *Curr. Biol.* **9,** R133–R136.
3. Moerner, W. E. (2002) A dozen years of single-molecule spectroscopy in physics, chemistry, and biophysics. *J. Phys. Chem. B* **106,** 910–927.
4. Haustein, E. and Schwille, P. (2004) Single-molecule spectroscopic methods. *Curr. Opin. Struct. Biol.* **14,** 531–540.
5. Michalet, X., Kapanidis, A. N., Laurence, T., et al. (2003) The power and prospects of fluorescence microscopies and spectroscopies. *Annu. Rev. Biophys. Biomol. Struct.* **32,** 161–182.
6. Tamarat, P., Maali, A., Lounis, B., and Orrit, M. (2000) Ten years of single-molecule spectroscopy. *J. Phys. Chem. A* **104,** 1–16.
7. Plakhotnik, T., Donley, E. A., and Wild, U. P. (1997) Single-molecule spectroscopy. *Annu. Rev. Phys. Chem.* **48,** 181–212.
8. Xie, X. S. (1996) Single-molecule spectroscopy and dynamics at room temperature. *Acc. Chem. Res.* **29,** 598–606.

9. Goodwin, P. M., Ambrose, W. P., and Keller, R. A. (1996) Single-molecule detection in liquids by laser-induced fluorescence. *Acc. Chem. Res.* **29,** 607–613.

10. Böhmer, M. and Enderlein, J. (2003) Fluorescence spectroscopy of single molecules under ambient conditions: methodology and technology. *Chem. Phys. Chem.* **4,** 793–808.

11. Haran, G. (2003) Single-molecule fluorescence spectroscopy of biomolecular folding. *J. Physics-Condensed Matter* **15,** R1219–R1317.

12. Förster, T. (1948) Zwischenmolekulare Energiewanderung und Fluoreszenz. *Annalen der Physik* **6,** 55–75.

13. Ha, T., Enderle, T., Ogletree, D. F., Chemla, D. S., Selvin, P. R., and Weiss, S. (1996) Probing the interaction between two single molecules: Fluorescence resonance energy transfer between a single donor and a single acceptor. *Proc. Natl. Acad. Sci. USA* **93,** 6264–6248.

14. Selvin, P. R. (2000) The renaissance of fluorescence resonance energy transfer. *Nature Struct. Biol.* **7,** 730–734.

15. Haas, E., Katchalskikatzir, E., and Steinberg, I. Z. (1978) Brownian-motion of ends of oligopeptide chains in solution as estimated by energy-transfer between chain ends. *Biopolymers* **17,** 11–31.

16. Vix, A. and Lami, H. (1995) Protein fluorescence decay: discrete components or distribution of lifetimes: really no way out of the dilemma. *Biophys. J.* **68,** 1145–1151.

17. Sakmann, B. and Neher, E. (1995) *Single Channel Recording,* Plenum Press, New York, NY.

18. Zhuang, X. and Rief, M. (2003) Single-molecule folding. *Curr. Opin. Struct. Biol.* **13,** 88–97.

19. Kellermayer, M. S., Smith, S. B., Granzier, H. L., and Bustamante, C. (1997) Folding-unfolding transitions in single titin molecules characterized with laser tweezers. *Science* **276,** 1112–1116.

20. Jia, Y. W., Talaga, D. S., Lau, W. L., Lu, H. S. M., DeGrado, W. F., and Hochstrasser, R. M. (1999) Folding dynamics of single GCN4 peptides by fluorescence resonant energy transfer confocal microscopy. *Chem. Phys.* **247,** 69–83.

21. Talaga, D. S., Lau, W. L., Roder, H., et al. (2000) Dynamics and folding of single two-stranded coiled-coil peptides studied by fluorescent energy transfer confocal microscopy. *Proc. Natl. Acad. Sci. USA* **97,** 13,021–13,026.

22. Deniz, A. A., Laurence, T. A., Beligere, G. S., et al. (2000) Single-molecule protein folding: Diffusion fluorescence resonance energy transfer studies of the denaturation of chymotrypsin inhibitor 2. *Proc. Natl. Acad. Sci. USA* **97,** 5179–5184.

23. Schuler, B., Lipman, E. A., and Eaton, W. A. (2002) Probing the free-energy surface for protein folding with single-molecule fluorescence spectroscopy. *Nature* **419,** 743–747.

24. Lipman, E. A., Schuler, B., Bakajin, O., and Eaton, W. A. (2003) Single-molecule measurement of protein folding kinetics. *Science* **301,** 1233–1235.

25. McCarney, E. R., Werner, J. H., Bernstein, S. L., et al. (2005) Site-specific dimensions across a highly denatured protein; a single molecule study. *J. Mol. Biol.* **352,** 672–678.

26. Groll, J., Amirgoulova, E. V., Ameringer, T., et al. (2004) Biofunctionalized, ultrathin coatings of cross-linked star-shaped poly(ethylene oxide) allow reversible folding of immobilized proteins. *J. Am. Chem. Soc.* **126,** 4234–4239.

27. Rhoades, E., Gussakovsky, E., and Haran, G. (2003) Watching proteins fold one molecule at a time. *Proc. Natl. Acad. Sci. USA* **100,** 3197–3202.

28. Rhoades, E., Cohen, M., Schuler, B., and Haran, G. (2004) Two-state folding observed in individual protein molecules. *J. Am. Chem. Soc.* **126,** 14,686–14,687.

29. Schuler, B. (2005) Single-molecule fluorescence spectroscopy of protein folding. *Chemphyschem.* **6,** 1206–1220.

30. Van Der Meer, BW, Coker, G. III, and Chen, S. Y. S. (1994) *Resonance energy transfer: theory and data.* New York, Weinheim, Cambridge: VCH Publishers, Inc.

31. Deniz, A. A., Laurence, T. A., Dahan, M., Chemla, D. S., Schultz, P. G., and Weiss, S. (2001) Ratiometric single-molecule studies of freely diffusing biomolecules. *Annu. Rev. Phys. Chem.* **52,** 233–253.

32. Eggeling, C., Berger, S., Brand, L., et al. (2001) Data registration and selective single-molecule analysis using multi- parameter fluorescence detection. *J. Biotechnol.* **86,** 163–180.

33. Schuler, B., Lipman, E. A., Steinbach, P. J., Kumke, M., and Eaton, W. A. (2005) Polyproline and the "spectroscopic ruler" revisited with single molecule fluorescence. *Proc. Natl. Acad. Sci. USA* **102,** 2754–2759.

34. Ha, T. (2001) Single-molecule fluorescence resonance energy transfer. *Methods* **25,** 78–86.

35. Moerner, W. E. and Fromm, D. P. (2003) Methods of single-molecule fluorescence spectroscopy and microscopy. *Rev. Sci. Instrum.* **74,** 3597–3619.

36. Wahl, M., Koberling, F., Patting, M., Rahn, H., and Erdmann, R. (2004) Time-resolved confocal fluorescence imaging and spectrocopy system with single molecule sensitivity and sub-micrometer resolution. *Curr. Pharm. Biotechnol.* **5,** 299–308.

37. Hohng, S., Joo, C., and Ha, T. (2004) Single-molecule three-color FRET. *Biophys. J.* **87,** 1328–1337.

38. Clamme, J. -P. and Deniz, A. A. (2005) Three-color single-molecule fluorescence resonance energy transfer. *Chem. Phys. Chem.* **6,** 74–77.

39. Rothwell, P. J., Berger, S., Kensch, O., et al. (2003) Multiparameter single-molecule fluorescence spectroscopy reveals heterogeneity of HIV-1 reverse transcriptase: primer/template complexes. *Proc. Natl. Acad. Sci. USA* **100,** 1655–1660.

40. Williams, R. M., Piston, D. W., and Webb, W. W. (1994) 2-photon molecular-excitation provides intrinsic 3-dimensional resolution for laser-based microscopy and microphotochemistry. *FASEB J.* **8,** 804–813.

41. Margittai, M., Widengren, J., Schweinberger, E., et al. (2003) Single-molecule fluorescence resonance energy transfer reveals a dynamic equilibrium between closed and open conformations of syntaxin 1. *Proc. Natl. Acad. Sci. USA* **100,** 15,516–15,521.

42. Mujumdar, R. B., Ernst, L. A., Mujumdar, S. R., Lewis, C. J., and Waggoner, A. S. (1993) Cyanine dye labeling reagents: sulfoindocyanine succinimidyl esters. *Bioconjug. Chem.* **4,** 105–111.

43. Panchuk-Voloshina, N., Haugland, R. P., Bishop-Stewart, J., et al. (1999) Alexa dyes, a series of new fluorescent dyes that yield exceptionally bright, photostable conjugates. *J. Histochem. Cytochem.* **47,** 1179–1188.

44. Murphy, C. J. (2002) Optical sensing with quantum dots. *Anal. Chem.* **74,** 520A–526A.

45. Chan, W. C., Maxwell, D. J., Gao, X., Bailey, R. E., Han, M., and Nie, S. (2002) Luminescent quantum dots for multiplexed biological detection and imaging. *Curr. Opin. Biotechnol.* **13,** 40–46.

46. Lippitz, M., Erker, W., Decker, H., van Holde, K. E., and Basche, T. (2002) Two-photon excitation microscopy of tryptophan-containing proteins. *Proc. Natl. Acad. Sci. USA* **99,** 2772–2777.

47. Shimomura, O. (2005) The discovery of aequorin and green fluorescent protein. *J. Microsc.* **217,** 1–15.

48. Sako, Y. and Uyemura, T. (2002) Total internal reflection fluorescence microscopy for single-molecule imaging in living cells. *Cell Struct. Funct.* **27,** 357–365.

49. Dawson, P. E. and Kent, S. B. (2000) Synthesis of native proteins by chemical ligation. *Annu. Rev. Biochem.* **69,** 923–960.

50. Ratner, V., Kahana, E., Eichler, M., and Haas, E. (2002) A general strategy for site-specific double labeling of globular proteins for kinetic FRET studies. *Bioconjug. Chem.* **13,** 1163–1170.

51. Kapanidis, A. N. and Weiss, S. (2002) Fluorescent probes and bioconjugation chemistries for single- molecule fluorescence analysis of biomolecules. *J. Chem. Phys.* **117,** 10,953–10,964.

52. David, R., Richter, M. P., and Beck-Sickinger, A. G. (2004) Expressed protein ligation. Method and applications. *Eur. J. Biochem.* **271,** 663–677.

53. Schuler, B. and Pannell, L. K. (2002) Specific labeling of polypeptides at amino-terminal cysteine residues using Cy5-benzyl thioester. *Bioconjug. Chem.* **13,** 1039–1043.

54. Yamaguchi, J., Nemoto, N., Sasaki, T., et al. (2001) Rapid functional analysis of protein-protein interactions by fluorescent C-terminal labeling and single-molecule imaging. *FEBS Lett.* **502,** 79–83.

55. Cropp, T. A. and Schultz, P. G. (2004) An expanding genetic code. *Trends Genet.* **20,** 625–630.

56. Hillisch, A., Lorenz, M., and Diekmann, S. (2001) Recent advances in FRET: distance determination in protein-DNA complexes. *Curr. Opin. Struct. Biol.* **11,** 201–207.

57. Norman, D. G., Grainger, R. J., Uhrin, D., and Lilley, D. M. (2000) Location of cyanine-3 on double-stranded DNA: importance for fluorescence resonance energy transfer studies. *Biochemistry* **39,** 6317–6324.

58. Clegg, R. M., Murchie, A. I., Zechel, A., and Lilley, D. M. (1993) Observing the helical geometry of double-stranded DNA in solution by fluorescence resonance energy transfer. *Proc. Natl. Acad. Sci. USA* **90,** 2994–2998.

59. Stryer, L. and Haugland, R. P. (1967) Energy transfer: a spectroscopic ruler. *Proc. Natl. Acad. Sci. USA* **58,** 719–726.

60. Cowan, P. M. and McGavin, S. (1955) Structure of poly-L-proline. *Nature* **176,** 501–503.
61. Schimmel, P. R. and Flory, P. J. (1967) Conformational energy and configurational statistics of poly-L-proline. *Proc. Natl. Acad. Sci. USA* **58,** 52–59.
62. Lakowicz, J. R. (1999) *Principles of Fluorescence Spectroscopy.* Kluwer Academic/Plenum Publishers, New York, NY.
63. Englander, S. W., Calhoun, D. B., and Englander, J. J. (1987) Biochemistry without oxygen. *Anal. Biochem.* **161,** 300–306.
64. Amirgoulova, E. V., Groll, J., Heyes, C. D., et al. (2004) Biofunctionalized polymer surfaces exhibiting minimal interaction towards immobilized proteins. *Chem. Phys. Chem.* **5,** 552–555.
65. Boukobza, E., Sonnenfeld, A., and Haran, G. (2001) Immobilization in surface-tethered lipid vesicles as a new tool for single biomolecule spectroscopy. *J. Phys. Chem. B* **105,** 12,165–12,170.
66. MacDonald, R. C., MacDonald, R. I., Menco, B. P., Takeshita, K., Subbarao, N. K., and Hu, L. R. (1991) Small-volume extrusion apparatus for preparation of large, unilamellar vesicles. *Biochim. Biophys. Acta* **1061,** 297–303.
67. Frieden, C., Chattopadhyay, K., and Elson, E. L. (2002) What fluorescence correlation spectroscopy can tell us about unfolded proteins. *Adv. Protein Chem.* **62,** 91–109.
68. Chattopadhyay, K., Elson, E. L., and Frieden, C. (2005) The kinetics of conformational fluctuations in an unfolded protein measured by fluorescence methods. *Proc. Natl. Acad. Sci. USA* **102,** 2385–2389.
69. Berglund, A. J., Doherty, A. C., and Mabuchi, H. (2002) Photon statistics and dynamics of fluorescence resonance energy transfer. *Phys. Rev. Lett.* **89,** 068101.
70. Gopich, I. V. and Szabo, A. (2005) Theory of photon statistics in single-molecule Förster resonance energy transfer. *J. Chem. Phys.* **122,** 14,707.
71. Kapanidis, A. N., Lee, N. K., Laurence, T. A., Doose, S., Margeat, E., and Weiss, S. (2004) Fluorescence-aided molecule sorting: Analysis of structure and interactions by alternating-laser excitation of single molecules. *Proc. Natl. Acad. Sci. USA* **101,** 8936–8941.
72. Kapanidis, A. N., Laurence, T. A., Lee, N. K., Margeat, E., Kong, X., and Weiss, S. (2005) Alternating-Laser Excitation of Single Molecules. *Acc. Chem. Res.* **38,** 523–533.
73. Cantor, C. R. and Schimmel, P. R. (1980) *Biophysical Chemistry*, W. H. Freeman and Company, San Francisco, CA.
74. Muller, B. K., Zaychikov, E., Brauchle, C., and Lamb, D. C. (2005) Pulsed Interleaved Excitation. *Biophys. J.* **89,** 3508–3522.
75. Kremer, W., Schuler, B., Harrieder, S., et al. (2001) Solution NMR structure of the cold-shock protein from the hyperthermophilic bacterium Thermotoga maritima. *Eur. J. Biochem.* **268,** 2527–2539.

9

Single Molecule Studies of Protein Folding Using Atomic Force Microscopy

Sean P. Ng, Lucy G. Randles, and Jane Clarke

Summary

Atomic force microscopy (AFM) offers new insights into the ability of proteins to resist mechanical force. The technique has been opened up by the availability of easy-to-use instruments that are commercially available, so that the technique no longer relies on the need to build instruments in the lab. Indeed it may become common for AFM instruments to sit beside stopped-flow apparatus in protein folding laboratories. In this chapter, we describe the instrument set-up, the preparation of suitable protein substrate, and the collection of data. Data selection and analysis are more complex than for conventional stopped-flow ensemble studies, but offer new insights into the function of proteins in vivo.

Key Words: Forced unfolding; AFM; Φ-value; energy landscape; dynamic force spectrum.

1. Introduction

Since its invention in 1986, the atomic force microscopy (AFM) has become one of the most widely used instruments for imaging surface topology of a wide range of systems, such as semiconductors, polymers, and biomaterials *(1)*. More recently, it has been used in dynamic force mode to extract information about the energy landscape of molecular bonds in a wide range of biomolecules *(2–4)*. This chapter describes how AFM experiments can be used to study protein unfolding.

AFM experiments have captured much recent attention in the study of protein folding. AFM uses an external force as a denaturant allowing the unfolding of individual protein domains to be observed. This method offers "new" insights into the folding mechanism, as well as the folding energy landscape, of many proteins under the influence of mechanical stress.

1. First, this method provides valuable information on forced unfolding of single protein molecules that has not been available previously. The results obtained from this

From: *Methods in Molecular Biology, vol. 350: Protein Folding Protocols*
Edited by: Y. Bai and R. Nussinov © Humana Press Inc., Totowa, NJ

method are, therefore, ideal for benchmarking the molecular dynamics simulations, which also study the behavior of single molecules under force extension.

2. Second, this technique introduces an excellent way to trigger unfolding of particular proteins that carry out biological functions under mechanical stress, e.g., fibronectin type III domains *(5–8)* and immunoglobulin domains *(9–12)*. Classical chemical or thermal denaturation cannot truly mimic the unfolding events of these mechanical proteins in vivo.

3. Third, it offers the possibility of probing alternative folding pathways of the energy landscape that may differ from those explored by chemical denaturants. Chemical denaturant has a nondirectional denaturing effect on proteins that leads to collapsed structures. In contrast, forced unfolding causes stretching of polypeptide chains in a well-defined directional reaction coordinate (N- to C-terminal extension). It is important to compare the pathways occurring with the two different unfolding methods. In a few AFM experiments, it has been shown that the mechanical stability and the unfolding pathway of a protein depend critically on the direction of the applied forced extension *(13,14)*.

4. Finally, it has been shown that it is possible to capture the presence of intermediate states *(15–18)* and to study refolding *(19)* and misfolding *(20)* through the combined use of protein engineering techniques with single molecule AFM. A fundamental requirement for these studies is the construction of an engineered polyprotein consisting of identical tandem repeats of the domain under study *(21)*.

With these possibilities in mind, this chapter was written as a "guide" to novices in the field, and will hopefully point out traps and pitfalls along the way.

1.1. Background Theory

1.1.1. Effects of Force on the Energy Landscape

Protein folding and unfolding reactions are, in many cases, simple two-state reactions and the rate constants for folding and unfolding can be considered to be determined by the height of the rate-determining energy barrier, or transition state, by analogy to simple chemical reactions *(22)*. In the absence of force, the unfolding and refolding rate constants (k_u^0 and k_f^0) in a simple two-state system are given by:

$$k_u^0 = \kappa \nu \exp\left(\frac{-\Delta G_{TS-N}}{k_B T}\right) \tag{1}$$

$$k_f^0 = \kappa \nu \exp\left(\frac{\Delta G_{D-TS}}{k_B T}\right) \tag{2}$$

where ΔG_{TS-N} and ΔG_{D-TS} are the activation energies for unfolding and refolding, k_B is the Boltzmann constant, T is the absolute temperature, κ is the transmission coefficient, and ν is the characteristic vibration frequency. **Figure 1** is a schematic diagram of the two-state energy landscape in the absence and presence

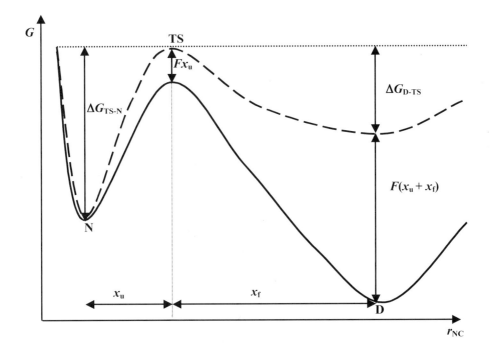

Fig. 1. A schematic representation of the two-state free energy landscape in the absence (broken curve) and with the perturbation of a mechanical force (solid curve). The distances along the unfolding reaction coordinate from the native (N) to transition state (TS) and from the TS to the denatured state (D) are given by x_u and x_f, respectively. The barrier for unfolding ΔG_{TS-N}, is lowered by Fx_u, whereas the barrier for refolding ΔG_{D-TS} is raised by Fx_f. The position of the TS is assumed to remain essentially stationary with respect to the increase in the applied force from the reference point N.

of an external force. In protein folding reactions the prefactor, $\kappa\nu$, is likely to be in the order of 10^6–10^7/s (discussed in **ref. 23**).

By applying an external mechanical force, F, the system is driven far from thermal equilibrium and the energy landscape of the protein is deformed (**Fig. 1**). In Bell's linear approximation *(24)*, the energy barrier for unfolding will be lowered by Fx_u, whereas the energy barrier for refolding will be raised by Fx_f, where the distances along the unfolding reaction coordinate from the native (N) to transition state (TS) and from TS to the denatured state (D) are given by x_u and x_f, respectively. The position of the TS is assumed to remain unchanged with respect to the increase in the applied force. The rate of unfolding and folding at any given force can then be calculated according to:

$$k_u(F) = \kappa\nu \exp\left(\frac{-(\Delta G_{TS-N} - Fx_u)}{k_B T}\right) = k_u^0 \exp\left(\frac{Fx_u}{k_B T}\right) \tag{3}$$

$$k_{\mathrm{f}}(F) = \kappa v \exp\left(\frac{\Delta G_{\mathrm{D-TS}} - F x_{\mathrm{f}}}{k_{\mathrm{B}}T}\right) = k_{\mathrm{f}}^0 \exp\left(\frac{-F x_{\mathrm{f}}}{k_{\mathrm{B}}T}\right) \qquad (4)$$

Because the system is driven far from equilibrium under force, only the unfolding rate is meaningful. Because x_{f} is large the probability of refolding is almost zero under experimental conditions and, therefore, the refolding rate is negligible.

1.1.2. Polymer Elasticity Modeling: Determining Protein Contour Length

When a polymer, such as a protein, is stretched by an applied force, the extension of the polymer causes an opposing entropic force. In general, small extensions require little force and the force–extension relationship follows a linear function where the polymer reacts like an ideal spring. However, for extensions larger than the contour length of the molecule, L (the contour length is the length of a linearly extended molecule without stretching the molecular backbone), the behavior of the protein becomes more complicated and its elasticity may be described by simple models, such as the worm-like chain (WLC) model (4,25).

The WLC uses L and the persistence length (P) of a molecule to model the response of the protein to force extension. Persistence length is a measure of the flexibility of a polymer. A polymer with smaller P is essentially more resistant to extension.

An analytical expression for the applied force (F) as a function of the polymer extension (x) is expressed by:

$$F(x) = \frac{k_{\mathrm{B}}T}{P}\left(\frac{1}{4(1 - x/L)^2} - \frac{1}{4} + \frac{x}{L}\right) \qquad (5)$$

In AFM mechanical unfolding studies, the WLC model is used to fit the force–extension relationship of a multiple module protein by following the unfolding of all of its domains. Fitting the model to the rising edge of each unfolding peak allows the protein length to be determined from the difference in the fitted contour length between adjacent peaks (4).

1.1.3. Unfolding Forces Depend on Pulling Speed

The most probable unfolding force at which a protein unfolds, $p(F)$, is approximated by an unfolding probability density equation (26)

$$p(F) = \frac{1}{r_{\mathrm{f}}} k_u(F) * \exp\left[-\frac{1}{r_{\mathrm{f}}} \int_0^F k_u(F')\partial F'\right] \qquad (6)$$

Increase in the unfolding speed (v)

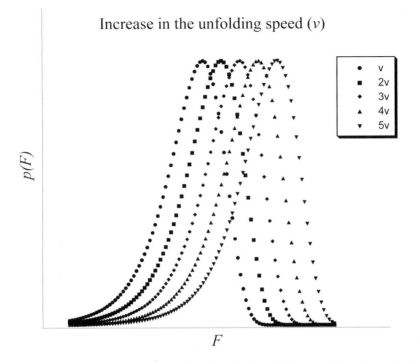

Fig. 2. Numerical simulations show the distribution of the unfolding forces is shifted to higher F with the increase of unfolding speed (v). The peak of the distribution corresponds to the most probably unfolding force (f^*).

where r_f is the loading rate and $k_u(F)$ is the unfolding rate constant at any given force F. Maximizing the distribution of the probability density (i.e.,

$$\left.\frac{\partial P(F)}{\partial F}\right|_{F=f^*} = 0)$$ derives the most probable unfolding force f^*(i.e., the mode of

the distribution)

$$f^* = \frac{k_B T}{x_u} \ln\left(\frac{x_u k_c v}{k_B T k_u^0}\right) \tag{7}$$

where v is the pulling speed and k_c is the spring constant of the cantilever (**Eq. 7** does not consider the contribution from the stiffness of the polymer being pulled and, therefore, $k_c v$ is equivalent to the loading rate). Increasing v results in a shift of the distribution to higher unfolding forces (**Eq. 7**; **Fig. 2**). This explains why there is a linear increase of unfolding force with the log of the pulling speed in all unfolding studies of proteins by AFM. In all studies, the width of the distribution depends on the thermal energy and the value of x_u (i.e., $\frac{k_B T}{x_u}$).

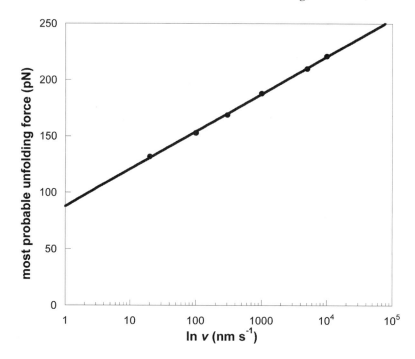

Fig. 3. The dynamic force spectrum (a plot of unfolding force vs pulling speed) of a polyprotein of TI I27 *(52)*. The speed dependence can be fitted to numerical expressions to extract the dynamic parameters x_u and k_u^0 of the protein.

The dynamic force spectrum is a plot of unfolding force vs logarithm of pulling speed (**Fig. 3**). Because the compliance of the system is largely governed by the compliance of the protein, and not the cantilever, in protein unfolding experiments the spring constant of the system, k_c, is unknown and dynamic force spectra cannot be analyzed directly. Kinetic parameters of a protein, such as the unfolding rate constant extrapolated to zero force (k_u^0) and x_u, can be extracted from the spectrum by fitting procedures discussed in **Subheading 3.5.**

2. Instrumentation

The schematic in **Fig. 4** illustrates the main features of an AFM. Because we are only considering the use of such an instrument to take force measurements in the z direction of a sample, we will not discuss any use of the microscope for imaging in the x–y plane.

2.1. Operation Modes of the Force Mode AFM

The force mode AFM can be operated in two different modes depending on the absence (open loop mode) or the presence (closed loop mode) of a feedback control to the piezo positioner.

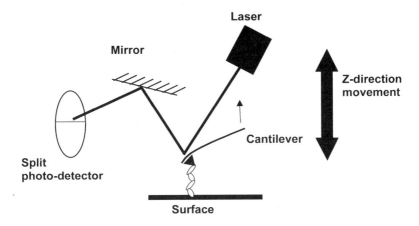

Fig. 4. The schematic diagram highlights the main features of an atomic force microscope. A laser beam is focused on the back of a cantilever. The laser beam reflected off the back of the cantilever is captured and is directed to the photodetector by a mirror. The z direction movement of the cantilever is controlled by a piezoelectric positioner.

In the open loop mode, the piezo positioner moves the laser–cantilever assembly down to a stationary substrate to pick up a sample and the sample is stretched as the laser–cantilever assembly retracts from the substrate. (Alternatively, the piezo positioner may move the substrate up and down while the cantilever–laser assembly remains unmoved depending on the AFM manufacturer specifications.)

In the closed loop mode (force-clamp), the piezo positioner compensates for the deflection of the cantilever by adjusting the tip-sample separation, such that the deflection of the cantilever always remains at a computer-controlled predetermined set value (e.g., at a constant pulling force). This is achieved by integrating a feedback system (i.e., PID controller) that simultaneously compares the signals generated on each deflection of the cantilever with the computer-controlled set point. The differences are amplified by the PID system and fed back to the piezo positioner.

2.2. Cantilevers

In an AFM experiment, the cantilever acts as a sensor for the local interaction between the tip and the sample. The deflection of the cantilever is detected by the laser deflection reflected from the backside of the cantilever by a photodetector. The cantilevers, with dimensions in the order of a hundred microns, are fabricated from silicon or silicon nitride using semiconductor micromachining techniques (**Fig. 5**) and are normally available in two shapes: rectangular (sensitive to lateral forces by having a greater degree of rotational freedom) and "V" shape (suitable for topographical imaging with minimized torsional motions). The backside of

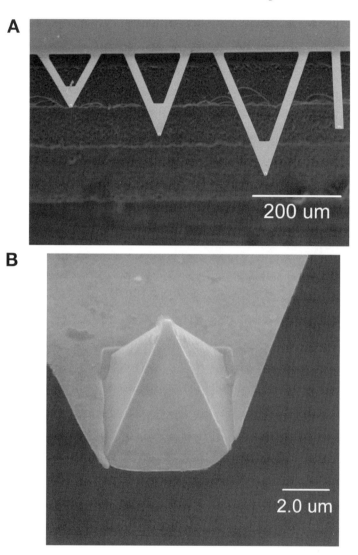

Fig. 5. Scanning electron micrograph images of (A) a rectangular and V-shaped cantilevers, and (B) the tip of the cantilever under higher magnification. (The scanning electron micrograph images are courtesy of Molecular Imaging, Inc.)

the cantilever (the face that is not in contact with the sample) is normally coated with a thin layer of gold to enhance reflectivity, especially in aqueous environments. Regardless of its shape, the bending of the cantilever is used to determine the force (F) exerted onto a sample by the application of Hooke's law

$$F = k_c d \qquad (8)$$

where k_c is the spring constant of the cantilever and d is the deflection of the cantilever in nanometers. k_c increases with cantilever thickness but decreases with cantilever length. Extremely small cantilevers (e.g., bio-levers) are, in principle, ideal for acquiring force data with reduced noise at high loading rates. However, they are not generally used in pulling experiments because of the added expense. Most of the pulling experiments use cantilevers with spring constants in the range of 10 to 100 pN/nm (*see* **Note 1**).

There are many types of cantilevers available commercially with different aspect ratios and geometries, each with a specific function mainly aimed for high-resolution imaging. However, the cantilever tips used for pulling experiments just provide a point for a molecule to attach and, therefore, do not require specific modifications. Unsharpened tips (radius of curvature ~50 nm) with "bare" silicon nitride are, in fact, found to have greater absorption toward sample molecules presumably owing to the presence of dangling bonds on the non-stoichiometric matrix of the SiN_x resulting from the chemical vapor deposition process. Some success has been achieved using modified cantilevers so that specific attachment to the protein at both the N- and C-termini is achieved *(27,28)*. To date, this is not common, and investigators appear to find the non-specific adsorption of protein to the cantilever (sometimes accompanied by nonspecific adsorption to the surface) to be sufficient for the purpose.

2.3. Optical Lever Detection

The most common method to detect cantilever deflections is the optical lever method. A laser beam is focused onto the end of the cantilever and the reflected beam is collected by a split photodiode detector. The difference in laser intensity between the sections of the detector quantifies the deflection of the cantilever. The deflection of the cantilever is used to calculate the force exerted onto the sample once the spring constant of the cantilever is determined (**Eq. 8**).

2.4. Spring Constant Determination

There are two common calibration methods to determine the spring constant of a cantilever: the added mass method *(29)* and the thermal fluctuation method *(30)*. Although the added mass method is considered to be more accurate, it can be destructive and is methodologically challenging and slow. The thermal fluctuation method, widely used in many commercial force spectroscopy, has been shown to yield results similar to added mass methods with less than approx 10% discrepancy *(31)*. This method approximates the cantilever as a simple harmonic oscillator fluctuating in response to thermal noise. The frequency and amplitude of these oscillations are related to the mechanical properties of the cantilever and, therefore, are sufficient to determine the spring constant. The advantage of this method is that the calibration can be done in liquid, it is easy to perform because the protocol is built into the software of commercial instruments, and it requires no further manipulation of the cantilever (nondestructive).

2.5. Vibration Isolation

The z direction measurements of an AFM are very sensitive to small displacements of the sample or the cantilever. As a result, vibration isolation is of great importance to obtain highly accurate force data. The vibrations may add a background noise to the measured signal.

There are several sources of mechanical vibrations, such as the vibrations from the building and the machine itself, which can be in the approx 15–200 Hz range. To dampen these types of vibrations, the AFM machine is mounted on a mechanically isolated platform (vibration isolation table). Acoustic vibrations in the approx 2–20 kHz range can be transmitted to the cantilever. Hence, it may be necessary to make the AFM measurements in a quiet room or to place the AFM systems under an acoustic proof cover. Finally, it is also important to have a stable control of the room temperature to avoid thermal drift to the cantilever or the piezo positioner.

2.6. Sample Holder: Fluid Cells

One of the appealing features of the AFM is its capability to operate under liquids using a suitable sample holder or fluid cells. All fluid cells, irrespective of their design, basically contain the aqueous sample and provide a stable optical path for the laser beam, which is reflected off the back of the cantilever. A schematic layout of a fluid cell is shown in **Fig. 6** with the open and sealed design. Open fluid cells (often simply a microscope slide) allow the buffers, reagents, or protein sample to be added directly to the cells with ease. The design of a closed fluid cell, however, prevents evaporation of the protein sample and allows flow of buffers through the cell during an experiment.

Recently, the study of protein forced unfolding under different temperatures *(32)* has been made possible with the combination of a fluid cell and a temperature controller. The commercially available temperature controller provides direct and symmetrical heating/cooling through its thermal element mounted on the fluid cell. The high precision and ultra sensitive thermal sensor of the temperature controller offers accurate control to the temperature of the sample with extremely low thermal drift (e.g., 0.1°C in a steady state and less than 0.5°C overshoot).

3. Methods
3.1. Multi-Modular Protein Constructs

Pulling experiments require the substrate to be attached at one end to a suitable surface, typically by engineering cysteine groups onto the C-terminal end of the protein that form covalent links with a gold surface. When the cantilever is

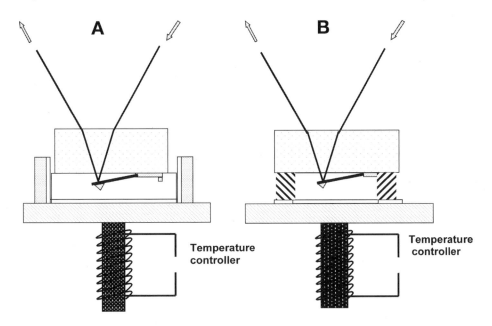

Fig. 6. Schematic illustration of liquid cells with (**A**) the open and (**B**) "O" ring sealed design. A temperature controller can be incorporated for heating/cooling of the cells.

brought toward the surface, it may attach to any part of the protein through nonspecific adsorption. This may be assisted by the presence of six histidine residues that are incorporated at the N-terminus after expression in a modified version of pRSETA vector (Invitrogen, Paisley, UK). Some experiments have been carried out on single protein domains; this is generally considered to be unattractive as tip–surface and protein–surface interactions may mask any unfolding peaks. This has been overcome by increasing the distance between the tip and surface by inserting long linkers to the ends of a single domain. More commonly, multimodular constructs are used; this increases the chance of picking up a reasonable length of protein and maximizes the data acquisition rate by allowing multiple force peaks to be obtained from a single approach–retraction cycle.

Naturally occurring chains of multi-domain proteins, such as titin or tenascin, can be used (*4,33*) or a tailor-made "polyprotein" can be constructed using molecular biology techniques (*34*). In this case, multiple repeats of a single domain or specific combinations of domains are cloned into a single expression vector. A common cloning system uses PCR amplification using primers that introduce unique restriction sites at the start and end of each domain allowing the modules to be inserted in a particular order. These versatile expression systems facilitate a single module to be mutated or replaced with another and also allow *in situ* sequencing (*16,21*). A well-tested system that has been extensively used in our laboratory consists of two sets of four TI I27 domains arranged in tandem that can easily be replaced by any other protein of choice and then ligated together to form an eight-module construct *(21)*. The advantage of this 4+4 system, is that each half of the protein can be sequenced, ensuring that PCR and cloning errors are detected. The initial TI I27 constructs are available on request from our laboratory. We have recently modified the published protocol to use fewer initial PCR steps and allow more efficient cloning. To prepare a "new" polyprotein gene using this protocol the steps are as follows:

1. A stock of "parent plasmids" 1–4 and 5–8 (**Fig. 7A**) are prepared. Modules 1–4 are contained within the multiple cloning site of a modified version of pRSET A (Invitrogen) that encodes an N-terminal His_6 tag, allowing rapid Ni-affinity chromatography. This cloning site is situated behind DNA encoding the T7 promoter. Modules 5–8 are in a holding vector.

2. The TI I27 module(s) that are to be replaced are removed from the parent plasmids by digestion with the appropriate restriction enzymes. The vector is gel-purified and treated with shrimp alkaline phosphatase (SAP) to dephosphorylate the vector DNA, thus preventing self-ligation.

3. All protein modules to be inserted are amplified using PCR with primers that incorporate specific restriction sites (**Fig. 7B**). These are initially cloned into individual T-vectors and sequenced. Note that the C-terminal primer for the C-terminal domain has to also encode two Cys residues and a stop codon after the end of the gene, before the *Eco*RI site.

4. After sequencing, the DNA sequences encoding the modules are cut with the appropriate restriction enzymes and purified from an agarose gel. Four protein modules can then be inserted in a single ligation reaction into the vector. Ligation reactions can be done on the "bench-top" in a few hours, but they are more reliable when carried out at 16°C overnight. Each 4-mer plasmid is sent for DNA sequencing.

5. The construct is then fully assembled by ligating modules 5–8 into the expression vector, cut with *Nhe*I and *Eco*RI (*see* **Note 2**).

An important point to be borne in mind: if the domain of interest is "too short" the inserted module may be destabilized in the new protein *(35–37)*, it is important to include a number (2–4) of extra amino acids at the start and end of each domain to ensure that inclusion in the polyprotein is intact. This may also prevent specific domain–domain interactions from interfering with the experiments (*see* **Note 2**).

3.2. Protein Expression and Purification

Once the multidomain construct has been produced, the expression vector is transformed into competent *Escherichia coli* cells that contain the gene for T7 polymerase, the product of which is highly specific for T7 promoters. Because

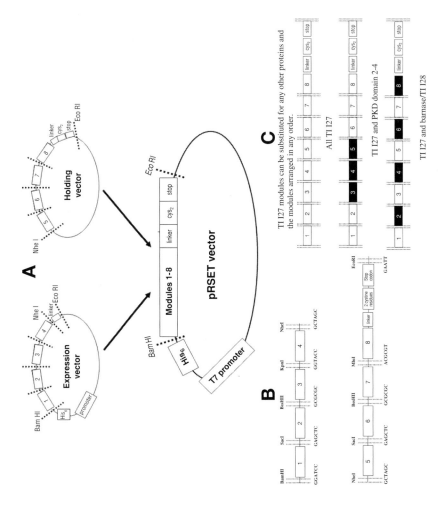

Fig. 7. (**A**) Schematic showing the constructs used to create multimeric proteins. Modules 1–4 are held in an expression vector (containing a T7 promoter site). Modules 5–8 are situated in a "holding" vector. The linker region in the expression vector is cleaved and modules 5–8 ligated into the expression vector to produce the 8-module construct. Restriction sites for each module are labelled in **B**. (**C**) Schematic to show how different domains can be inserted into the construct in any order using the unique restriction sites (*39,40,53*).

of this specificity, expression from T7 promoters is very high. We have always had good expression using *E. coli* C41 cells (Avidis, Saint Beauzire, France) *(38)*. Cells are grown in rich nutrient media to an OD_{600nm} of 0.4–0.6. Protein expression is induced with the addition of IPTG (isopropyl-β-D-thiogalactopyranoside) to initiate expression of the T7 polymerase. After induction, cells are grown overnight and then harvested by centrifugation. Soluble protein is released either by sonication or cell cracking and subsequently purified on a nickel column and then by size fractionation chromatography.

In order to be analyzed, the protein must be soluble and folded. It is possible to resolubilize a protein by fully unfolding in a chemical denaturant and then slowly decreasing the denaturant to allow refolding to occur, but this is a difficult procedure in a polyprotein where the local concentration of protein is always high (domains are attached to each other, and this cannot be altered by dilution).

If the protein is insoluble as a polyprotein construct (or indeed if it is only weakly soluble as a single domain), these protein modules may be expressed as polyproteins using flanking wild-type TI I27 domains (**Fig. 7C**) *(39,40)*. This also has the advantage that the TI I27 domains provide an internal standard for AFM experiments, whereby force peaks from the inserted protein can be distinguished from TI I27 by either the force profile or by the distance between peaks.

Proteins should be stored in dilute solutions in buffer at 4°C with sodium azide (0.01%) to prevent bacterial growth. Polyproteins can be stable for months under these conditions (*see* **Note 3**).

3.2.1. Characterizing the Polyprotein

It is surprising that most investigators do not do simple biophysical experiments to check that the domains are folded and "well behaved" in the polyprotein construct. Yet it has been shown that the proteins can indeed be influenced by the presence of neighbors in a multimodular construct *(41)*. To check if a protein is folded, it is necessary to use a spectroscopic technique. This can be by following tryptophan fluorescence between 300 and 400 nm after excitation at 280 nm or by circular dichroism between 200 and 250 nm. The resultant trace should be compared to a sample that has been unfolded by denaturant (8 *M* guanidinium chloride). The two traces should be significantly different if the protein is folded. If possible, the traces should be compared with sample traces of the individual domains. Simple unfolding experiments can be used to optimize solution conditions (pH, salt content) for the AFM experiments.

We have also shown that it is possible to use NMR to check the structural integrity of proteins in multimodular constructs *(17,39)*. It is also easy to use stopped-flow techniques to check whether the folding and unfolding kinetics are affected by inclusion in a polyprotein *(17,39,41)*.

3.3. Data Acquisition

3.3.1. Protein Adsorption

In a typical AFM pulling experiment, 200–500 µL of polyprotein solution (OD_{280} ~0.5) is spread (about 2 cm^2) onto a surface of freshly evaporated gold and incubated for approx 15–30 min to allow protein molecules to adsorb onto the gold surface. After this step, any loosely bound protein molecules are washed off and the protein on the surface is covered with a fresh layer of buffer. It is important to regulate the concentration of protein on the surface so that it is dilute enough that, in general, only one protein molecule is picked up at a time, but concentrated enough that there are enough protein molecules on the surface for a reasonable number of successful attachments. It has been estimated *(42)* that a concentration of protein that gives 1 in 10 successful pick-ups is ideal. The optimum protein concentration and absorption time can only be determined empirically.

3.3.2. Force–Extension Trace

The polyprotein is picked up by lowering the cantilever to the gold surface and is extended when the cantilever retracts from the substrate. The approach–retract cycles produce force–extension traces particular to specific "polyproteins." The saw-tooth pattern of a force trace, shown in **Fig. 8**, shows the series of events in a pulling cycle of a cantilever.

The retraction trace (superimposed with the approach trace) serves as a force baseline to measure the unfolding peak of the polyprotein, provided there is no drift in the approach and retract cycles. The force to unfold a module is taken to be the maximum height measured from the force baseline of each unfolding peak. Force traces are collected at different pulling speeds ranging from 100 to 2500 nm/s. At each pulling speed, at least approx 30–40 "clean and long" traces containing at least four unfolding events are required to construct a force histogram.

3.4. Data Analysis 1: Analyzing Unfolding Traces

3.4.1. Criteria for Choosing Force Traces and Peaks

If only all force traces collected were "ideal," like that shown in **Fig. 8**. Unfortunately, most force traces collected are far from ideal (*see* **Note 4** for poor and bad force traces). The most difficult part of the single molecule measurements of protein unfolding is the choice of which force traces to analyze from the many obtained from the AFM measurements. To date, there is no common standard to distinguish "good" and "bad" traces. On one hand, excluding nonideal traces avoids any uncertainty in deciding the mode of the distribution. On the other hand, having a small sample size results in an inadequate estimate of the shape or mode of the distribution. In view of this, a number of criteria for

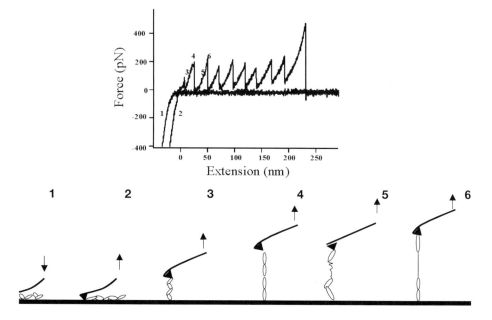

Fig. 8. The sequential events of a pulling cycle in a force–extension trace. The cantilever approaches and presses against the substrate (1) and begins to leave the substrate with an attached polyprotein (2). At both of these positions, the cantilever is bent upward and negative forces are observed in the trace. The cantilever elongates the polyprotein (3) by exerting forces onto the polyprotein until it is fully extended (4). The cantilever is bent downward at positions 3 and 4. One of the domains unfolds causing the cantilever recovers to its resting position (5). The cantilever continues to extend the polyprotein until another domain unfolds (6).

choosing acceptable force traces and deciding which peaks in the trace to measure are essential. These criteria include the following (*see also* **Note 5**):

1. Each trace should have peaks that are evenly spaced, accounting for the difference in length between the folded and unfolded domains. A precise measurement of the protein length can be achieved by fitting the WLC model to the force–extension curve.
2. Each trace should have at least three clearly resolved peaks so that the inter-peak distances can be evaluated.
3. The first peak can only be counted if the preceding irregular force peak is pulled-off from the surface and the peak base returns to the baseline.
4. The final overlap of the approach and the retraction traces should be close enough to represent only a small drift in the vertical displacement of the cantilever.
5. For analysis of the unfolding force for an eight-module protein molecule, the first and the last peaks are routinely disregarded and at least approx 30–40 clear traces are recorded at each pulling speed for analysis (*39*).

3.4.2. Contour Length

In an AFM experiment, the presence of the evenly spaced "saw-tooth pattern" of force peaks is the "fingerprint" of a modular protein. The WLC model *(4,25)* can be used to fit the force–extension relationship of the protein following the unfolding of each of the domains. In the pulling studies of a polyprotein of TI I27, fitting the rising edge of each peak to the WLC model with a fixed persistence length ($P = 3.5$ Å at 298 K) derives the contour length of the polyprotein with an increase of approx 29 nm each time when a domain unfolds. This value corresponds to a stretch of a folded domain comprising of 89 amino acids *(4)*.

3.4.3. Dynamic Force Spectrum

There are a number of ways of displaying force unfolding data. The first parameter to be extracted is the most probable unfolding force at any given retraction velocity. Because outliers in force traces can distort the mean value, it is the modal force that is most representative. Simply, the observed forces can be fit as a histogram. Determination of the modal force can, however, be affected by the bin widths and boundaries selected. An alternative method that we have found to be more reliable is as follows: to generate a force histogram from a set of force data, each of the forces is independently fitted to a Gaussian distribution, assuming each measurement is degraded by a specific level of noise. The width of the distribution corresponds to the noise level of the instrument (e.g., 20 pN owing to the thermal fluctuation of the cantilever). The summation of these Gaussians creates the force histogram; and the mode of the distribution corresponds to the most probable unfolding force of the protein.

The modes determined at a range of pulling speeds are then used to show the dependence of unfolding force on pulling speed in a dynamic force spectrum (*see* **Fig. 3**). Kinetic parameters such as x_u and k_u^0 can be extracted from the dynamic force spectrum.

3.5. Data Analysis 2: Determining k_u^0 and x_u

The kinetic parameters, k_u^0 and x_u, that define the mechanical properties of the single protein molecules can be derived from a dynamic force spectrum (unfolding force vs log pulling speed) by fitting the speed dependence to the appropriate analytical expressions that characterize their dynamic behavior and varying the pairwise combination values of k_u^0 and x_u to yield the best agreement between experimental and simulated data. Unfortunately, the force spectrum cannot be fitted to a simple equation because the spring constant of the system is unknown. Instead, two methods are generally accepted for determining k_u^0 and x_u from the experimental data. These models are the Monte Carlo fitting procedure and the numerical fitting procedure. These have been described in detail elsewhere *(4,43,44)*. Importantly, the value of k_u^0 is highly dependent on getting a correct value of x_u *(45)*. This means that for any single protein the value of k_u^0 cannot be determined with great confidence. This can be overcome, however, when looking at mutants of a single protein when changes in k_u^0 can be determined accurately, because a single shared value of x_u can be used *(16,46)* (*see* **Note 6**).

3.6. Measuring Φ–Values of Mutant Proteins by AFM

Φ-Value analysis is the only existing method to determine the structure of a transition state at the resolution of individual amino acid residues *(22)*. The transition state represents the highest point on an energy landscape along a folding pathway. Knowledge of these structures can, thus, identify the rate-limiting factors in the folding of a protein and may be used to re-engineer the folding rate of a protein in a rational way.

The analysis is performed by making a series of single-point mutations throughout the protein to act as reporters on the local environment. These mutations should be conservative, preferably nondisruptive, deletions of a small number of methylene groups. **Figure 9** shows the free energy diagrams of the energy perturbation of a mutation when Φ is 0 and 1, respectively.

This method has now been extended for analyzing the transition state structure and the folding pathways of a protein under the influence of an external force using AFM *(47)*. In the analysis, the dynamic force spectrum of a mutant is directly compared with that of wild-type to define its Φ-value. It is important to note that only mutants with the same x_u as the wild-type are eligible for the comparison. Mutants with different x_u may have different folding pathways and, therefore, should be analyzed separately (*see* **Note 7**).

The transition state Φ-value is determined by comparing the effect of mutation on the native state (N) directly with the effect of the mutation on the TS, and can be determined from a comparison of unfolding rates between a wild-type and mutant using **Eqs. 9** and **10**.

$$\Phi = 1 - \left(\frac{\Delta\Delta G_{TS-N}}{\Delta\Delta G_{D-N}} \right) \tag{9}$$

where $\Delta\Delta G_{D-N}$ is the change in free energy of the protein on mutation and

$$\Delta\Delta G_{TS-N} = -RT \ln \left(\frac{k_u^{0,wt}}{k_u^{0,mut}} \right) \tag{10}$$

where $k_u^{0,wt}$ and $k_u^{0,mut}$ are the unfolding rate constants of wild-type and mutant proteins at zero force, respectively.

In mechanical Φ-value analysis, $\Delta\Delta G_{D-N}$ is determined from equilibrium denaturation studies on the wild-type and mutant polyproteins and $k_u^{0,wt}$ and $k_u^{0,mut}$ can be obtained from the analysis of the force spectra, either by assuming

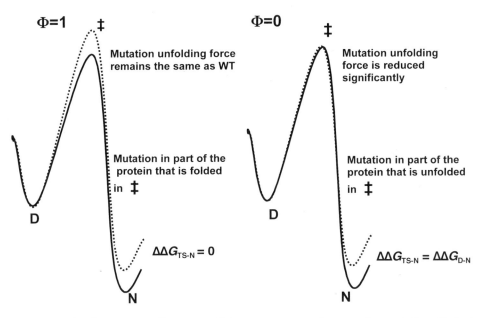

Fig. 9. The free energy diagrams of the energy perturbation of a mutation (dotted curve) in a two-state reaction. Φ is defined as $\Phi = 1 - \dfrac{\Delta\Delta G_{TS-N}}{\Delta\Delta G_{D-N}}$. When $\Phi = 1$, the unfolding forces remain unchanged upon a mutation. The barrier to unfolding is conserved upon a mutation (i.e., $\Delta\Delta G_{D-N} = 0$). The mutation is in part of the protein that is fully structured at the transition state (TS). When $\Phi = 0$, the unfolding forces reduce significantly upon a mutation. All free energy lost upon a mutation is reflected in the loss of unfolding kinetics (i.e., $\Delta\Delta G_{D-N} = \Delta\Delta G_{TS-N}$). The mutation is in part of the protein that is fully unfolded at the TS.

a common (mean) x_u (where justified from the data) and calculating the values of k_u^0 individually, or more simply from direct analysis of the force spectra fitted to a common slope.

3.6.1. Determining $\Delta\Delta G_{TS-N}$ Directly From Comparison of Mutant and Wild-Type Force Spectra

The unfolding forces for the mutant and wild-type proteins at the same pulling speed can be used to determine $\Delta\Delta G_{TS-N}$ directly using

$$\Delta\Delta G_{TS-N} = (f^{wt} - f^{mut})A\bar{x}_u \qquad (11)$$

where f^{wt} and f^{mut} are the most probable unfolding force for the wild-type at a given pulling speed, A is Avogadro's number, and $\bar{x}_u$ is the mean x_u of wild-type and the mutant protein.

Alternatively the pulling speeds at which the wild-type and mutant proteins unfold at the same force can be compared:

$$\Delta\Delta G_{TS-N} = -RT \ln\left(\frac{\nu^{wt}}{\nu^{mut}}\right) \tag{12}$$

3.6.2. Defining the Upper Boundary ($\Phi = 1$) and the Lower Boundary ($\Phi = 0$) in the Dynamic Force Spectrum of a Mutant

If the unfolding forces for the mutant are the same as that for wild-type, then its Φ-value is equal to 1. This implies that the mutation is in part of the protein that is as structured in the transition state as it is in the native state. The transition state is unfolded to the same extent as the native state by the mutation, so the height of the barrier to unfolding for the mutant remains the same as wild-type.

If $\Phi = 0$ then mutation is in a region of the protein that is completely unfolded in the transition state, i.e., $\Delta\Delta G_{TS-N} = \Delta\Delta G_{D-N}$. The force boundary that determines this limit can be calculated from **Eq. 11**, knowing the unfolding forces of wide-type, and assuming that $\Delta\Delta G_{TS-N} = \Delta\Delta G_{D-N}$ (as determined from the equilibrium denaturation experiments).

Figure 10 shows the data for wild-type and two mutants of the all-β protein TNfn3. One mutant, I20A, has unfolding forces that are the same as wild-type at all pulling speeds, the second, I8A, has unfolding forces that fall between the two limiting Φ-values and has been calculated to have a Φ-value of 0.6.

By making a number of mutations throughout the protein it is possible to "map" the transition state for unfolding experimentally allowing direct comparison with simulations *(16,46)*.

4. Notes

1. Factors affecting the force distribution. In a single molecule experiment, the unfolding event is dependent on the intrinsic thermal distribution of the protein molecules where the width of the distribution corresponds to the force scale ($\frac{k_B T}{x_u}$) of the unfolding forces measured from AFM traces.

Fig. 10. *(Opposite page)*   The comparison of the modal unfolding force for wild-type (filled circles and broken line) and two mutants (filled squares and continuous line) of the all-β protein TNfn3. Mutation I20A, has unfolding forces that are the same as wild-type at all pulling speeds, whereas mutation I8A, has unfolding forces that fall between the two limiting Φ-values and has been calculated to have a Φ-value of 0.6. The bottom dashed line in each panel shows the unfolding force that would be predicted if the Φ-value were 0 according to **Eqs. 11** and **12**. If the mutant protein unfolds at the same force as wild-type, that is equivalent to $\Phi = 1$. All data are fitted to the globally determined pulling speed dependence (slope).

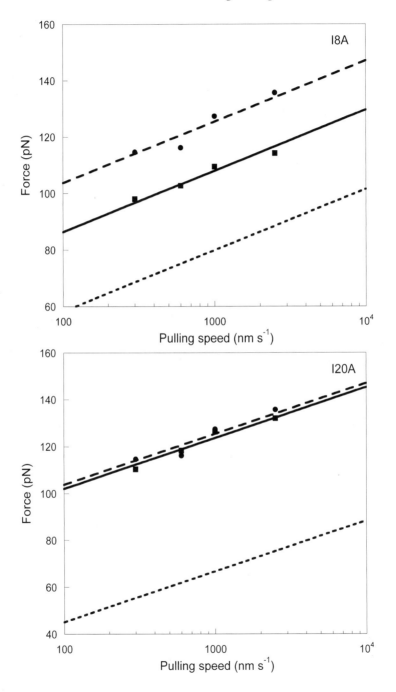

In the AFM experiments, thermal fluctuations of the cantilever affect the measured forces. A soft cantilever is sensitive enough to detect small variations in force but is susceptible to thermal fluctuations in position. Stiff cantilevers reduce thermal noise but cause larger fluctuations to the measured force *(49)*.

The calibration method itself is subject to a certain degree of error. However, the effect of this error can be reduced by averaging over a large number of measurements. For each new data set, the cantilever should be replaced. The error reduces significantly with the collection of more datasets.

2. Difficulty in cloning techniques. Primers encoding restriction sites must be sufficiently long to allow good overlap with the template protein sequence (typically by 3–4 codons). Amplification is carried out with *Taq* DNA polymerase, which automatically adds a single-A tail to the 3′ end of the newly synthesized DNA strand. This allows simple cloning of this DNA into a vector containing an overhanging T motif. This vector should be as fresh as possible as, with time, the T motif gradually deteriorates and cloning efficiency decreases.

Digestion with restriction enzymes works best when they are fresh and have been exposed to minimal cycles of freeze-thaw. Cut DNA is purified on agarose gels and digestion is immediately evident by the presence of correct size bands on the gel. If there is a low yield of DNA, as shown by a weak band, either the DNA can be used as a template for another PCR reaction or the amount of elution buffer in the gel extraction procedure can be reduced to effectively concentrate the DNA.

Ligase efficiency is particularly affected by freeze-thaw. Using fresh ligase often eliminates the majority of cloning problems that may otherwise be encountered. To increase the chance of successful ligations, reactions are best carried out overnight at 16°C. Transformation into competent cells may also be problematic. Ideally cells should be freshly prepared and must be kept cold. Media plates on which cells are grown should also be fresh.

Ligations can also be affected by SAP efficiency. SAP is needed to prevent the vector from religating back to itself and to allow the new DNA to be inserted. Ligations where this enzyme has not worked typically produce very large numbers of bacterial colonies when plated, none of which contain the gene of interest. As with ligase, fresh SAP should ideally be used and freeze-thaw cycles should be minimized.

To help identify problems during the cloning procedure it is vital to run controls at every stage, particularly during the ligation reactions.

3. Problems encountered in protein production. Multimeric proteins often have more problems associated with their expression than their monomeric counterparts. After inducing cells it is best to leave them to express protein at a temperature of no greater than 26°C to ensure that the protein remains as stable as possible. Once the cells are grown and harvested, it is important to keep them as cold as possible while protein is released. Sonication, although being a good method of releasing proteins, is prone to heating up the sample although this can be reduced if carried out in a glass container in ice. A better method, particularly for less stable proteins, is cell cracking which, because the machinery itself can be kept in a cold room, keeps the sample cooler and, therefore, is less likely to harm the protein.

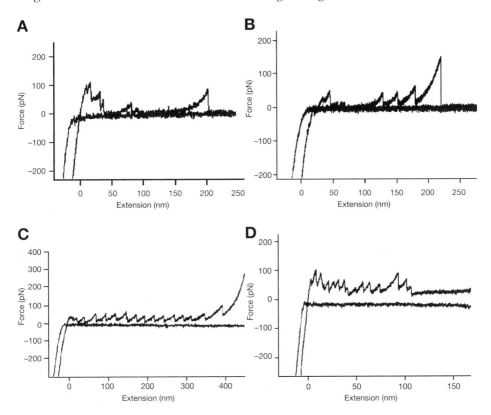

Fig. 11. Examples of "bad traces" that are difficult to interpret and therefore, they are excluded in our analysis. Traces showing (**A**) an almost completely unfolded protein being extended (**B**) a misfolded or partially unfolded protein being pulled (**C**) two proteins that have become attached to the atomic force microscopy tip, and (**D**) the pulling of aggregated proteins.

Multimers often form aggregates during purification. In order to isolate only the 8-mer form, size exclusion chromatography is carried out followed by running on a polyacrylamide gel to check if the product is of the correct size. To minimize this aggregation, multimers should always be stored in a suitable buffer in a dilute form and before use should be passed through a filter. Subsequent analysis by AFM should be done using fresh, filtered protein solutions.

4. Ambiguity in the force traces and choice of cantilever. The force extension curve in the AFM experiments should ideally consist of unfolding peaks equal to the number of domains in the polyprotein. However, this is rarely the case. Indeed, very few peaks (or even no peaks) or ambiguous force traces are normally observed. These traces often come from the attachment of more than one protein or protein in denatured or aggregated forms to the tip of the cantilever (**Fig. 11**). This may cause great difficulty in interpreting the data. To avoid further uncertainty, all these

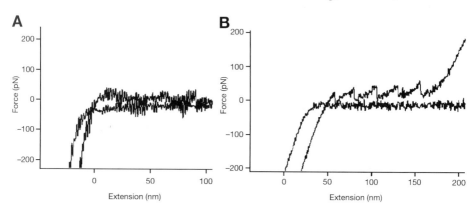

Fig. 12. Traces using normal cantilever and bio lever (reduced noise) for mechanically weaker protein. The unfolding events of mechanically weaker proteins are almost smeared out or become invisible in the traces when using normal cantilever. However, with the use of bio lever, the unfolding events become observable.

traces are rejected. This results in a small sample size and requires considerable time to sample clean and unambiguous traces. To acquire a good quality data set, one has to suffer this inevitable agony, owing to the nature of the single molecule measurements.

It is interesting to note that, in recent AFM studies of mechanically weak proteins, the force traces acquired using the conventional AFM tip and bio lever show significant differences (**Fig. 12**). By using bio lever, it is apparent that the noise level in the traces is greatly reduced. However, there are some drawbacks in the use of bio lever, and among them are the added expense and the small surface area for protein absorption.

5. The mechanical resistance of a folded domain is dependent on the unfolding history of the polyprotein. In principle, the sequence of peaks is important in a force–extension curve (*48*). This results from the fact that there are two competing factors affecting the unfolding of domains of a polyprotein. As more domains become unfolded, the probability of another domain unfolding is reduced and consequently, the unfolding force for the later peaks in a trace is increased. On the other hand, the later peaks in a trace are subjected to a lower loading rate (resulting from the increase in system compliance) as more domains unfold and this causes a reduction in the observed unfolding force. Ideally, unfolding forces should be analyzed in accord with the sequence of peaks in the force traces. However, in order to maximize the amount of data in our studies, the data from all peaks of a trace are pooled regardless of the positions of the peaks in a trace. The justification is that the error owing to the dependence of the unfolding force on the sequence of peaks in a trace is much less than the fluctuation in an unfolding force because of thermal noise of the cantilever (i.e., 10–20 pN).

6. Problems in the constant speed and the constant force pulling experiments. Constant speed measurements: in a constant speed (open loop mode) unfolding

experiment, it is not possible to determine the rate constant directly from the force measurements. Indeed, numerical fitting of the experimental data to the analytical expressions or use of a Monte Carlo simulation approach is required to extract the kinetic parameters x_u and k_u^0. However, it has been shown that (47) a small change in slope (i.e., x_u) results in a significant error in k_u^0. To increase the accuracy of the fitted k_u^0, measurements at a wide range of pulling speeds are required. Present commercially available AFMs have a pulling speed range of approx 10 to 10,000 nm/s. However, many experimental results show that the most reliable data collection range falls within 300–5000 nm/s. This is because at very high pulling speeds (>5000 nm/s) viscous drag retards the defection of the cantilever (49) and piezo positioner response time increases error in measurements; at very low pulling speeds (<100 nm/s) thermal drift in the AFM mechanical loop affects the reliability of the measurements.

It must be noted that it is almost always possible to have more than one pairwise combination of x_u and k_u^0 that fits the experimental data well. Although the values of x_u may be insignificantly different, the corresponding values of k_u^0 can be appreciably different (45). This is because the two kinetic parameters x_u and k_u^0 are strongly coupled to each other in the fitting regardless of using the Monte Carlo or numerical fitting methods. This uncertainty in fitting is analogous to the uncertainly in determining the unfolding rate constant in conventional protein folding studies with long extrapolation to 0 *M* denaturant concentrations.

Constant force measurements: to minimize the uncertainty in the fitted kinetic parameters and to avoid the laborious task of the numerical extrapolation in the dynamic force spectrum, the kinetic parameters should ideally be measured using a force clamp (closed loop mode). It has been shown that k_u^0 and k_f^0 can be derived directly from the constant force measurements (19).

However, the refolding of a protein can only be observed at a force lower than the force to unfold it. In such circumstances, it is important to remember that the refolding and unfolding may not follow the same pathway (16,46). As a matter of fact, pulling experiments occur far from equilibrium. In addition, the refolding measurements at very low force (e.g., in the absence of force) are likely to be masked by the thermal fluctuation of the cantilever (~20 pN).

Other limitations such as the deadtime between the feedback control and the piezo positioner, and the thermal drift in the long mechanical loops may add also significant error in the measurements.

7. The presence of intermediates. Methods to detect intermediate states under force: many proteins unfold via intermediate states. These intermediate states have been observed and validated in the AFM experiments by means of the following methods:

   a. The intermediate state may manifest itself as a "hump" (18), or appear as a secondary peak from each unfolding event (15,50). In the latter case histograms of contour length increments and unfolding forces should show two distinct distributions.
   b. Site-directed mutagenesis of certain residues may be used to detect and investigate the intermediate states under force (15,17,46,50).

c. The presence of high-energy intermediate states may not be observed directly from the AFM experiments. Nevertheless, the existence of these intermediates may be implied by the change in x_u in the force spectrum upon mutation *(16,51)*.

If unfolding occurs via an intermediate then the observed peaks in force traces correspond to forces to unfold the protein from the intermediate state.

Intermediate states complicate the mechanical Φ-value analysis: the estimation of the mechanical Φ-values can be complicated by the presence of force-stabilized intermediate states *(16,17)*. In the analysis where N is the ground state, the effect of the mutation on the TS, derived from the force measurements, is normalized by $\Delta\Delta G_{D-N}$, the effect of the mutation on the native state (N). In the AFM experiments, $\Delta\Delta G_{D-N}$ is determined from conventional equilibrium denaturation experiments and is assumed to be the same as the $\Delta\Delta G_{D-N}$ of the protein under force.

In the presence of a force-stabilized intermediate, the ground state is shifted to I. $\Delta\Delta G_{TS-I}$ should, therefore, be normalized against $\Delta\Delta G_{D-I}$, the effect of the mutation on the ground state where the ground state is I. $\Delta\Delta G_{TS-I}$ is still measured from the AFM experiments but $\Delta\Delta G_{D-I}$ is essentially not assessable by the conventional method if the intermediate states is only populated under force. In such case other indirect approaches such as the molecular dynamics simulations, or the application of protein engineering techniques to construct an intermediate model protein, or the combination of both methods will have to be sought to estimate $\Delta\Delta G_{D-I}$ *(15–17,46)*.

Acknowledgments

This work was supported by the Wellcome Trust. S.P.N. holds a scholarship from the Agency of Science and Technology and Research (A*STAR) of the Singapore Government, L. G. R. holds a Biotechnology and Biological Sciences Research Council (BBSRC) studentship, and J. C. is a Wellcome Trust Senior Research Fellow.

References

1. Binnig, G., Quate, C. F., and Gerber, C. (1986) Atomic force microscope. *Phys. Rev. Lett.* **56,** 930–933.
2. Lee, G. U., Chrisey, L. A., and Colton, R. J. (1994) Direct measurement of the forces between complementary strands of DNA. *Science* **266,** 771–773.
3. Marszalek, P. E., Li, H., and Fernandez, J. M. (2001) Fingerprinting polysaccharides with single-molecule atomic force microscopy. *Nature Biotechnol.* **19,** 258–262.
4. Rief, M., Gautel, M., Oesterhelt, F., Fernandez, J. M., and Gaub, H. E. (1997) Reversible unfolding of individual titin immunoglobulin domains by AFM. *Science* **276,** 1109–1112.
5. Halliday, N. L. and Tomasek, J. J. (1995) Mechanical properties of the extracellular matrix influence fibronectin fibril assembly in vitro. *Expt. Cell Res.* **217,** 109–117.

6. Hynes, R. O. (1999) The dynamic dialogue between cells and matrices: implications of fibronectin's elasticity. *Proc. Natl Acad. Sci. USA* **96,** 2588–2590.

7. Sharma, A., Askari, J. A., Humphries, M. J., Jones, E. Y., and Stuart, D. I. (1999) Crystal structure of a heparin- and integrin-binding segment of human fibronectin. *EMBO J.* **18,** 1468–1479.

8. Sechler, J. L., Rao, H., Cumiskey, A. M., et al. (2001) A novel fibronectin binding site required for fibronectin fibril growth during matrix assembly. *J. Cell Biol.* **154,** 1081–1088.

9. Maruyama, K. (1997) Connectin/titin, giant elastic protein of muscle. *FASEB J.* **11,** 341–345.

10. Labeit, S. and Kolmerer, B. (1995) Titins: giant proteins in charge of muscle ultrastructure and elasticity. *Science* **270,** 293–296.

11. Wang, K., McCarter, R., Wright, J., Beverly, J., and Ramirez-Mitchell, R. (1991) Regulation of skeletal muscle stiffness and elasticity by titin isoforms: a test of the segmental extension model of resting tension. *Proc. Natl Acad. Sci. USA* **88,** 7101–7105.

12. Wang, K., McClure, J., and Tu, A. (1979) Titin: major myofibrillar components of striated muscle. *Proc. Natl Acad. Sci. USA* **76,** 3698–3702.

13. Carrion-Vazquez, M., Li, H., Lu, H., Marszalek, P. E., Oberhauser, A. F., and Fernandez, J. M. (2003). The mechanical stability of ubiquitin is linkage dependent. *Nature Struct. Biol.* **10,** 738–743.

14. Brockwell, D. J., Paci, E., Zinober, R. C., et al. (2003) Pulling geometry defines the mechanical resistance of a beta-sheet protein. *Nature Struct. Biol.* **10,** 731–737.

15. Li, L., Huang, H. H., Badilla, C. L., and Fernandez, J. M. (2005) Mechanical unfolding intermediates observed by single-molecule force spectroscopy in a fibronectin type III module. *J. Mol. Biol.* **345,** 817–826.

16. Best, R. B., Fowler, S. B., Herrera, J. L., Steward, A., Paci, E., and Clarke, J. (2003) Mechanical unfolding of a titin Ig domain: structure of transition state revealed by combining atomic force microscopy, protein engineering and molecular dynamics simulations. *J. Mol. Biol.* **330,** 867–877.

17. Fowler, S. B., Best, R. B., Toca Herrera, J. L., et al. (2002) Mechanical unfolding of a titin Ig domain: structure of unfolding intermediate revealed by combining AFM, molecular dynamics simulations, NMR and protein engineering. *J. Mol. Biol.* **322,** 841–849.

18. Marszalek, P. E., Lu, H., Li, H., et al. (1999) Mechanical unfolding intermediates in titin modules. *Nature* **402,** 100–103.

19. Fernandez, J. M. and Li, H. (2004) Force-clamp spectroscopy monitors the folding trajectory of a single protein. *Science* **303,** 1674–1678.

20. Oberhauser, A. F., Marszalek, P. E., Carrion-Vazquez, M., and Fernandez, J. M. (1999) Single protein misfolding events captured by atomic force microscopy. *Nature Struct. Biol.* **6,** 1025–1028.

21. Steward, A., Toca-Herrera, J. L., and Clarke, J. (2002) Versatile cloning system for construction of multimeric proteins for use in atomic force microscopy. *Protein Sci.* **11,** 2179–2183.

22. Fersht, A. R., Matouschek, A., and Serrano, L. (1992) The folding of an enzyme. I. Theory of protein engineering analysis of stability and pathway of protein folding. *J. Mol. Biol.* **224,** 771–782.
23. Pain, R. H. (2000) *Mechanisms of Protein Folding*, Oxford University Press.
24. Bell, G. I. (1978) Models for the specific adhesion of cells to cells. *Science* **200,** 618–627.
25. Bustamante, C., Marko, J. F., Siggia, E. D., and Smith, S. (1994) Entropic elasticity of lambda-phage DNA. *Science* **265,** 1599, 1600.
26. Evans, E. and Ritchie, K. (1997) Dynamic strength of molecular adhesion bonds. *Biophys. J.* **72,** 1541–1555.
27. Gamsjaeger, R., Wimmer, B., Kahr, H., et al. (2004) Oriented binding of the His(6)-tagged carboxyl-tail of the L-type Ca2+ channel alpha(1)-subunit to a new NTA-functionalized self-assembled monolayer. *Langmuir* **20,** 5885–5893.
28. Hinterdorfer, P., Baumgartner, W., Gruber, H. J., Schilcher, K., and Schindler, H. (1996) Detection and localization of individual antibody-antigen recognition events by atomic force microscopy. *Proc. Natl Acad. Sci. USA* **93,** 3477–3481.
29. Cleveland, J. P., Manne, S., Bocek, D., and Hansma, P. K. (1993) A nondestructive method for determining the spring constant of cantilevers for scanning force microscopy. *Rev. Sci. Instrum.* **64,** 403–405.
30. Hutter, J. L. and Bechhoefer, J. (1993) Calibration of atomic force microscope tips. *Rev. Sci. Instrum.* **64,** 1868–1873.
31. Proksch, R. (2004) Nondestructive added mass spring calibration with the MFP-3D. *Technical Note of Asylum Research.* www.AsylumResearch.com. Last accessed March 17, 2006.
32. Law, R., Liao, G., Harper, S., Yang, G., Speicher, D. W., and Discher, D. E. (2003) Pathway shifts and thermal softening in temperature-coupled forced unfolding of spectrin domains. *Biophys. J.* **85,** 3286–3293.
33. Oberhauser, A. F., Marszalek, P. E., Erickson, H. P., and Fernandez, J. M. (1998) The molecular elasticity of the extracellular matrix protein tenascin. *Nature* **393,** 181–185.
34. Carrion-Vazquez, M., Oberhauser, A. F., Fowler, S. B., et al. (1999) Mechanical and chemical unfolding of a single protein: a comparison. *Proc. Natl Acad. Sci. USA* **96,** 3694–3699.
35. Politou, A. S., Gautel, M., Joseph, C., and Pastore, A. (1994) Immunoglobulin-type domains of titin are stabilized by amino-terminal extension. *FEBS Lett.* **352,** 27–31.
36. Pfuhl, M., Improta, S., Politou, A. S., and Pastore, A. (1997) When a module is also a domain: the role of the N terminus in the stability and the dynamics of immunoglobulin domains from titin. *J. Mol. Biol.* **265,** 242–256.
37. Hamill, S. J., Meekhof, A. E., and Clarke, J. (1998) The effect of boundary selection on the stability and folding of the third fibronectin type III domain from human tenascin. *Biochemistry* **37,** 8071–8079.
38. Miroux, B. and Walker, J. E. (1996) Over-production of proteins in Escherichia coli: mutant hosts that allow synthesis of some membrane proteins and globular proteins at high levels. *J. Mol. Biol.* **260,** 289–298.

39. Best, R. B., Li, B., Steward, A., Daggett, V., and Clarke, J. (2001) Can non-mechanical proteins withstand force? Stretching barnase by atomic force microscopy and molecular dynamics simulation. *Biophys. J.* **81,** 2344–2356.

40. Forman, J. R., Qamar, S., Paci, E., Sandford, R. N., and Clarke, J. (2005) The remarkable mechanical strength of polycystin-1 supports a direct role in mechanotransduction. *J. Mol. Biol.* **349,** 861–871.

41. Rounsevell, R. W. S., Steward, A., and Clarke, J. (2005) Biophysical investigations of engineered polyproteins:implications for force data. *Biophys. J.* **88,** 2022–2029.

42. Evans, E. and Williams, P. M. (2002) *Dynamic Force Spectroscopy I: Single Bonds in Les Houches-Ecole d'Ete de Physique Theorique.* Springer and Verlag, Germany.

43. Rief, M., Fernandez, J. M., and Gaub, H. E. (1998) Elastically coupled two-level systems as a model for biopolymer extensibility. *Phys. Rev. Lett.* **81,** 4764–4767.

44. Clarke, J. and Williams, P. M. (2005) Unfolding induced by mechanical force. In: *Protein Folding Handbook Part 1* (Kiefhaber, T. and Buchner, J., ed.), Wiley-VCH Verlag GmbH and Co., Weinheim, Germany, pp. 1111–1142.

45. Best, R. B., Brockwell, D. J., Toca-Herrera, J. L., et al. (2003) Force mode atomic force microscopy as a tool for protein folding studies. *Anal. Chim. Acta,* **479,** 87–105.

46. Ng, S. P., Rounsevell, R. W., Steward, A., et al. (2005) Mechanical unfolding of TNfn3: the unfolding pathway of a fnIII domain probed by protein engineering, AFM and MD simulation. *J. Mol. Biol.* **350,** 776–789.

47. Best, R. B., Fowler, S. B., Toca-Herrera, J. L., and Clarke, J. (2002) A simple method for probing the mechanical unfolding pathway of proteins in detail. *Proc. Natl Acad. Sci. USA* **99,** 12,143–12,148.

48. Zinober, R. C., Brockwell, D. J., Beddard, G. S., et al. (2002) Mechanically unfolding proteins: the effect of unfolding history and the supramolecular scaffold. *Protein Sci.* **11,** 2759–2765.

49. Evans, E. (2001) Probing the relation between force—lifetime—and chemistry in single molecular bonds. *Annu. Rev. Biophys. Biomol. Struct.* **30,** 105–128.

50. Schwaiger, I., Kardinal, A., Schleicher, M., Noegel, A. A., and Rief, M. (2004) A mechanical unfolding intermediate in an actin-crosslinking protein. *Nature Struct. Mol. Biol.* **11,** 81–85.

51. Williams, P. M., Fowler, S. B., Best, R. B., et al. (2003) Hidden complexity in the mechanical properties of titin. *Nature* **422,** 446–449.

52. Rounsevell, R., Forman, J. R., and Clarke, J. (2004) Atomic force microscopy: mechanical unfolding of proteins. *Methods* **34,** 100–111.

53. Li, H., Oberhauser, A. F., Fowler, S. B., Clarke, J., and Fernandez, J. M. (2000) Atomic force microscopy reveals the mechanical design of a modular protein. *Proc. Natl Acad. Sci. USA* **97,** 6527–6531.

10

Using Triplet–Triplet Energy Transfer to Measure Conformational Dynamics in Polypeptide Chains

Beat Fierz, Karin Joder, Florian Krieger, and Thomas Kiefhaber

Summary

Intrachain diffusion processes play an important role in protein folding and function. In this chapter we discuss the application of triplet–triplet energy transfer to directly measure rate constants for intrachain contact formation in polypeptide chains. The photochemistry of triplet–triplet energy transfer is described, experimental prerequisites of the method are discussed, and a detailed description of the experimental protocols and data analysis is given.

Key Words: Triplet–triplet energy transfer; chain dynamics; protein folding; electron transfer; laserflash photolysis.

1. Introduction

Dynamics of polypeptide chains play an essential role in folding and function of proteins and peptides. They represent the underlying motions in the transitions between different protein conformations and they determine the rate at which a folding polypeptide chain can explore conformational space. Conformational search within a polypeptide chain is limited by intrachain diffusion processes, i.e., by the rate at which two points along the chain can form an interaction. The knowledge of the rates of intrachain contact formation in polypeptide chains and their dependence on amino acid sequence, chain length, and local structure formation is therefore essential for an understanding of the dynamics of protein folding and for the characterization of the free energy barriers of the folding reactions. In addition, intrachain diffusion provides an upper limit for the speed at which a protein can reach its native state just like free diffusion provides an upper limit for the rate constant of bimolecular reactions.

In the last years, several experimental systems have been developed to measure contact formation between specific sites on a polypeptide chain. The dynamics of

From: *Methods in Molecular Biology, vol. 350: Protein Folding Protocols*
Edited by: Y. Bai and R. Nussinov © Humana Press Inc., Totowa, NJ

unfolded polypeptide chains have first been studied using fluorescence resonance energy transfer (FRET) from an energy donor to an energy acceptor group *(1)*. FRET was shown to be a powerful tool to determine donor–acceptor distances *(2)*, but time-resolved FRET kinetics also contain major contributions from diffusion of the two FRET probes relative to each other *(1,3)*. Determination of the diffusion constants allowed the first estimates of the time-scale of chain diffusion processes in polypeptide chains, but it remained unclear whether the same diffusion constants represent the dynamics for formation of van der Waals contact between two groups. In addition, analysis of the FRET data critically depends on the shape of the donor–acceptor distribution function, which had to be introduced in the data analysis and, thus, did not allow model-free analysis of the data.

Rates of intrachain diffusion have later been estimated from the rate constant for intramolecular bond formation in guanidinium chloride (GdmCl)-unfolded cytochrome c *(4)*. The reaction of a methionine with a heme group separated by 50–60 amino acids along the chain gave time constants around 35–40 µs. Extrapolating these data to shorter distances gave an estimate of 1 µs for the time constant of contact formation at a distance of 10 peptide bonds. However, it turned out that the recombination reaction of heme with methionine is too slow to directly measure chain dynamics and the resulting rate constants did not directly reflect chain dynamics.

These early studies on intrachain diffusion did not yield absolute rate constants for contact formation, but they were important to trigger the development of more direct and model-free methods to study chain diffusion. Recently, several experimental systems using triplet states to directly monitor contact formation in model peptides and protein fragments have been introduced. In the following, we will present systems using triplet–triplet energy transfer (TTET) and triplet quenching to directly measure chain dynamics. We will briefly introduce the photochemistry of the processes involved and will discuss the experimental requirements for direct model-free measurements of chain dynamics in detail.

2. Using TTET to Measure Peptide Dynamics

Figure 1 illustrates the concept of TTET experiments applied to measurements of protein dynamics. A triplet donor and a triplet acceptor group are introduced at specific points in polypeptide chains. The donor is selectively excited by a laserflash and typically undergoes fast intersystem crossing into the triplet state (*see* **Fig. 2**). Upon intrachain diffusion the triplet donor and triplet acceptor groups meet and the triplet state is transferred to the acceptor. Because TTET is a two-electron exchange process (Dexter mechanism), it has a very strong distance dependence and usually requires van der Waals contact to allow for electron transfer *(5)*. This should be contrasted to FRET, which occurs through dipole–dipole interactions and, thus, allows energy transfer over larger distances.

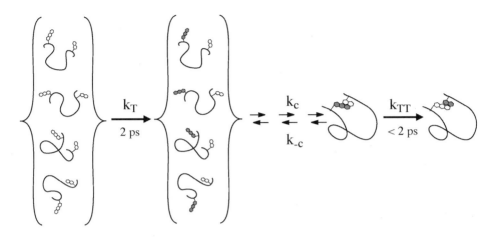

Fig. 1. Schematic representation of the triplet–triplet energy transfer experiments. The experiments allow determination of the absolute rate constant for contact formation (k_c) in the ensemble of unfolded conformation, if formation of the triplet state (k_T) and the transfer process (k_{TT}) are much faster than chain dynamics (k_{TT}, $k_t \gg k_c$, k_{-c}; *see* **Subheading 2.1.**). The rate constants given for k_T and k_{TT} were taken from **ref. 6**.

A prerequisite for efficient and rapid transfer is that the acceptor group has a significantly lower triplet energy than the donor group (**Fig. 2**). A well-suited pair for TTET measurements are xanthone derivates as a triplet donor and naphthalene as a triplet acceptor group (**Fig. 3**) The triplet states of these molecules have specific absorbance bands, which can easily be monitored to measure the decay of the donor triplet states and the concomitant increase of acceptor triplet states (**Fig. 4**). The time constant for formation of xanthone triplet states is around 2 ps *(6)*. TTET between xanthone and naphthalene is faster than 2 ps and has a bimolecular transfer rate constant of 4×10^9 $M^{-1}s^{-1}$ *(6,7)*, which is the value expected for a diffusion-controlled reaction between small molecules in water (*see* **Subheading 2.2.**). Owing to this fast photochemistry, TTET between xanthone and naphthalene allows measurements of absolute rate constants for diffusion processes slower than 5–10 ps (*see* **Subheading 2.1.**). The upper limit of the experimental time window accessible by TTET is set by the intrinsic lifetime of the donor, which is around 20–30 μs for the xanthone triplet state in water. Xanthone has a high quantum yield for intersystem crossing ($\phi_{ISC} = 0.99$, $\varepsilon \approx$ 4000 $M^{-1}cm^{-1}$) and the triplet state has a strong absorbance band with a maximum around 590 nm in water ($\varepsilon_{590}^T \approx 10{,}000$ $M^{-1}cm^{-1}$). This allows single pulse measurements at rather low peptide concentrations (10–50 μM). The low concentrations applied in the experiments also rule out contributions from intermolecular transfer reactions, which would have half times higher than 50 μs in this concentration range *(7,8)*. Also contributions from through-bond transfer

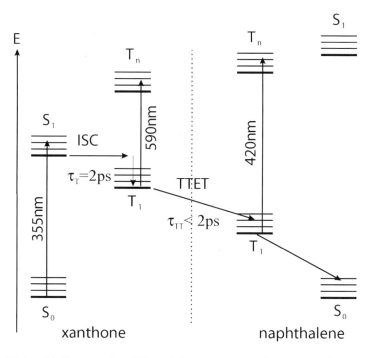

Fig. 2. Jablonski diagram for triplet–triplet energy transfer (TTET) from xanthone to naphthalene. Rate constants for triplet formation (k_T) and TTET (k_{TT}) were measured by laserflash photolysis using femtosecond pulses *(6)*.

processes can be neglected, because this can not occur over distances beyond eight bonds *(9,10)* and even the shortest peptides used in TTET studies had donor and acceptor separated by 11 bonds.

2.1. Kinetics of TTET Coupled to Polypeptide Chain Dynamics

The coupling of electron transfer to chain dynamics displayed in **Fig. 1** can be kinetically described by a three-state reaction if the excitation of the donor is fast compared with chain dynamics (k_c, k_{-c}) and electron transfer (k_{TT}).

$$O \underset{k_{-c}}{\overset{k_c}{\rightleftharpoons}} C \overset{k_{TT}}{\longrightarrow} C^* \tag{1}$$

O and C represent open chain conformations (no contact) and contact conformations of the chain, respectively. k_c and k_{-c} are apparent rate constant for contact formation and breakage, respectively. If the excited state quenching or transfer (k_{TT}) reaction is on the same time-scale or slower than breaking of the contact (k_{-c}), the measured rate constants (λ_i) are functions of all microscopic rate constants.

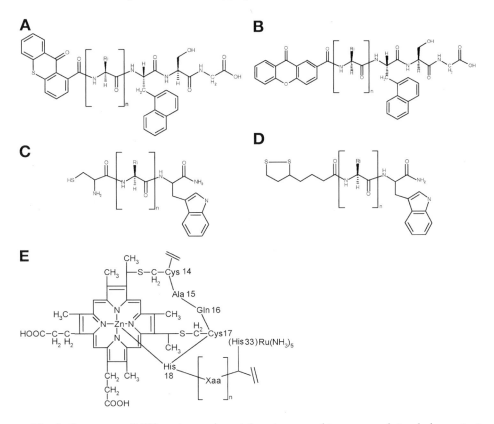

Fig. 3. Structures of different experimental systems used to measure intrachain contact formation. Results from these systems are summarized in **Table 1**. (**A**) Thioxanthone- and naphthylalanine-labeled poly(Gly-Ser) peptides to measure contact formation by triplet–triplet energy transfer (TTET) *(8)*. (**B**) Xanthone- and naphthylalanine-labeled peptides to measure contact formation by TTET in various homopolymer chains and natural sequences with up to 57 amino acids between donor and acceptor *(7,20,21,23)*. (**C,D**) Tryptophan and cysteine or lipoate-labeled peptides to measure contact formation by Trp triplet quenching in various short peptide chains *(19,28,29)*. (**E**) Quenching of Zn-porphyrin triplets by Ru-His complexes in unfolded cytochrome c *(27)*.

$$\lambda_{1,2} = \frac{B \pm \sqrt{B^2 - 4C}}{2}$$
$$B = k_c + k_{-c} + k_{TT}$$
$$C = k_c \cdot k_{TT}$$

(2)

If the population of contact conformations is low ($k_{-c} \gg k_c$) the kinetics will be single exponential and the observed rate constant corresponds to the smaller

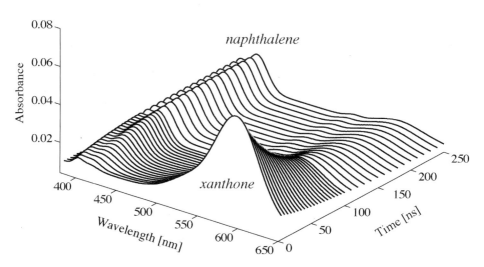

Fig. 4. Time-dependent change in the absorbance spectrum of a Xan-(Gly-Ser)$_{14}$-NAla-Ser-Gly peptide after a 4-ns laser flash at 355 nm. The decay in the intensity of the xanthone triplet absorbance band around 590 nm is accompanied by a corresponding increase in the naphthalene triplet absorbance band around 420 nm (adapted from **ref. 7**).

eigenvalue (λ_1) in **Eq. 2**. Because this is in agreement with experimental results, we will only consider this scenario in the following. We can further simplify **Eq. 2** if the time scales of electron transfer and chain dynamics are well separated. In the regime of fast electron transfer compared with chain dynamics ($k_{TT} \gg k_c, k_{-c}$) **Eq. 2** can be approximated by:

$$\lambda_1 = k_c \tag{3}$$

This is the desired case, because the observed kinetics directly reflect the dynamics of contact formation. In the regime of fast formation and breakage of the contact compared to electron transfer ($k_{TT} \ll k_c, k_{-c}$) **Eq. 2** can be approximated by:

$$\lambda_1 = \frac{k_c}{k_c + k_{-c}} \cdot k_{TT} \cong \frac{k_c}{k_{-c}} \cdot k_{TT} = K_c \cdot k_{TT} \tag{4}$$

where K_c reflects the ratio of contact conformations (C) over open chain conformations (O). In this limit the chain dynamics cannot be measured but the fraction of closed conformations can be determined. It should be kept in mind that these simplifications only hold if the formation of excited states is fast compared with the following reactions. If the excitation is on the nanosecond time-scale or slower, the solutions of a linear four-state model have to be used to analyze the kinetics (*11,12*).

2.2. Test for Suitability of a Donor–Acceptor Pair to Directly Measure Chain Dynamics

2.2.1. Characterization of the Photochemistry

The considerations discussed in **Subheading 2.1.** show that it is crucial to characterize the rate constants for the photochemical processes in the applied system to be able to interpret the kinetic data. The photochemistry of the donor excitation and of the electron transfer process are usually characterized by investigating the isolated labels. Because the photochemical processes involved in TTET are often fast, the kinetics should be studied on the femtosecond to the nanosecond time-scale to gain complete information on all photochemical processes in the system *(6)* (**Fig. 2**).

2.2.2. Test for Diffusion-Controlled Reactions

To test whether each donor–acceptor contact leads to transfer, it should be examined whether the reaction is diffusion controlled. This can be done by studying the bimolecular transfer process from the donor to acceptor groups and determining its rate constant, its temperature dependence, and its viscosity dependence.

2.2.2.1. Determination of the Bimolecular Transfer Rate Constant

The bimolecular rate constant for energy transfer (k_q) can be measured using the free labels in solution under pseudo first-order conditions. In these experiments the concentration of the quencher or acceptor [Q] should be at least 10 times higher than the concentration of the donor to be in the pseudo first-order regime. Because [Q] is approximately constant during the experiment, the apparent first-order rate constant (k) under pseudo first-order conditions is given by *(13)*

$$k = k_q \cdot [Q] \tag{5}$$

Thus, the bimolecular transfer constant (k_q) can be obtained by varying the quencher concentration and analyzing the data according to the Stern–Volmer equation *(14)* (*see* **Fig. 5**).

$$k = k_0 + k_q[Q] \tag{6}$$

Here, k_0 denotes the rate constant for triplet decay of the donor in absence of quencher (acceptor) and k the rate constant in presence of the acceptor or quencher. For diffusion-controlled reactions of small molecules in water, k_q is around 4–6×10^9 s^{-1}.

2.2.2.2. Viscosity Dependence of the Transfer Reaction

For a diffusion-controlled reaction, the viscosity dependence of k_q is inversely proportional to solvent viscosity ($k \sim 1/\eta$). To test for the effect of solvent viscosity

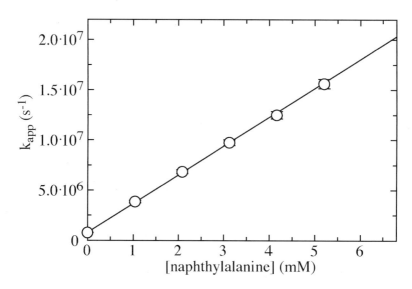

Fig. 5. Pseudo first-order measurements of bimolecular triplet–triplet energy transfer from xanthonic acid to naphthylalanine in water. Xanthone concentration was 30 μM. The slope gives a bimolecular transfer constant (k_q) of 3×10^9 $M^{-1}s^{-1}$. The data were discussed in **ref. 7** (adapted from **ref. 22**).

on k_1 the viscosity is usually varied by adding cosolutes like ethylene glycol or glycerol. To determine the final macroscopic viscosity, instruments like the falling-ball viscosimeter or the Ubbelohde viscosimeter are used. The molecular size of the cosolute used in a viscosity study is an important factor. The macroscopic viscosity of a glycerol/water mixture and a solution of polyethlyleneglycol might be the same, although the microscopic properties of these solutions are very different *(15)*. In the latter case, the dissolved reactant molecules only feel a fraction of the macroscopic viscosity. Thus, it is advisable to use viscous cosolutes of small molecular weight. Glycerol/water mixtures have proven to be most useful because the viscosity can be varied from 1 cP (pure water) over three orders of magnitude by addition of glycerol. It is imperative to control the temperature when working with high concentrations of glycerol because the viscosity is strongly temperature dependent. Note that viscous cosolutes can interfere with the reaction that is studied. In protein folding studies, the polyols used to vary the viscosity tend to stabilize the native state of proteins *(16)* making an analysis of the kinetic data difficult. However, for measurements of chain dynamics in unfolded polypeptide chains, this effect should not interfere with the measurements as long as the cosolutes do not induce a structure. Specific effects of the polyols on chain conformations/dynamics can be tested by comparing the results from different cosolutes.

2.2.2.3. Determination of the Activation Energy

Diffusion-controlled reactions typically have activation energies close to zero. The lack of an activation barrier in a diffusion-controlled quenching reaction can be verified by determining the temperature dependence of k_q. The diffusion coefficients of the reactants are, however, temperature dependent, mainly through the effect of temperature on solvent viscosity according to the Stokes–Einstein equation.

$$D = \frac{k_B T}{6 \pi r \eta} \tag{7}$$

After correcting for the change in viscosity of the solvent with temperature, the slope in an Arrhenius plot should not be higher than $k_B T$. As any photophysical reaction has an intrinsic rate constant, the reaction is only diffusion controlled to a certain maximal concentration of quencher molecules. After that the photochemistry of the quenching or transfer process becomes rate limiting. This provides an upper limit of the rate constants that can be measured, even if the system is diffusion controlled at lower quencher concentrations. For the xanthone/naphthalene system TTET has been shown to occur faster than 2 ps and thus this system is suitable to obtain absolute time constants for all processes slower than 5–10 ps *(6)*.

2.2.3. Testing Interference From Other Amino Acids

To test for possible interference from amino acid side chains with the TTET reaction, the effect of various amino acids on the donor triplet lifetimes must be measured. **Table 1** shows bimolecular quenching and TTET rate constants for the interaction of various amino acids with xanthone measured under pseudo first-order conditions. The thioether group of methionine ($k_q = [2.0 \pm 0.1] \times 10^9 \ M^{-1} \ s^{-1}$) and the deprononated imidazole ring of histidine ($k_q = [1.8 \pm 0.1] \times 10^9 \ M^{-1} \ s^{-1}$) quench xanthone triplets very efficiently with a rate constant close to the diffusion limit. The other amino acid side chains quench xanthone triplets either very inefficiently (Cys, His$^+$, N-terminus) or not at all (Ala, Arg, Asn, Asp, Gly, Lys, Ser, Phe). The aromatic amino acids tryptophan, tyrosine, and phenylalanine are possible acceptors for xanthone triplets in TTET reactions. **Table 1** shows that TTET between xanthone and tryptophan ($k_{TTET} = (3.0 \pm 0.1) \times 10^9 \ M^{-1} \ s^{-1}$) and tyrosine ($k_{TTET} = [2.5 \pm 0.1] \times 10^9 \ M^{-1} \ s^{-1}$) occur in diffusion-controlled reactions with virtually the same bimolecular rate constants as observed for TTET from xanthone to naphthylalanine ($k_{TTET} = [2.8 \pm 0.1] \times 10^9 \ M^{-1} \ s^{-1}$). However, TTET from xanthone to tryptophan and tyrosine are complex reactions with at least two observable rate constants. For both amino acids, TTET is accompanied by radical formation *(17,18)*, which explains the complex kinetics and makes them not suitable for

Table 1
Interaction of Different Amino Acids With the
Triplet State of Xanthone[a]

Amino acid	$k_q(M^{-1}s^{-1})^b$	Conditions
Naphthyl acetic acid[c]	$(4.0 \pm 0.1) \times 10^9$	water, pH 7.0
Trp[c]	$(3.0 \pm 0.1) \times 10^{9e}$	water, pH 7.0
NAla[c]	$(2.8 \pm 0.1) \times 10^9$	water, pH 7.0
Tyr[c]	$(2.5 \pm 0.1) \times 10^{9e}$	water, pH 7.0
Met[d]	$(2.0 \pm 0.1) \times 10^9$	water, pH 7.0
His[d]	$(1.8 \pm 0.1) \times 10^{9e}$	0.1 M KKP, pH 8.0
His[+d]	$(2.8 \pm 0.2) \times 10^7$	0.1 M NaOAc, pH 4.0
Cys[d]	$(5.1 \pm 0.2) \times 10^7$	water, pH 7.0
N-Terminal NH^{3+d}	$(2.0 \pm 0.5) \times 10^6$	water, pH 7.0

[a]No effect on the xanthone triplet lifetime was observed for Ala, Arg+, Asn, Gly, Lys+, Phe, Ser, Asp, Asp−.
[b]Bimolecular quenching constants were measured under pseudo first-order conditions as described in **Subheading 2.2.2.1.**
[c]Triplet–triplet energy transfer.
[d]Triplet quenching.
[e]Radical formation as side reaction.

the use as TTET acceptors in polypeptide chains. These results show that mainly methionine, tryptophan, and tyrosine interfere with TTET from xanthone to naphthalene in intrachain diffusion experiments. Histidine containing sequences can be measured with this donor/acceptor pair if the pH of the solution is below 5.5.

2.3. Experimental Procedure to Measure TTET

2.3.1. Instrumentation

The instrumentation required to perform electron transfer experiments on contact formation kinetics consists of a high-energy light source to produce excited states and of a detection mechanism (**Fig. 6**). A pulsed laser is used to produce triplet donor states in TTET or triplet quenching experiments. The duration of the light pulse has to be very short, i.e., shorter than the time-scale of the reaction of interest. Additionally, the excitation pulse has to provide enough energy to excite the major portion of the molecules in the sample in order to generate a large signal. A pulsed Nd:YAG laser with a pulse width of <5 ns and a pulse energy of approx 100 mJ is well suited for these purposes. Transient absorption is used to detect the triplet states in TTET or triplet quenching experiments. A pulsed flash lamp generates enough light for the absorption measurements. The lamp intensity typically

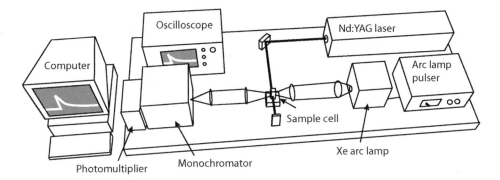

Fig. 6. Schematic representation of a laserflash setup used to measure electron transfer reactions. (Adapted from **ref. 22**.)

stays at a constant plateau value for several hundred microseconds, which is sufficient for TTET measurements. Using a monochromator allows the monitoring of single wavelengths. Transient spectra can be reconstructed from measurements at different wavelengths or by using a charge-coupled device (CCD) camera.

2.3.2. Data Analysis

For data analysis transient spectra may be analyzed and fitted globally to different mechanisms (*see* **Fig. 4**). This allows testing for side reactions, such as radical formation, which occurs in some triplet systems and interferes with the actual transfer process *(19)*. If a system has been photochemically characterized, the kinetics are typically measured at the wavelengths of maximum signal change of donor and acceptor groups. In the case of the xanthone/naphthalene pair, kinetics are measured at 590 and 420 nm (**Fig. 3**). Typical peptide concentrations required for this donor/acceptor pair are 30–50 μM in a 1-cm cuvet. To test for possible contributions from bimolecular processes, TTET should be measured in a mixture of peptides containing either the donor or the acceptor label *(7,8)*.

For all unstructured peptides investigated up-to-date single exponential kinetics of intrachain contact formation on the nanosecond time-scale were observed (**Fig. 7**) *(7,20–22)*. The only exceptions were proline-containing peptides. In this case, the dynamics of molecules containing a *cis* or a *trans* Xaa–Pro peptide bond gave rise to double exponential kinetics *(21)*.

In addition to the determination of the rate constants, the amplitude of the observed reaction should be analyzed to test for possible fast processes occurring in the dead time of the laserflash experiments. These may be caused by conformations, which have donor and acceptor in close proximity at the time point of the laserflash.

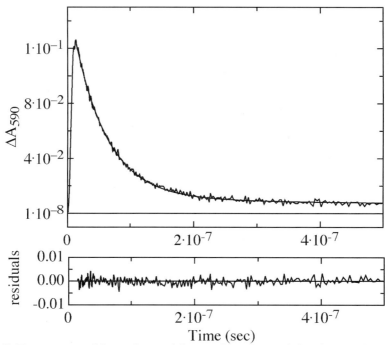

Fig. 7. Time course of formation and decay of xanthone triplets in parvalbumin loop fragment 85–102 after a 4-ns laser flash at t=0. The change of xanthone triplet absorbance is measured at 590 nm. The dynamics of contact formation can be described by a single exponential with a time constant of 54 ± 3 ns. (Data are taken from **ref. *20*.**)

3. Comparison of Results From Different Experimental Systems

Recently, several experimental systems using TTET or triplet quenching to monitor contact formation in model peptides and protein fragments have been introduced (*see* **Table 2**). The main advantage of energy transfer systems is the ability to monitor the population of both the donor and the acceptor populations during the experiment (*see* **Fig. 4**). Bieri et al. *(8)* and Krieger et al. *(7,20)* used TTET between thioxanthone or xanthone derivatives and naphthalene as discussed in **Subheadings 2.1.–2.4.** (*see* **Fig. 3A,B**) to directly measure rate constants for intrachain contact formation in synthetic polypeptide chains. In the initial studies a derivate of thioxanthone was used as triplet donor and the nonnatural amino acid naphthylalanine (NAla) as triplet acceptor *(8)*. Because of the sensitivity of the triplet energy of the donor on solvent polarity, the measurements had to be carried out in ethanol. TTET detected single exponential kinetics in all peptides with time constants of 20 ns for the shortest loops in poly(glycine-serine)-based polypeptide chains *(8)*. Based on these experiments, the minimum time constant for intrachain contact formation was shown to be around 20 ns for

Table 2
Comparison of End-to-End Contact Formation Rate Constants Observed in Different Systems

k_{app} (s^{-1})	Loop size (n)[a]	Method	Labels[b]	Sequence	Conditions	Reference
$5 \times 10^7 - 1.2 \times 10^7$	3–9	TTET	thioxanthone/NAla	(Gly-Ser)$_x$	EtOH	(8)
9.1×10^6	10	triplet quenching	Trp/Cys	(Ala-Gly-Gln)$_3$	H$_2$O, phosphate	(19)
1.1×10^7	10	triplet quenching	Trp/cystine	(Ala-Gly-Gln)$_3$	H$_2$O, phosphate	(19)
1.7×10^7	10	triplet quenching	Trp/lipoate	(Ala-Gly-Gln)$_3$	H$_2$O, phosphate	(19)
6.2×10^4	10	triplet quenching	Trp/lipoate	Pro$_9$	H$_2$O, phosphate	(19)
2.4×10^7	10	triplet quenching	Trp/lipoate	(Ala)$_2$Arg(Ala)$_4$ArgAla	H$_2$O, phosphate	(19)
$1.8 \times 10^8 - 6.5 \times 10^6$	3–57	TTET	xanthone/NAla	(Gly-Ser)$_x$	H$_2$O, 22.5°C	(7)
$8 \times 10^7 - 3.4 \times 10^7$	3–12	TTET	xanthone/NAla	(Ser)$_x$	H$_2$O, 22.5°C	(7)
$2.5 \times 10^8 - 2.0 \times 10^7$	4	TTET	xanthone/NAla	Ser-Xaa-Ser[c]	H$_2$O, 22.5°C	(7)
4.0×10^6	15	triplet quenching	Zn-porphyrin/Ru	unfolded cytochrome c	5.4 M GmdCl, 22°C	(27)
2.0×10^7	17	TTET	xanthone/NAla	carp parvalbumin	H$_2$O, 22.5°C	(20)

[a]n is the number of peptide bonds between the reacting groups.
[b]The structures of the labels are shown in **Fig. 3**.
[c]Xaa = Gly, Ala, Ser, Gly, Arg, His, Ile, Pro. The highest (*cis* Pro) and lowest values (*trans* Pro) for the observed rate constants are given.
TTET, triplet–triplet energy transfer experiments.

formation of short and flexible loops. The thioxanthone used in the initial experiments as triplet donor was later replaced by xanthone *(7)*, which has higher triplet energy than thioxanthone and thus allowed measurements in water (*see* **Fig. 3B**). These results showed two- to threefold faster rate constants for contact formation compared with the same peptides in ethanol *(7)*, setting the time constant for intrachain diffusion in short flexible chains in water of 5–10 ns. The faster kinetics compared to the thioxanthone/NAla system could be attributed to the effect of ethanol *(7)*, which slows down chain dynamics. The xanthone/naphthalene system was further used to test the effect of chain length, amino acid sequence, and denaturants on chain dynamics in model peptides and in peptides derived from natural proteins *(7,20–23)*. These results allowed setting an upper limit for the rate of formation of local structures early in the folding process (**Fig. 8**). A disadvantage of TTET from xanthone to NAla in its application to natural proteins is that Tyr, Trp, and Met interact with the xanthone triplet state either by TTET or by triplet quenching (*see* **Subheading 2.2.3.**) and, thus, should not be present in the studied polypeptide chains *(20)*.

Lapidus et al. *(19)* used a related system to measure chain dynamics in short peptides. In this approach contact formation was measured by quenching of tryptophan triplet states by cysteine (*see* **Fig. 3C**). Tryptophan can be selectively excited by a laserflash ($\phi_{ISC} = 0.18$, $\varepsilon_{266} \approx 3500\ M^{-1}cm^{-1}$, $\varepsilon^T_{460} \approx 5000\ M^{-1}cm^{-1}$) and its triplet decay can be monitored by absorbance spectroscopy *(24)*. The use of naturally occurring amino acids is an advantage of this system. Thus, the groups can be introduced at any position in peptides and proteins. As in the case of TTET from xanthone to naphthalene, some amino acids interfere with the measurements (e.g., Tyr, Met) because they interact with tryptophan triplets and, thus, should not be present in the studied polypeptide chains *(25)*. Major disadvantages of the Trp/Cys system are, (1) that the formation of tryptophan triplets is slow ($\tau = 3$ ns) *(17)*. (2) That triplet quenching is accompanied by the formation of S· radicals *(19)* and (3) that the quenching process is not diffusion controlled *(26)*. These properties reduce the time window available for the kinetic measurements and do not allow direct and model-free analysis of the quenching time traces. In addition, its low quantum yield makes the detection of the triplet states and the data analysis difficult, especially because the kinetics are obscured by radical absorbance bands. In addition to cysteine quenching, two other systems were presented by the same group *(19)* with cystine or the cyclic disulfide lipoate serving as quencher (**Fig. 3D**). The advantage of using lipoate is that the quenching kinetics are much closer to the diffusion limit. Thus, the rates measured with this system are generally faster. Still, all systems used for quenching of tryptophan triplets gave significantly slower kinetics of intrachain contact formation compared to TTET from xanthone to NAla in the

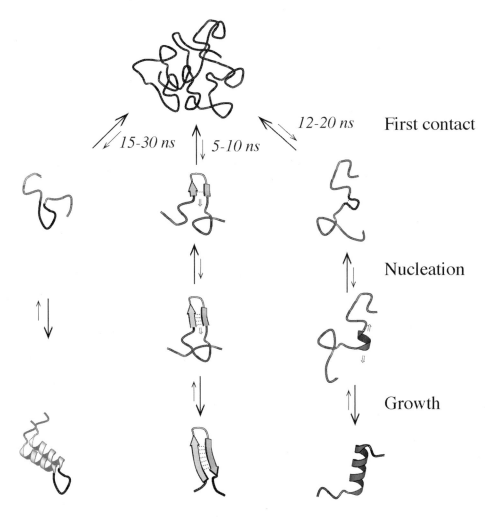

Fig. 8. Schematic representation of the time constants for the first steps in formation of loops, β-hairpins, and α-helices during protein folding derived from the data measured by triplet–triplet energy transfer in water. (Adapted from **ref. 7**.)

same or in similar sequences (**Table 3**). This indicates that tryptophan triplet quenching does not allow measurements of chain dynamics on the absolute time-scale *(26)*.

Another recent experimental approach investigated intrachain contact formation in unfolded cytochrome c using electron transfer from a triplet excited Zn-porphyrine group to a Ru complex, which was bound to a specific histidine residue (His 33; **Fig. 3E**). Because electron transfer is fast and close to the

Table 3
Comparison of End-to-End Contact Formation Rates Observed in Different Systems Measured or Extrapolated to $n \approx 10$

k_{app} (s^{-1})	Loop size (n)	Method	Labels	Sequence	Conditions[a]	Reference
1.2×10^7	9	TTET	thioxanthone/NAla	$(GS)_4$	EtOH	(8)
9.1×10^6	10	triplet quenching	Trp/Cys	$(AGQ)_3$	H_2O, phosphate buffer	(19)
1.1×10^7	10	triplet quenching	Trp/cystine	$(AGQ)_3$	H_2O, phosphate buffer	(19)
1.7×10^7	10	triplet quenching	Trp/lipoate	$(AGQ)_3$	H_2O, phosphate buffer	(19)
8.3×10^7	9	TTET	xanthone/NAla	$(GS)_4$	H_2O, 22.5°C	(7)
4.1×10^7	10	TTET	xanthone/NAla	$(Ser)_9$	H_2O, 22.5°C	(7)
1×10^7	10	triplet quenching	Zn-porphyrin/Ru	unfolded cytochrome c	5.4 M GmdCl, 22°C	(27)

[a]Solvent was H_2O unless indicated. The data measured in concentrated GdmCl solutions or EtOH were not corrected for solvent effects. This would lead to a significant increase in k_c.
TTET, triplet–triplet energy transfer experiments.

diffusion limit, these experiments should also yield absolute rate constants for chain diffusion. Contact formation in the 15 amino acid loop from cytochrome c was observed with a time constant of 250 ns *(27)* in the presence of 5.4 *M* GdmCl. This is significantly faster than the dynamics in unfolded cytochrome c reported by Hagen et al. *(4)* under similar conditions. However, the dynamics measured by Chang et al. agree well with TTET measurements *(7,8,20)* when the kinetics are extrapolated from 5.4 *M* GdmCl to water *(22)* (**Table 2**).

A more detailed comparison of the results obtained from the different experimental systems and a discussion of other methods applied to measure chain dynamics is given in **ref. *22***.

4. Conclusions

Several experimental systems have recently been developed to measure rate constants for intrachain diffusion processes. An essential prerequisite of methods applied to measure absolute rate constants for intrachain diffusion is that photochemistry is faster than the formation and breakage of the intramolecular contact between donor and acceptor or between donor and quencher. Presently the xanthone/naphthalene and the Zn-porphyrine/Ru system fulfill these requirements and allow model-free measurements of intrachain diffusion reactions.

References

1. Haas, E., Katchalski-Katzir, E., and Steinberg, I. Z. (1978) Brownian motion at the ends of oligopeptid chains as estimated by energy transfer between chain ends. *Biopolymers* **17,** 11–31.
2. Stryer, L. and Haugland, R. P. (1967) Energy transfer: a spectroscopic ruler. *Proc. Natl. Acad. Sci. USA* **58,** 719–726.
3. Beechem, J. M. and Haas, E. (1989) Simultaneous determination of intramolecular distance distributions and conformational dynamics by global analysis of energy transfer measurements. *Biophys. J.* **55,** 1225–1236.
4. Hagen, S. J., Hofrichter, J., Szabo, A., and Eaton, W. A. (1996) Diffusion-limited contact formation in unfolded cytochrome c: estimating the maximum rate of protein folding. *Proc. Natl. Acad. Sci. USA* **93,** 11,615–11,617.
5. Klessinger, M. and Michl, J. (1995) *Excited States and Photochemistry of Organic Molecules,* VCH, Weinheim, Germany.
6. Satzger, H., Schmidt, B., Root, C., et al. (2004) Ultrafast quenching of the xanthone triplet by energy transfer: new insight into the intersystem crossing kinetics. *J. Phys. Chem. A* **108,** 10,072–10,079.
7. Krieger, F., Fierz, B., Bieri, O., Drewello, M., and Kiefhaber, T. (2003) Dynamics of unfolded polypeptide chains as model for the earliest steps in protein folding. *J. Mol. Biol.* **332,** 265–274.
8. Bieri, O., Wirz, J., Hellrung, B., Schutkowski, M., Drewello, M., and Kiefhaber, T. (1999) The speed limit for protein folding measured by triplet-triplet energy transfer. *Proc. Natl. Acad. Sci. USA* **96,** 9597–9601.

9. Closs, G. L., Johnson, M. D., Miller, J. R., and Piotrowiak, P. (1989) A connection between intramolecular long-range electron, hole and triplet energy transfer. *J. Am. Chem. Soc.* **111,** 3751–3753.
10. Wagner, P. J. and Klán, P. (1999) Intramolecular triplet energy transfer in flexible molecules: electronic, dynamic, and structural aspects. *J. Am. Chem. Soc.* **121,** 9626–9635.
11. Szabo, Z. G. (1969) Kinetic characterization of complex reaction systems. In: *Comprehensive Chemical Kinetics,* (Bamford, C. H. and Tipper, C. F. H., eds.) Elsevier Publishing Company, Amsterdam, the Netherlands, pp. 1–80.
12. Kiefhaber, T., Kohler, H. H., and Schmid, F. X. (1992) Kinetic coupling between protein folding and prolyl isomerization. I. Theoretical models. *J. Mol. Biol.* **224,** 217–229.
13. Moore, J. W. and Pearson, R. G. (1981) *Kinetics and Mechanisms,* John Wiley and Sons, New York, NY.
14. Stern, O. and Volmer, M. (1919) Über die Abklingungszeit der Fluoreszenz. *Physik. Z.* **20,** 183–188.
15. Kleinert, T., Doster, W., Leyser, H., Petry, W., Schwarz, V., and Settles, M. (1998) Solvent composition and viscosity effects on the kinetics of CO binding to horse myoglobin. *Biochemistry* **37,** 717–733.
16. Timasheff, S. N. (2002) Protein hydration, thermodynamic binding and prefential hydration. *Biochemistry* **41,** 13,473–13,482.
17. Bent, D. V. and Hayon, E. (1975) Excited state chemistry of aromatic amino acids and related peptides: III. Tryptophan. *J. Am. Chem. Soc.* **97,** 2612–2619.
18. Bent, D. V. and Hayano, T. (1975) Excited state chemistry of aromatic amino acids and related peptides: I. Tyrosine. *J. Am. Chem. Soc.* **97,** 2599–2606.
19. Lapidus, L. J., Eaton, W. A., and Hofrichter, J. (2000) Measuring the rate of intramolecular contact formation in polypeptides. *Proc. Natl. Acad. Sci. USA* **97,** 7220–7225.
20. Krieger, F., Fierz, B., Axthelm, F., Joder, K., Meyer, D., and Kiefhaber, T. (2004) Intrachain diffusion in a protein loop fragement from carp parvalbumin. *Chemical Physics* **307,** 209–215.
21. Krieger, F., Möglich, A., and Kiefhaber, T. (2005) Effect of proline and glycine residues on the dynamics and barriers of loop formation in polypeptide chains. *J. Am. Chem. Soc.* **127,** 3346–3352.
22. Fierz, B. and Kiefhaber, T. (2004) Dynamics of unfolded polypeptide chains. In: *Protein Folding Handbook* (Buchner, J. and Kiefhaber, T., eds.) Wiley-VCH Verlag GmbH and Co. KGaA, Weinheim, Germany, pp. 805–851.
23. Möglich, A., Krieger, F., and Kiefhaber, T. (2004) Molecular basis of the effect of urea and guanidinium chloride on the dynamics of unfolded proteins. *J. Mol. Biol.* **345,** 153–162.
24. Volkert, W. A., Kuntz, R. R., Ghiron, C. A., Evans, R. F., Santus, R., and Bazin, M. (1977) Flash photolysis of tryptophan and N-acetyl-L-tryptophanamide; the effect of bromide on transient yields. *Photochem. Photobiol.* **26,** 3–9.
25. Gonnelli, M. and Strambini, G. B. (1995) Phosphorescence lifetime of tryptophan in proteins. *Biochemistry* **34,** 13,847–13,857.

26. Yeh, I. C. and Hummer, G. (2002) Peptide loop-closure kinetics from microsecond molecular dynamics simulations in explicit solvent. *J. Am. Chem. Soc.* **124,** 6563–6568.
27. Chang, I. -J., Lee, J. C., Winkler, J. R., and Gray, H. B. (2003) The protein-folding speed limit: intrachain diffusion times set by electron-transfer rates in denatured Ru(NH3)5(His-33)-Zn-cytochrome c. *Proc. Natl. Acad. Sci. USA* **100,** 3838–3840.
28. Lapidus, L. J., Eaton, W. A., and Hofrichter, J. (2001) Dynamics of intramolecular contact fromation in polypeptides: distance dependence of quenching rates in a room-temperature glass. *Phys. Rev. Lett.* **87,** 258101-1–258101-4.
29. Lapidus, L. J., Steinbach, P. J., Eaton, W. A., Szabo, A., and Hofrichter, J. (2002) Effects of chain stiffness on the dynamics of loop formation in polypeptides. Appendix: Testing a 1-diensional diffusion model for peptide dynamics. *J. Phys. Chem. B* **106,** 11,628–11,640.

11

A Hierarchical Protein Folding Scheme Based on the Building Block Folding Model

Nurit Haspel, Gilad Wainreb, Yuval Inbar, Hui-Hsu (Gavin) Tsai, Chung-Jung Tsai, Haim J. Wolfson, and Ruth Nussinov

Summary

The building block protein folding model states that the native protein structure is the product of a combinatorial assembly of relatively structurally independent contiguous parts of the protein that possess a hydrophobic core, i.e., building blocks (BBs). According to this model, our group proposed a three-stage scheme for a feasible time-wise semi ab-intio protein structure prediction. Given a protein sequence, at the first stage of the prediction scheme, we propose cutting the sequence into structurally assigned BBs. Next, we perform a combinatorial assembly and attempt to predict the relative three-dimensional arrangement of the BBs. In the third stage, we refine and rank the assemblies. The scheme has proven to be very promising in reducing the complexity of the protein folding problem and gaining insight into the protein folding process. In this chapter, we describe the different stages of the scheme and discuss a possible application of the model to protein design.

Key Words: Protein folding; building blocks; protein superimposition; combinatorial assembly; clustering; parallel tempering.

1. Introduction

How a one-dimensional (1D) polypeptide chain folds into a three-dimensional (3D) structure is still an unsolved problem, despite the improvements made in the methodologies. All the information needed to specify the protein 3D structure is contained within its amino acid sequence, and given suitable conditions, proteins will spontaneously fold into their native states. Several models have been proposed for protein folding. This work is based on the hierarchical model, which postulates that the unit from which a fold is constructed is the outcome of a combinatorial assembly of a set of folding units. The assemblies, in turn, associate to form intramolecular domains. The hydrophobic folding units (HFUs) possess relatively

From: *Methods in Molecular Biology, vol. 350: Protein Folding Protocols*
Edited by: Y. Bai and R. Nussinov © Humana Press Inc., Totowa, NJ

strong hydrophobic cores, and their hydrophobic interactions with their surroundings, or with other units, are weaker. They are compact and may consist of noncontiguous segments on the amino acid chain *(1)*.

According to the building block model, an HFU consists of contiguous segments of the protein chain, which are defined as building blocks (BBs). According to this model, if a BB is removed from the protein chain, the most highly populated conformation of the resulting peptide in solution would very likely be similar to that of the BB when it is embedded in the native protein structure, even though it may happen that an alternative conformation is selected in the combinatorial assembly process. The formation of any BB in a given sequence can be described and guided by microfunnel-like energy landscapes. The mutual recognition of BBs resembles a fusion of two microfunnel-like landscapes. At the bottom of a subfunnel-like landscape resides a compact, stable HFU. It has been shown that the BB model provides a coherent rationale for the kinetics of a two- and three-state protein folding process *(1)*. Tsai et al. *(2,3)* devised an empirical fragment size-independent stability scoring function that measures the relative conformational stability of protein fragments and favors folding units that are compact objects, highly isolated, and highly hydrophobic modules. A comparison of the data from experimental-limited proteolysis with those of the computational cuttings to the BBs showed a high correspondence between the two methods *(4)*. Further research *(5)* has also shown that the locations of hinges in protein structures tend to fall between BB elements, therefore, enabling the trial-association process of the BBs.

Based on the BB folding model, our group has developed a protein prediction scheme depending solely on the protein sequence. The biological role of a protein is determined by its function, which is, in turn, largely determined by its structure. Although an increasing number of structures are determined experimentally at an accelerated rate, it is simply not possible to determine all the protein structures from experiments, and the necessity for computational prediction methods is becoming even more evident. Our three-stage scheme protein prediction technique *(6)* follows the protein folding process as outlined by the BB folding model and aims to reduce the complexity of the protein folding process by dividing it into subproblems and efficiently solving each one, making the entire problem more tractable. According to this scheme, we first cut the sequence into structurally assigned BBs. Next, we perform a combinatorial assembly of the assigned BBs and attempt to predict the relative 3D arrangement of the BBs. In the third stage, we refine and rank the assemblies. This stage is currently under development. Here, we also introduce an application for our scheme, which can be utilized to design new proteins based on a sequence *(7)*. **Figure 1** shows a flowchart of the three stages of the scheme.

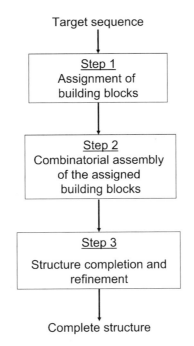

Fig. 1. An outline of the three stages of the building block (BB)-based folding scheme: the overall scheme is composed of three steps. The first step involves cutting the target sequence into BBs and assigning their conformations. In the second step, the BBs are assembled combinatorially. In the third step, the structure is refined to yield the predicted conformation. (Adapted from **ref. 15**.)

Our scheme is justified if the BBs can be referred to as stand-alone units, regardless of their structural context. Our results suggest that BBs are indeed, with high probability, independent protein fragments. Furthermore, we show that we are able to detect and assess protein fragments with high precision.

2. Methods

2.1. The BB Cutting Algorithm

This algorithm *(2,3)* uses a scoring function that measures the relative conformational stability of a candidate BB fragment.

1. We define a BB stability score:

$$Score^{BB}(Z,H,I) = \frac{(Z^1_{avg} - Z)}{Z^1_{dev}} + \frac{(H^1_{avg} - H)}{H^1_{dev}} + \frac{(I^1_{avg} - I)}{I^1_{dev}}$$

$$+ \frac{(Z^2_{avg} - Z)}{Z^2_{dev}} + \frac{(H^2_{avg} - H)}{H^2_{dev}} + \frac{(I^2_{avg} - I)}{I^2_{dev}}$$

    a. The three ingredients in the scoring function are the following:

      i. Isolation (I): the degree of isolation of a unit is based on the solvent accessible surface area (ASA) originally buried in the interior of the protein and exposed to the solvent after cutting.

      ii. Compactness (Z): the ASA of a structure, divided by its minimum possible surface area, which is the surface area of a sphere with an equal volume to the structure.

      iii. Hydrophobicity (H): the fraction of the buried nonpolar ASA with respect to the total nonpolar ASA.

2. Each of the score components is calculated as the deviation from the average value of known protein structures. The average and standard deviation of these quantities were calculated from a nonredundant dataset of 930 representative single-chain proteins from the Protein Data Bank (PDB) (*8*) (*see* **Note 1**).

3. Candidates with high local scores are classified as local minima on the scoring surface.

4. A region whose stability is not high enough to constitute a "valid" BB, and whose addition to a neighboring, sequentially linked BB does not increase the stability of that BB, is defined as "unassigned."

5. This process continues iteratively until the BB can no longer be dissected (*see* **Note 2**).

6. The resulting tree outlines the most probable folding routes. The different levels are referred to as cutting levels. **Figure 2** shows an example of the BB assignment of adenylate kinase from Baker's yeast.

2.2. Creation of the BB Structural Template Library

The way to implement the first stage of the prediction scheme and identify BBs in a given protein sequence is done by using a BB structural template library (*9,10*). In order to relate each BB to a characteristic profile that represents its typical sequence, i.e., BB structural template library, we cluster BBs from a BB library according to their structural similarities. The BB library consists of approx 68,000 protein segments, exhibiting a relatively high conformational stability as measured by the scoring function proposed by Tsai et al. (*3*) and assembled from a structurally nonredundant protein dataset. To relate a BB with a structural template we developed a pairwise superimposition tool, coined SSGS (secondary structure guided superimposition tool) (*11*) to efficiently detect structural similarity among the BBs and cluster them accordingly (*see* **Notes 3** and **4**). **Figure 3** shows an example of a BB cluster.

2.2.1. Clustering of the BBs

1. Iteratively compare every BB to the representatives of existing clusters.

2. If a match is found between the candidate BB and a cluster representative, the BB is assigned to the cluster.

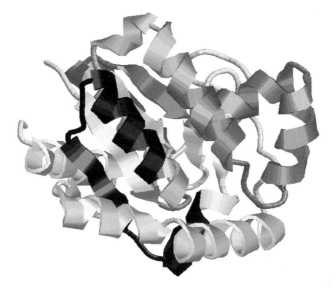

Fig. 2. An example of building block (BB) cutting. Each BB is displayed in a different shade of gray. The protein is adenylate kinase from aker's yeast (PDB: 1aky), cutting level 2.

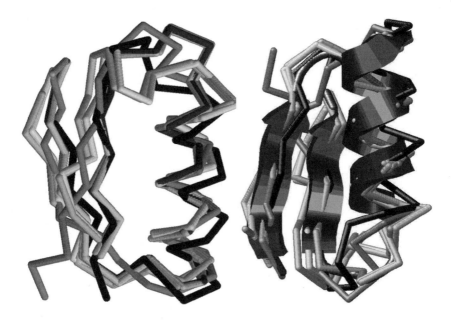

Fig. 3. Building block (BB) cluster structural multiple alignment from two views. The representative of this cluster is the BB consisting of residues 2–35 of the human estrogenic 17 β-hydroxysteroid dehydrogenase (PDB = 1bhs, cartoon chain), which is considered to be the average structure for the cluster.

3. If none of the representatives of the clusters match the BB, a new cluster is created with this BB as its representative. This way, the cluster representative is the first BB that opened that cluster (*see* **Note 5**).

In order to match each BB to the representatives of the existing clusters, we use a pairwise protein matching algorithm, developed especially for this purpose. The algorithm is described in detail in **Subheading 2.2.2.**

2.2.2. Secondary Structure Guided Superimposition Tool

The motivation for this algorithm is to perform a protein pairwise superimposition that is permissive and allows a loose matching between the compared proteins in assigned loop regions. In addition, the algorithm should be efficient because we should make many structural comparisons of BBs in a feasible time. This tool was developed based on the observation that secondary structure elements, such as α-helices and β-sheets, are hydrogen bonded and fluctuate less than loop regions. Therefore, it aligns the Cα atoms of two proteins, whereas considering their secondary structure assignment, allowing a more permissive alignment of loop regions. The stages of the algorithm are as described in **Subheadings 2.2.2.1.–2.2.2.3.**

2.2.2.1. TRANSFORMATION SET CREATION

The input to this stage consists of the coordinates of the Cα atoms of two proteins, A and B. We shall regard protein A as the target and protein B as the model, with M and N residues, respectively. To create a transformation set we use the pose clustering method (*12*) that performs object recognition by determining hypothetical object poses and finds clusters of the poses in the space of object positions. The transformation set creation stage of the algorithm consists of the following stages:

1. Preprocessing: for every noncollinear triplet of the model's Cα atoms (i,j,k), compute the lengths of the vertices of the triangle (i,j,k) and use the lengths as a 3D key to place the indices (i,j,k) in a 3D hash table.
2. Recognition: iteratively choose noncollinear triplets of the target's Cα atoms (A,B,C) and compute the corresponding vertices' lengths.
3. Use the lengths as a query key to access the hash table to find possible instances of the model. The query returns all of the model's hashed triangles whose distance from the query key is bounded by a given resolution factor f (*see* **Note 6**).
4. For every target triplet T = (A,B,C) compute the transformation needed to transform the triplet to each of the query returned model triplets, thus, creating a redundant transformation set. These transformations place a target and model triplet in the same plane and coincides their baricenters (*see* **Note 7**).

2.2.2.2. CLUSTERING OF THE TRANSFORMATIONS

We cluster the transformations according to the root-mean-square deviation (RMSD) between them to create a nonredundant set. The transformation clusters that have the largest number of members will be tested in the next stage, hence, avoiding the examination of all the possible transformations. We set a linear time complexity by using a geometric hash table *(13,14)* during the search for a matching cluster for the candidate transformation. The transformation clustering procedure is as follows:

1. Prior to the clustering, we select a triplet of target Cα atoms Q = (A,B,C) and form three geometric hash tables a,b,c.
2. During the clustering procedure, given a transformation T_j $(0 \leq j \leq \infty)$, we apply T_j on Q and create *Q = (*A, *B, *C).
3. We search the geometric hash tables (a, b, c) with *A, *B, *C, respectively, for Cα atoms E_a, E_b, and E_c that are in radius, r, from the query key. Denote $T_{Ea}{}^i$, $T_{Eb}{}^i$, $T_{Ec}{}^i$ as the transformations that were used to transform A, B, and C's Cα atoms to $E_a{}^i$, $E_b{}^i$, and $E_c{}^i$, respectively.
4. The intersection U of the transformation groups $T_{Ea}{}^i$, $T_{Eb}{}^i$, $T_{Ec}{}^i$ contain only transformations that appeared in all these transformation groups. Therefore, U contains possible matching transformations to T_j.
5. If U_f is a transformation in U and the RMSD between $Q*U_f$ and $Q*T_j$ is smaller than the required threshold, we add T_j as a member to the cluster containing U_f.
6. If none of the transformations match the candidate transformation, we add a new cluster whose representative is T_j, and insert *Q to the hash tables as follows: A* into hash table a, B* into hash table b, and insert C* into hash table c.

2.2.2.3. THE ALIGNMENT CREATION AND EVALUATION

Both the alignment and its evaluation are performed by a dynamic programming recursion:

1. Creation of the spatial neighboring relations matrix: prior to the dynamic programming recursion, we preprocess the proteins and translate the input into a spatial neighboring relations matrix (SNRM), R (M × N). The input is the Cα atoms coordinates of the model protein and the target protein, with lengths of M and N, respectively.
2. We index the Cα atoms according to their location on the backbone, starting from the N-terminal side, and set all the values in R to a mismatch.
3. For every $C\alpha_j$ $(0 \leq j \leq N)$ atom of the model, we compute its Euclidian distance to $C\alpha_i$ $(0 \leq j \leq M)$ atom of the target.
4. If the distance is smaller than a radius r and |i − j| *Maximal shift value*, we set R(i,j) to a match between $C\alpha_i$ and $C\alpha_j$. Accordingly, if $C\alpha_i$ and $C\alpha_j$ are spatially distant from each other, we set R(i,j) as a mismatch (*see* **Note 8**).
5. The affine gap and mismatch score depend on the cell the gap starts from, the size

of the gap, and whether the cell the gap starts from is a mismatch or a match (*see* **Note 9**).

6. The secondary structure nature is introduced into the algorithm by further differentiating between cell types according to the secondary structure assignment of their aligned Cα atoms: structures and loops. Residues that belong to β-sheets or α-helices are considered as a "structure." The residues that do not belong to β-sheets or α-helixes are considered as a "loop" (*see* **Notes 10** and **11**).

7. The secondary structure dependency of the alignment is manifested in the topology factor, which is the payoff given by aligning two cells one after another depending on (1) the secondary structure assignment of the aligned residues in the cell, (2) whether the cell is a match or a mismatch, and (3) the route progress situations being either opening a mismatched region, continuing a matched region, continuing a mismatched region, or closing a mismatched region. The final payoff for aligning two cells will be the average of the payoffs for opening a mismatch region in a loop-to-loop residues alignment and a structure-to-loop residues alignment (*see* **Note 12**).

8. The dynamic programming: we create three matrices: $Opt_{(M+1,N+1)}$, $Left_{(M+1,N+1,6)}$, and $Up_{(M+1,N+1,6)}$. The Opt matrix holds the value for an optimal alignment that ends in the current cell. The matrices Up and Left are composed of six values, each for a different cell type for the optimal path that ends with a gap leading to the current cell, depending on the topology of the aligned residues and the match or mismatch.

9. At every stage of the dynamic programming we compute 13 values, as the outcome of our differentiation between six cells types (determined by the match/mismatch state and the secondary structure alignment) from which a gap region can start. At every stage of the recursion, we save the highest payoffs for these gap types (the gap type is determined by the cell it starts from).

10. As the recursion progresses, we check the current cell type and determine whether opening a gap at the current cell type achieves a higher payoff (*see* **Note 13**).

11. Trace back: given matrices Opt, Left, and Up, find the type of the highest value among the cells Opt (M,N), Left(M,N), and Up(M,N). According to the matrix that contains the highest value, we set the next cell in the alignment.

2.3. The BB Assignment Algorithm

The clusters of BBs, grouped by their structures and by their sequences, constitute the input to a graph theoretic sequence assignment algorithm *(15)* (*see* **Note 14**).

The stages of the algorithm are as follows:

1. Alignment of the target sequence against the BB sequence database: for a given target sequence whose BB composition we wish to assign, we carry out a sequence comparison with the database containing the representatives of all BB clusters, using BLAST *(16)* with default parameters, allowing gaps (*see* **Note 15**).

2. Constructing a weighted, directed acyclic graph:

   a. If a sequence similarity above a given threshold is found (that is, a BB in the database is found to match an area of the target protein sequence), this BB is represented as a weighted graph vertex.

   b. The weighting scheme contains two components: the BLAST match score and the BB stability score. A candidate BB is considered only if the match spans the entire BB (allowing a 10-residue gap at each side) and if the match length is at least 15 residues long.

   c. The BB's score is: –(BLAST score + 3 * stability score) (*see* **Note 16**). Using this scheme, the more negative the BB total score, the more "suitable" it is.

   d. The edges: a directed edge connects two vertices if they are sequentially adjacent, and if they adhere to the rules followed in the generation of the BB from the native structures (no more than a 7-residue overlap, and not over 15 residues apart). The edge connecting the vertices is assigned the average weight of the two vertices.

   e. Adding fictitious start and end vertices to the graph: a zero-weight edge connects each of these to a vertex that is up to a distance of 15 residues. These either follow (the starting source vertex) or are prior to (the ending target vertex) (*see* **Note 17**).

3. Finding a BB assignment to the target sequence using a graph theoretic algorithm that finds the shortest path from the start to the end vertices (*see* **Note 18**). A path is a consecutive set of vertices that leads from the starting source vertex to the ending target vertex. The path represents a possible assignment of BBs to the sequence.

4. Among the obtained paths, the highest scoring ones are retained. **Figure 4** presents an illustration of the algorithm.

2.4. The Combinatorial Assembly of the BBs

Given a protein sequence with a set of BBs assigned to it, we assemble the BBs on one another combinatorially to determine the overall fold *(6)*. We assume that the same physical forces that drive the protein–protein interactions are also the forces that assemble the protein structural units into the overall structure of the protein. Therefore, we choose to approach the combinatorial assembly by multiple pairwise docking. We define a valid complex as a spatial organization of the BBs such that they do not penetrate each other more than a predefined threshold. Our solution is inspired by the way one solves a jigsaw puzzle. We gradually get the ultimate overall structure by fitting pairs of substructural units.

The input of the algorithm consists of two ordered sets. A set of N structural units (SUs) of the target protein and a set of N –1 positive integers that represent the lengths of the linkers, i.e., the unassigned regions of the protein sequence between the SUs. The output is a list of ranked valid complexes.

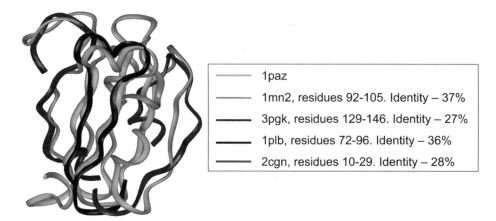

Fig. 4. An example of the building block assignment algorithm. The model protein is pseudoazurin (PDB 1paz). (Adapted from **ref. *15*.**)

Given N SUs, the algorithm detects and ranks putative complexes of all the N units. The goal of the algorithm is to maximize the interface between the SUs within the complexes and validate the constraints (*see* **Notes 19** and **20**). **Figure 5** shows an example of the combinatorial assembly algorithm applied to citrate synthase.

The stages of the algorithm are as follows:

1. All pairs docking: we apply a geometric docking algorithm to each ordered pair of subunits. The docking algorithm output candidate transformations that once applied on the second SU we get a hypothetic complex. The transformations are scored by the algorithm (*see* **Note 21**).
2. Assuming that there are N SUs, we keep for each of the N(N − 1)/2 pairs a set of the best K scoring transformations (*see* **Note 22**).
3. We perform the assembly of the SUs by combining different subsets of the candidate pairwise transformations that were generated in the all pairs docking stage.
   a. We use the following reduction: let G be a graph of N vertices, where each vertex represents a different SU.
   b. Each pairwise transformation that docks two SUs is represented by an edge between the corresponding vertices.
   c. G is a complete graph with K parallel edges. By a simple generalization we get that a complex of N SUs that was generated by the input transformations is represented by a spanning tree of G and any spanning tree of G represents a different complex. A complete graph with N vertices contains N^{N-2} possible spanning trees (*17*). In our case, we have K parallel edges between each pair of vertices and therefore G has $N^{N-2}K^{N-1}$ spanning trees.
4. We developed a polynomial time–complexity algorithm that suggests a heuristic solution to the problem, because it is not feasible to examine all the possible trees even for small N and K.

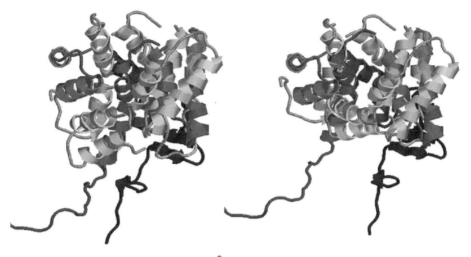

Model arrangement (4.5Å)   Native arrangement

Fig. 5. An example of combinatorial assembly of the 13 building blocks (BBs) of citrate synthase. (**Left**) Arrangement of BB obtained by the algorithm. (**Right**) The native BB arrangement.

5. The method is based on three basic concepts: (1) the gradual generation of trees/complexes; (2) the restriction of the topology of the spanning trees/complexes; and (3) the greedy selection of subtrees/subcomplexes.

   a. The algorithm has N stages. In each stage we generate the trees of size i. In the first stage the complexes are therefore the SUs themselves. In the Nth stage we get the spanning trees that represent the final complexes.

   b. In the ith stage we generate trees of size i by connecting two disjoint trees of smaller size with one of the input edges. Because each valid tree/complex has at least one edge whose removal creates two disjoint subtrees, all the trees may be generated in such manner (*see* **Note 23**).

   c. Instead of keeping all the subtrees for each subset of consecutive SUs we keep only the D subtrees with the best score (*see* **Note 24**).

6. The final scoring: we first score the final complexes and perform an iterative clustering to minimize the final set of solutions.

   a. The scoring function weighs both the geometric fit between the subunits and the nonpolar surface of the subunits that get buried in the complex. A good geometrical fit should include a wide interface area and a small rate of penetrations.

   b. In order to measure the nonpolar buried surface area, we mark the surface of uncharged atoms and check if this marked surface becomes involved in an interface with another SU in the evaluated complex.

c. The clustering algorithm consists of three stages. First, we cluster by a hash table where the hash key is a bit set that represent the SU contacts that exist in the complex. The bit set is of size $N(N-1)/2$, a bit for each pair of SUs. A bit is set on if and only if the corresponding pair of subunits has some interface between them in the complex. This enables us a rough clustering in O(N).

d. In the second stage we compare between SUs with the same hash key. As the distance function, we calculate the RMSD between the Cα atoms once the complexes are optimally superimposed. Complexes that are within the RMSD clustering parameter (*see* **Note 25**) are put into the same cluster.

e. To speed up the process even further we first calculate the RMSD between the centers of mass of the SUs, and only if this distance is smaller than 1.5 times the RMSD threshold, we use a finer all-Cα function. In each cluster, only the complex with the highest score is kept as a representative.

f. In the third stage, we perform the same RMSD clustering routine as the second stage, only this time we compare all the representative complexes, ignoring the contact maps.

2.5. Protein Design Using the BB Algorithms

In this section we show a possible application to our model by proposing a protocol for protein design *(7)*. The input is a given target protein:

1. The protein is cut into its BBs.
2. Each of the BBs is compared with all the structures in the PDB.
3. During the comparison, candidates are searched according to the following terms:

    a. Small RMSD from the target BB, low sequence identity between the candidate BB and the one in the target structure, and a high similarity in the binary (hydrophobic/hydrophilic) pattern (*see* **Note 26**).

    b. The difference in the hydrophobicity scales should be small (*see* **Note 27**).

    c. The fragment ends should match well structurally, as otherwise it may introduce a strain when we ligate consecutive fragments in the designed protein.

    d. For small fragments of the target protein, which do not constitute valid BBs (unassigned BBs, with small sizes [under 15 residues], and their addition to a joined BB does not improve the score) no searches are made. These are retained as they appear in the target protein.

4. Selected BBs are substituted and ligated.
5. To validate the construct, the candidate-designed proteins are submitted to implicit simulations.
6. Stable BBs are submitted to long time explicit solvent simulations, and their stabilities are tested by MMGB/SA free energy calculations.
7. "Nonproteins" are constructed by selecting BBs with similar structures with an inverted pattern of hydrophobic and hydrophilic residues (*see* **Note 28**).
8. The engineered proteins should be submitted to validation by experiment.

3. Notes

1. Terms with superscript 1 were determined with respect to fragment size and those with superscript 2 as a function of the fraction of the fragment size to the whole protein.

2. In the cutting procedure, a seven-residue overlap is allowed between BBs. The minimum size of a BB is defined as 15 residues.

3. In order to make the clustering more efficient we use several heuristics: the difference between the two compared BB sizes should not be greater than 20% of the size of the smaller of the two BBs. In addition, we compare only BBs whose secondary structure assignments are similar.

4. Other sequence order-independent algorithms may achieve matches with lower RMSD, but this algorithm prefers the matching of secondary structure elements over loop assignment. In the case of BBs, this is preferable because secondary structure elements are usually better preserved and are more significant to functionality than loops.

5. When this clustering method is used, there may be clusters whose members resemble the representative but not one another. However, this is the fastest method.

6. The underlying metric is defined as follows: let $Tr = (i,j,k)$ be a triangle in the hash table, and let $T = (A,B,C)$ be the query triangle. Denote the lengths of the vertices of Tr by $(u1,u2,u3)$, where $u1 = |i{\rightarrow}j|$, $u2=|j{\rightarrow}k|$ and $u3=|k{\rightarrow}i|$ and the lengths of the vertices of T, similarly, by $(v1,v2,v3)$. Then $d(Tr, T)= \max_{1\leq i\leq 3} |u_i-v_i|$.

7. To reduce the number of comparisons, two BBs are compared only if the following holds: the $C\alpha$ atoms that make the model triplet had the same secondary structure assignment as the corresponding $C\alpha$ atoms in the target triplet or their neighboring residues. Only triplets whose $C\alpha$ indices difference is smaller than a third of the length of the larger protein are considered as relevant triplets. In addition, we only compare atoms in which the difference in the respective indices along the proteins is no more than one-third of the length of the larger protein. We account for sequence order and seek only matches that transform the N-terminal of the target to that of the model. In larger BBs we also add randomization to reduce the number of comparisons.

8. We discard a transformation if the number of matched cells is lower than the maximal shift value, because these transformations cannot yield a satisfying alignment with sufficient matches. We set the maximal shift value to a third of the longest protein length, and the radius r to 3 Å.

9. The calculation of the value that represents the best alignment that ends in some cell of the matrix is the maximum product of a comparison between five values: the leftward and upward cells each contribute two values gap (where each gap starts from a different cell) and the diagonal alignment value.

10. We assess the secondary structural assignment by the DSSP program (*18*).

11. This definition of the secondary structure assignment gives rise to three mismatch cell types (loop-to-loop mismatch cell, loop-to-structure mismatch cell, and a

structure-to-structure mismatch cell) and three types of match cells (loop-to-loop match cell, loop-to-structure match cell, and a structure-to-structure match cell).

12. The payoff ratio is designed in a way that gives preference to structure match over loop match and also prefers few long matches over many short matches.

13. At every cell we compute 12 values that contain the optimal payoff for a path that ends with a gap leading to the current cell. The thirteenth is the value for the optimal path to the current cell.

14. The advantage of using a graph algorithm instead of a simple search and assignment scheme is that the algorithm performs the assignment more efficiently, that is, in polynomial time rather than testing all the possible combinations of BB covers which can be exponential.

15. This is a one-against-all pairwise alignment, not a multiple alignment.

16. The factor of three is there because the stability score is usually smaller than the BLAST score by a factor of at least three. This factor gives the two measurements roughly equal weights in the scoring function. The minus sign leads to negative weights, so that the best path will be the "shortest," that is, with the smallest weight. Although the BLAST score is always positive, a larger stability score reflects a more stable BB.

17. The resulting graph is directed and acyclic. In such a graph, there exists an algorithm that finds the shortest path between any pair of known vertices in a short time.

18. The algorithm is the Single Source Shortest Path algorithm *(19)*, which, given a directed, weighted, noncyclic graph with a source vertex and a sink vertex, finds the shortest (minimum weight) path from the source to the sink.

19. An interface between two translated SUs, A and B, is defined as the maximal fraction of the two surfaces such that any point that belongs to the interface that is originated from one SU surface is within a 2.5 Å distance from the surface of the second SU.

20. Given the two molecules and their accessible surfaces, the penetration value between the molecules is the maximum penetration of the accessible surface of one SU into the interior of the other SU. A point within the interior of a 3D body penetrates by d if d is its distance from the body surface.

21. The docking algorithm that was specifically developed for this purpose is based on local feature matching *(20–22)*. Its main advantage is that it avoids a brute force search of the orientation space and thus achieves high performances in terms of time and space. Notice that one can use for this stage any other docking algorithm that produces a ranked set of candidate transformations.

22. K is usually set to few hundreds so the right orientation would not be missed.

23. There are two advantages to this approach: the first is that when we check whether the resolved tree is valid we need only validate the intermediate subtrees constraints, because each subtree has been validated earlier. The second is that many trees share common subtrees and this way we perform the computations that involve their common parts only once. Instead of generating all the trees in each stage we only construct trees that consist of consecutive vertices/SUs.

24. The score of each tree is the sum of the scores of the transformations that are represented by its edges. This restriction, together with the restricted topology, bounds the algorithm complexity to $O(KD^2 \log DN^4 + KD^2N^6)$.

25. By default, the RMSD threshold parameter is set to 4 Å.
26. Currently we use an RMSD threshold of 2.5 Å, 35% or less sequence identity, and over 70% similarity in the hydrophobicity pattern. ·
27. Currently, a candidate fragment should have less than 0.75 of the expected hydrophobic scale, based on the native fragment. The two hydrophobic parameters are related, but the latter is more accurate, and consequently is given a larger weight.
28. These "nonprotein" constructs should in principle be unstable, as their hydrophobic residues will be largely exposed and the polar/charged largely buried. These constructs serve as additional tests of the stability of our "valid" engineered proteins.

Acknowledgments

The authors thank Maxim Shatsky and Oranit Dror for their help and comments. The computation times are provided by the National Cancer Institute's Frederick Advanced Biomedical Supercomputing Center. The research of R. Nussinov in Israel has been supported in part by the "Center of Excellence in Geometric Computing and its Applications" funded by the Israel Science Foundation (administered by the Israel Academy of Sciences), and by the Adams Brain Center. This project has been funded in whole or in part with Federal funds from the National Cancer Institute, National Institutes of Health (NIH), under contract number NO1-CO-12400. This research was supported (in part) by the Intramural Research Program of the NIH, National Cancer Institute, Center for Cancer Research. The content of this publication does not necessarily reflect the view or policies of the Department of Health and Human Services, nor does mention of trade names, commercial products, or organization imply endorsement by the US government.

References

1. Struthers, M. D., Cheng, R. P., and Imperiali, B. (1996) Design of a monomeric 23-residue polypeptide with defined tertiary structure. *Science* **271**, 342–345.
2. Tsai, C. J. and Nussinov, R. (2001) The building block folding model and the kinetics of protein folding. *Protein Eng.* **14**, 723–733.
3. Tsai, C. J., Maizel, J. V., and Nussinov, R. (2000) Anatomy of protein structures: Visualizing how a one-dimensional protein chain folds into a three-dimensional shape. *Proc Natl Acad Sci USA* **97**, 12,038–12,043.
4. Tsai, C. J., Polverino de Laureto, P., Fontana, A., and Nussinov, R. (2002) Comparison of protein fragments identified by limited proteolysis and by computational cutting of proteins. *Protein Sci.* **11**, 1753–1770.
5. Sinha, N., Tsai, C. J., and Nussinov, R. (2001) Building blocks, hinge-bending motions and protein topology. *J Biomol Struct Dyn.* **19**, 369–380.
6. Inbar, Y., Benyamini, H., Nussinov, R., and Wolfson, H. J. (2005) Prediction of multimolecular assemblies by multiple docking. *J. Mol. Biol.* **349**, 435–447.

7. Tsai, H. H., Tsai, C. J., Ma, B., and Nussinov, R. (2004) In silico protein design by combinatorial assembly of protein building blocks. *Protein Sci.* **13,** 2753–2765.
8. Bernstein, F. C., Koetzle, T. F., Williams, G. J. B., et al. (1977) The Protein Data Bank: a computer-based archival file for macromolecular structures. *J. Mol. Biol.* **112,** 535–542.
9. Haspel, N., Tsai, C. J., Wolfson, H., and Nussinov, R. (2003) Hierarchical protein folding pathways: a computational study of protein fragments. *Protein* **51,** 203–215.
10. Wainreb, G. (2005) Templating the building blocks and secondary-structure guided superimposition tool. Master's thesis. Faculty of medicine, Tel Aviv University, Tel-Aviv, Israel.
11. Wainreb, G., Haspel, N., Wolfson, H., and Nussinov, R. (2006) A permissive secondary structure-guided superimposition tool for clustering of protein fragments toward protein structure prediction via fragment assembly. *Bioinformatics*, in press.
12. Ballard, D. H. (1981) Generalizing the Hough transform to detect arbitrary shapes. *Pattern Recognit.* **13,** 111–122.
13. Lamdan, Y. and Wolfson, H. J. (1988) Geometric hashing: a general and efficient model-based recognition scheme. In: *Second International Conference on Computer Vision (Tampa, FL, December* 5–8, 1988), Washington, D.C., IEEE Computer Society Press, pp. 238–249.
14. Wolfson, H. J. (1990) Model-based recognition by geometric hashing. *In Proc. 1st. Euro. Conf. on Comput. Vis* April 23–27, 1990, Antibes, France. 526–536.
15. Haspel, N., Tsai, C. -J., Wolfson, H., and Nussinov, R. (2003) Reducing the computational complexity of protein folding via fragment folding and assembly. *Protein Sci.* **12,** 1177–1187.
16. Altschul, S. F., Gish, W., Miller, W., Myers, E. W., and Lipman, D. J. (1990) Basic local alignment search tool. *J. Mol. Biol.* **215,** 403–410.
17. Cayley, A. (1889) A theorem on trees. *Quart. J. Math.* **23,** 276–378.
18. Kabsch, W. and Sander, C. (1983) Dictionary of protein secondary structure: pattern recognition of hydrogen-bonded and geometrical features. *Biopolymers* **22,** 2577–2637.
19. Cormen, T., Leiserson, C., and Rivest, R. (1990) *Introduction to Algorithms,* MIT Press, Cambridge, MA.
20. Norel, R., Lin, S. L., Wolfson, H., and Nussinov, R. (1995) Molecular surface complementarity at protein-protein interfaces: the critical role played by surface normals at well placed, sparse points in docking. *J. Mol. Biol.* **252,** 263–273.
21. Polak, V. (2003) Budda: backbone unbound docking application. Master's thesis School of Comp. Sci., Tel-Aviv University, Israel.
22. Duhovny, D., Nussinov, R., and Wolfson, H. (2002) Efficient unbound docking of rigid molecules. In: *Workshop on Algorithms in Bioinformatics. Lecture Notes in Computer Science 2452*, (Guigo, R. and Gusfield, D., eds.), Springer Verlag, Rome, Italy, pp. 185–200.

12

Replica Exchange Molecular Dynamics Method for Protein Folding Simulation

Ruhong Zhou

Summary

Understanding protein folding is one of the most challenging problems remaining in molecular biology. In this chapter, a highly parallel replica exchange molecular dynamics (REMD) method and its application to protein folding are described. The REMD method couples molecular dynamics trajectories with a temperature exchange Monte Carlo process for efficient sampling of the conformational space. A series of replicas are run in parallel at temperatures ranging from the desired temperature to a high temperature at which the replica can easily surmount the energy barriers. From time to time the configurations of neighboring replicas are exchanged and this exchange is accepted or rejected based on a Metropolis acceptance criterion that guarantees the detailed balance. Two example protein systems, one α-helix and one β-hairpin, are used as case studies to demonstrate the power of the algorithm. Up to 64 replicas of solvated protein systems are simulated in parallel over a wide range of temperatures. The simulation results show that the combined trajectories in temperature and configurational space allow a replica to overcome free energy barriers present at low temperatures. These large-scale simulations also reveal detailed results on folding mechanisms, intermediate state structures, thermodynamic properties, and the temperature dependences for both protein systems.

Key Words: Replica exchange method; parallel tempering; molecular dynamics; Monte Carlo; protein folding; free energy landscape; conformation space sampling.

1. Introduction

The ability of protein molecules to fold or self-assemble into their highly structured functional states is one of the most remarkable evolutionary achievements of biology. Understanding this protein folding process is of great current interest *(1–6)*. The interest is not only in obtaining the final fold (generally referred to as structure prediction problem) but also in understanding the folding mechanism and folding kinetics. Large-scale computer simulations are playing a growing

From: *Methods in Molecular Biology, vol. 350: Protein Folding Protocols*
Edited by: Y. Bai and R. Nussinov © Humana Press Inc., Totowa, NJ

role in supplementing experiment and filling in some of the gaps in our knowledge about folding mechanisms. However, how to efficiently sample the complex conformational space of protein folding is a great challenge *(1–3,5,7)*. The free energy landscape of protein folding is believed to be at least partially rugged. At room temperature, protein systems get trapped in many local minima. This trapping limits the capacity to effectively sample the conformational space. Many methods have been proposed to enhance the conformation space sampling, including J-Walking *(8,9)*, S-Walking *(10)*, q-jumping based on Tsallis statistics *(11)*, multicanonical sampling *(12)*, replica exchange method (also called parallel tempering) *(13,14)*, simulated tempering *(15,16)*, catalytic tempering *(17)*, deformation of potential energy surface with the diffusion equation method *(18,19)*, and so on. For a comprehensive review, readers are directed to the paper by Straub and Berne *(20)*. Despite the enormous efforts from many groups, it is still difficult to perform realistic all-atom folding simulations for normal size proteins, which often take microseconds to milliseconds to fold. The only microsecond simulation with an all-atom model and explicit solvent was done by Duan and Kollman on a 36-residue α-helical villin headpiece protein *(7)*. Thus, more efficient simulation methods and more powerful supercomputers such as IBMs BlueGene machine (http://www.research.ibm.com/ bluegene), or super PC-clusters such as folding@home (http://folding.stanford.edu), are in great demand to tackle this problem.

This chapter will go over the replica exchange method (REM) *(13,14)* and its coupling to molecular dynamics (MD)—the replica exchange molecular dynamics (REMD) method *(21–25)*. In the REM method, replicas are run in parallel at a sequence of temperatures ranging from the desired temperature to a high temperature at which the replica can easily surmount the energy barriers. From time to time the configurations of neighboring replicas are exchanged and this exchange is accepted by a Metropolis acceptance criterion that guarantees the detailed balance. Thus, REM is essentially a Monte Carlo (MC) method. Because the high temperature replica can traverse high-energy barriers, there is a mechanism for the low temperature replicas to overcome the quasi ergodicity they would otherwise encounter in a single temperature replica. The replicas can be generated by MC, by hybrid Monte Carlo (HMC), or by MD with velocity rescaling. Okamoto et al. *(21)* have developed a temperature rescaling scheme for coupling MD with REM. These large-scale simulations can not only study the protein folding mechanism, but also provide extensive data for force field and solvation model assessment and further improvement.

Two example small protein systems, one α-helix (Ace-A5[AAARA]3A-Nme) and one β-hairpin (Ace-GEWTYDDATKTFTVTE-Nme) are used as case studies to illustrate the power of the parallel algorithm in this chapter. Larger protein systems have also been studied with the REMD method, and interested readers can refer to previous publications for details *(24,26,30)*. Here, we select these two simpler systems partly because larger proteins are mainly composed of these two major secondary structures. Understanding the folding of these structural elements is believed to be the foundation of understanding the folding of larger and more complex proteins. It was also pointed out *(22,31)* that despite of their structural simplicity, these peptides still display many of the complexities associated with the folding free energy landscape. Thus, they are ideal systems to investigate the folding mechanism and competing roles of various interactions.

It should be pointed out that this chapter is not aimed to be a complete overview of the subject, but instead focuses on picking a few simple and specific examples, some of our own, to illustrate the parallel REMD method and its application in protein folding.

2. Methods

2.1. Replica Exchange Method

Suppose there is a molecular system of N atoms with masses m_k ($k = 1, 2,..., N$) and coordinates and momenta $q \equiv \{q_1, q_2, ..., q_N\}$ and $p \equiv \{p_1, p_2,..., p_N\}$, the Hamiltonian H (p, q) of the system can be expressed as:

$$H(p,q) = \sum_{k=1}^{N} \frac{p_k^2}{2m_k} + U(q),$$ (1)

where $U(q)$ is the potential energy of the N atom system. In the canonical ensemble at temperature T, each state $x \equiv (p, q)$ with the Hamiltonian H (p, q) is weighted by the Boltzmann factor:

$$\rho(x;T) = \frac{1}{Z} \exp^{[-\beta H(p,q)]}$$ (2)

where $\beta = 1/k_B T$ (k_B is the Boltzmann constant), and Z is the partition function $Z = \int \exp^{[-\beta H(p,q)]} dpdq$. The generalized ensemble for REM consists of M noninteracting replicas of the original system at M different temperatures T_m ($m = 1, 2,..., M$). The replicas are arranged such that there is always exactly one replica at each temperature. Then there is a one-to-one correspondence between replicas and temperatures; the label i ($i = 1, 2,..., M$) for replicas is a permutation of the label m ($m = 1, 2,..., M$) for temperatures and vice versa,

$$i = i(m) \equiv f(m),$$
$$m = m(i) \equiv f^{-1}(i),$$ (3)

where $f(m)$ is a permutation function of m, and $f^{-1}(i)$ is its inverse.

The meta state X of this generalized ensemble will be a collection of all the M sets of coordinates $q^{[i]}$ and momenta $p^{[i]}$ of the N atoms in replica i at temperature T_m: $x_m^{[i]} = (p^{[i]}, q^{[i]})_m$:

$$X = \left(x_1^{[i(1)]}, ..., x_M^{[i(M)]} \right) = \left(x_{m(1)}^{[1]}, ..., x_{m(M)}^{[M]} \right) \tag{4}$$

where the superscript and the subscript in $x_m^{[i]}$ label the replica and the temperature indices, respectively, which have a one-to-one correspondence. Because the replicas are noninteracting, the weight factor for the state X in this generalized ensemble is given by the product of Boltzmann factors for each replica or each T:

$$\rho_{REM}(X) = \exp\left\{ -\sum_{i=1}^{M} \beta_{m(i)} H(p^{[i]}, q^{[i]}) \right\} = \exp\left\{ -\sum_{m=1}^{M} \beta_m H(p^{[i(m)]}, q^{[i(m)]}) \right\}, \tag{5}$$

where $i(m)$ and $m(i)$ are the permutation functions defined in **Eq. 3**. Now suppose a pair of replicas is exchanged. For generality, we assume the pair being swapped is (i, j), which are at temperatures (T_m, T_n), respectively,

$$X = \left(..., x_m^{[i]}, ..., x_n^{[j]}, ... \right) \rightarrow X' = \left(..., x_m^{[j]'}, ..., x_n^{[i]'}, ... \right). \tag{6}$$

The indices i, j and m, n are related by the permutation function. Upon the exchange, the permutation function will be updated, let us rename it f':

$$\begin{aligned} i = f(m) &\rightarrow j = f'(m), \\ j = f(n) &\rightarrow i = f'(n). \end{aligned} \tag{7}$$

The previously mentioned exchange of replicas can be rewritten in more detail as,

$$\begin{aligned} x_m^{[i]} = \left(p^{[i]}, q^{[i]} \right)_m &\rightarrow x_m^{[j]'} = \left(p^{[j]'}, q^{[j]} \right)_m, \\ x_n^{[j]} = \left(p^{[j]}, q^{[j]} \right)_n &\rightarrow x_n^{[i]'} = \left(p^{[i]'}, q^{[i]} \right)_n, \end{aligned} \tag{8}$$

where the new momenta $p^{[i]'}$ and $p^{[j]'}$ might be rescaled in REMD, as shown next. It is easy to see that this process of exchanging a pair of replicas (i, j) is equivalent to exchanging the two corresponding temperatures T_m and T_n:

$$\begin{aligned} x_m^{[i]} = \left(p^{[i]}, q^{[i]} \right)m &\rightarrow x_n^{[i]'} = \left(p^{[i]'}, q^{[i]} \right)_n, \\ x_n^{[j]} = \left(p^{[j]}, q^{[j]} \right)_n &\rightarrow x_m^{[j]'} = \left(p^{[j]'}, q^{[j]} \right)_m. \end{aligned} \tag{9}$$

This mathematical equivalence is very useful in practice, because it can be used to reduce the communication costs in REM, i.e., rather than exchanging the two full sets of coordinates and momenta, one can just swap the two temperatures for the two replicas and then update the permutation function (*see* **Note 1**). In the original implementations of the REM *(13,14)*, Monte Carlo algorithms were used, thus only the coordinates q and the potential energy $U(q)$ need to be taken into account (*see* **Note 2**). In order for this exchange process to generate the equilibrium canonical distribution functions, it is necessary and sufficient to impose the detailed balance condition on the transition probability $\Pi\,(X \to X')$ from meta state X to X',

$$\rho_{REM}(X)\Pi(X \to X') = \rho_{REM}(X')\Pi(X' \to X). \tag{10}$$

From **Eqs. 1, 5**, and **10**, one can easily derive that

$$\frac{\Pi(X \to X')}{\Pi(X' \to X)} = \frac{\rho_{REM}(X')}{\rho_{REM}(X)} = \exp(-\Delta). \tag{11}$$

where

$$\Delta = (\beta_m - \beta_n)\left(U(q^{[j]}) - U(q^{[i]})\right). \tag{12}$$

In the Monte Carlo method, the Hamiltonian $H(p, q)$ reduces to the potential energy $U(q)$ only. The previously detailed balance condition can be easily satisfied, for instance, by the usual Metropolis criterion,

$$\Pi(X \to X') \equiv \Pi\left(x_m^{[i]} x_n^{[j]}\right) = \begin{cases} 1, & \text{for } \Delta \leq 0, \\ \exp(-\Delta), & \text{for } \Delta > 0. \end{cases} \tag{13}$$

2.2. Coupling to MD

For MD simulations, both the potential energy and kinetic energy are present in the Hamiltonian, thus, Okamoto et al. *(21)* introduced a momenta rescaling scheme to simplify the detailed balance condition,

$$p_n^{[i]'} = \sqrt{\frac{T_n}{T_m}} p_m^{[i]},$$

$$p_m^{[j]'} = \sqrt{\frac{T_m}{T_n}} p_n^{[j]}. \tag{14}$$

With the **Eq. 14** velocity rescaling scheme, the detailed balance equation can be reduced into:

$$\Delta = (\beta_m - \beta_n)\left(H(x^{[j]}) - H(x^{[i]})\right) = (\beta_m - \beta_n)\left(U(q^{[j]}) - U(q^{[i]})\right). \tag{15}$$

Note that because of the velocity rescaling in **Eq. 14** the kinetic energy terms are canceled out in the previously detailed balance condition, and that the same

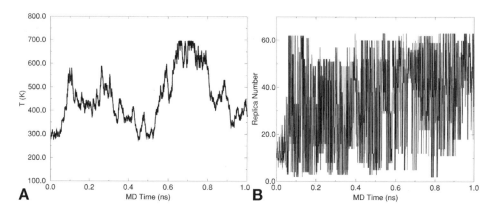

Fig. 1. The replica exchange trajectory from the β-hairpin simulation: **(A)** the temperature trajectory for one replica (started at 310 K); **(B)** the replica trajectory in temperature 310 K. Both show the replica exchange method is "surfing" the temperature space effectively. (Reproduced from **ref. 32**, with permission from Elsevier.)

criterion, **Eq. 15**, which was originally derived for Monte Carlo algorithm, is recovered (*see* **Note 3**). It should also be noted that this detailed balance criterion is exactly the same as in Jump Walking methods *(9,10)*.

Thus, the REMD can be summarized as the following two-step algorithm:

1. Each replica i ($i = 1,2,…,M$), which is in a canonical ensemble of the fixed temperature T_m ($m = 1,2,…,M$) is simulated *simultaneously* and *independently* for certain MD steps.
2. Pick some pairs of replicas, for example $xm[i]$ and $xn[i]$, and exchange the replicas with the probability $T(xm[i] \mid xn[j])$ as defined in **Eq. 13**, and then go back to **step 1**.

In **step 2**, only the replicas in neighboring temperatures are attempted for exchanges because the acceptance ratio of the exchange decreases exponentially with the difference of the two temperatures. Note that whenever a replica exchange is accepted in **step 2**, the permutation functions in **Eq. 3** must be updated. Thus, in REMD method, a random walk in the "temperature" space is realized for each replica, which in turn induces a random walk in potential energy space. This alleviates the problem of being trapped in states of energy local minimum.

2.3. Optimal Temperature Sequences in REMD

The optimal temperature distributions in the replica exchange method can be obtained by running a few trial replicas with short MD simulations. The temperature gap can thus be determined by monitoring the acceptance ratio desired between neighboring temperatures. The rest of the temperature list can

A   **B**

Helix   Hairpin

Fig. 2. The ribbon view of the α-helix (built by program IMPACT37) and β-hairpin (from PDB 2gb1.pdb, residue 41–56). The graphs are generated by molecular visualization program RasMol *(38)*.

usually be interpolated, because the optimal temperature distribution should be roughly exponential assuming the heat capacity is relatively a constant *(23)*(*see* **Note 4**). Using the β-hairpin as an example, the optimal temperature sequence is found to be 270, 274, 278, 282, 287, 291, 295, 300, 305, 310, 314, 318, 323, 328, 333, 338, 343, 348, 354, 359, 365, 370, 376, 381, 387, 393, 399, 405, 411, 417, 424, 430, 437, 443, 450, 457, 464, 471, 478, 485, 492, 500, 507, 515, 523, 531, 539, 547, 555, 563, 572, 581, 589, 598, 607, 617, 626, 635, 645, 655, 665, 675, 685, and 695 K, with an acceptance ratio of about 30%. The temperature gaps between these replicas range from 4 to 10 K.

Figure 1A shows the temperature trajectory for one replica started at 310 K for the β-hairpin. It is clear that this replica walks freely in the temperature space. A similar graph, **Fig. 1B**, monitors the replica trajectory at one particular temperature 310 K. It shows that various replicas visit this temperature randomly. These two plots basically show the time trajectory of the permutation function and its inverse in **Eq. 3** as we discussed in the REMD methodology section. The results indicate that our temperature series is reasonably optimized for this system with sufficiently high acceptance ratios for replica exchanges (*see* **Note 5**).

3. Case Study

Two example protein systems, one α-helix and one β-hairpin, as shown in **Fig. 2**, are used as case studies to illustrate the power of REMD. The α-helix system is a 21-residue peptide with a large propensity of forming α-helical structures in water at room temperature. It contains three Arg residues in its Ace-A$_5$(AAARA)$_3$A-Nme sequence (called Fs peptide, where Fs stands for folded short, and Ace and Nme are capping groups), and has been widely described in the experimental literature *(33–36)*. The Fs peptide is then solvated in a 43.7 × 43.7 × 43.7 Å^3 water box containing 2660 TIP3P water molecules

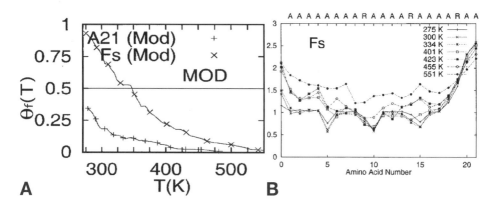

Fig. 3. (A) Comparison of the helical content of A21 and Fs peptides as a function of temperature obtained from a modified AMBER94 force field. (B) Water coordination to the backbone carbonyl oxygen atoms along the peptide sequence. The coordination number is about 1, except when the carbonyl group is four amino acids before an Arg side chain. The Arg side chain shields the carbonyl oxygen atom from exposure to water. (Reproduced with permission from **ref. 22**.)

(22). The solvated systems are subjected to 500 steps of steepest descent energy minimization and a 100-ps MD simulation at constant pressure and temperature (1 atm, 300 K). Up to 48 replicas are simulated with temperatures ranging from 275 to 551 K and an acceptance ratio of $\approx 20\%$. The initial configurations for REMD are selected from a 1-ns simulation at 700 K starting from an α-helical configuration and random velocities. The helical content of the starting configurations ranges from 15 to 73%. A modified AMBER94 force field is used with a cutoff of 8.0 Å for nonbonded interactions. Nonbonded pair lists were updated every 10 integration steps. The integration step in all simulations is 0.002 ps. The system is coupled to an external heat bath with relaxation time of 0.1 ps. All bonds involving hydrogen atoms are constrained by using SHAKE with a tolerance of 0.0005 Å. Exchanges are attempted every 125 integration steps (0.25 ps) and the total simulation time is 8 ns/replica.

The β-hairpin system, on the other hand, is taken from the C-terminus (residues 41–56) of protein G (PDB entry 2gb1). It has 16 residues with a residue sequence of Ace-GEWTYDDATKTFTVTE-Nme. The initial hairpin structure is solvated in explicit SPC water, with 1361 water molecules and also 3 Na$^+$ ions for neutralizing the molecular system, which results in a total of 4342 atoms in each replica. A total of 64 replicas are simulated for the explicit solvent systems with temperatures spanning from 270 to 695 K. The OPLSAA force field *(39)* is used with the periodic boundary condition. The particle–particle particle mesh Ewald (P3ME) method *(40)* is used for the long-range electrostatic interactions. A time step of 4.0 fs (outer timestep) is used for all temperatures with the P3ME/RESPA4143 algorithm. A standard equilibration protocol is used for each replica. It starts with a conjugate gradient minimization. Then a 100-ps MD equilibration is followed at each temperature. The final configurations at each temperature are then used as the starting points for the replica simulations and data are collected after another 100-ps REMD simulations. Each replica is run for 3.0 ns for data collection, with replica exchanges attempted every 0.4 ps *(23)*.

3.1. Folding of an α-Helix

Despite the large number of experimental studies conducted for α-helical peptides, there is still much debate concerning the propensity of Ala residues to stabilize α-helices. Garcia and coworkers have applied the REMD method to the 21-residue Fs peptide folding. **Figure 3A** shows a comparison of the helix content profiles as a function of temperature for the Fs and A21 peptides. A pure Ala 21-residue peptide (A21) has also been simulated for comparison. A residue is classified whether as helical or not based on the backbone (Φ, Ψ) angles, helical if the $\Phi = -60 \pm 30°$, and $\Psi = -47 \pm 30°$, and nonhelical otherwise. The simulations obtain a 90% helical content at 275 K for the Fs peptide, with a folding transition temperature $T_{1/2}$ of 345 K. The results seem to agree well with experimental data for Fs *(33–36)*. As for comparison, only 34% helical content is obtained for A21 at 275 K. The thermodynamic properties, such as enthalpy (ΔH) and entropy (ΔS) can also be determined from these temperature-dependent data. This is another benefit of the REMD method, because it not only helps the lower temperature replica to cross the free energy barrier, but also obtains the equilibrium properties at each temperature. Assuming a linear temperature dependence on the specific heat, and taking the unfolded state as reference, the authors obtained $\Delta H(273K) = -4.6$ kJ/mol, $\Delta S(273K) = -15$ J/K-mol, and $\Delta Cp(273K) = -60$ J/molK *(22)*.

Simulations reveal higher stability of the Fs peptide over A21. For example, at 350 K Fs peptide is 5.8 kJ/mol more stable than the A21 peptide *(22)*. This difference in stability has to come from the Arg side chains. To test the hypothesis that chain desolvation might be responsible for this stabilization *(44)* and to provide a molecular description of the shielding, the authors studied the coordination of water to the peptide backbone. **Figure 3B** shows the water coordination number for the Fs peptide backbone carbonyl oxygen. Carbonyl oxygen atoms involved in backbone hydrogen bonding, on average, have one coordinated water molecule. End carbonyl oxygen atoms not participating in hydrogen bonds have two coordinated water molecules. At low temperature, the peptide is in the α-helical conformation and has an average coordination number about one. At high temperature, the peptide is in the coil conformation and shows a coordination greater than 1.5 on average. The A21 peptide shows exactly these expected results (data not shown, *see* **ref. 22** for details), i.e., at low temperatures the water coordination number is about 1.0 for backbone carbonyl oxygen atoms. However, the Fs peptide, on the other hand, shows a coordination number of 0.5 at three positions along the sequence at low temperatures, as shown in **Fig. 3B**. The atoms with low coordination correspond to carbonyl oxygens four residues before Arg side chains, which are shielded from water by the Arg side chain. There are two separate configurations, which yield coordination numbers of 1 and 0, respectively. The shielded carbonyl oxygen atoms have zero coordinated water molecules. On average, they are approximately equally populated, yielding an average coordination number of 0.5.

This analysis indicates that the additional stabilization observed for the Fs peptide relative to the A21 peptide is produced by the partial shielding of the backbone hydrogen bonds from water, provided by the Arg side chains. The enhanced stability of shielded hydrogen bonds can be explained in terms of the competition for backbone hydrogen bonds between water molecules and backbone donors and acceptors. Thermal fluctuations can cause local opening and closing of backbone CO...NH hydrogen bonds. When the local environment is shielded from access to water, the hydrogen bond-breaking event is energetically unfavorable *(45)*. The destabilization of the opened CO...NH hydrogen bonds by side chain shielding results in the stabilization of the shielded hydrogen bond conformation, which contributes to the overall stability of helical conformations. The Arg side chain partially shields the carbonyl oxygen of the fourth amino acid upstream from the Arg.

3.2. Folding of a β-Hairpin

The second example, the C-terminus β-hairpin of protein G, has recently received much attention from both experimental and theoretical fronts *(31,46–54)*. Its fast folding speed (folds in 6 ms) and reasonable stability in aqueous solution make it a system of choice for studying the β-hairpin folding in isolation. Again, we use the highly parallel REMD to study the free energy landscape and folding mechanism of this β-hairpin.

The free energy or potential of mean force (PMF) is often calculated by the histogramming analysis *(51,55)*.

$$P(X) = \frac{1}{Z} \exp\left(-\beta W(X)\right), \tag{16}$$

and

$$W(X_2) - W(X_1) = -kT \log\left(\frac{P(X_2)}{P(X_1)}\right), \tag{17}$$

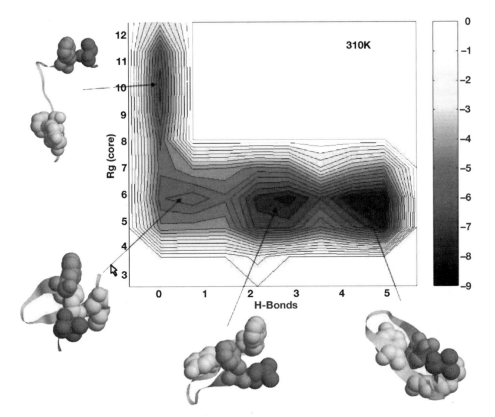

Fig. 4. Folding free energy contour map of the β-hairpin vs the two reaction coordinates, the number of β-sheet H-bonds $N^β_{HB}$ and the radius gyration of the hydrophobic core residues Rg^{core}. The representative structures are shown for each energy state. The hydrophobic core residues Trp43, Tyr45, Phe52, and Val54 are shown in space-fill mode, while all other residues are shown as ribbons. The free energy is in units of kT, and contours are spaced at intervals of 0.5 kT. (Reproduced from **ref. 32**, with permission from Elsevier.)

where $P(X)$ is the normalized probability obtained from a histogram analysis as a function of X. X is any set of reaction coordinates or any parameters describing the conformational space. Thus, $W(X_2) - W(X_1)$ gives the relative free energy. Here we use the number of strand hydrogen bonds ($N^β_{HB}$) and the radius of gyration of the hydrophobic core (Rg^{core}) as the reaction coordinates for the free energy contour map. $N^β_{HB}$ is denied as the number of native β-strand backbone hydrogen bonds excluding the two at the turn of the hairpin (five out of total seven such hydrogen bonds) *(23)*. Rg^{core} is the radius of gyration of the side chain atoms on the four hydrophobic residues, Trp43, Tyr45, Phe52, and Val54.

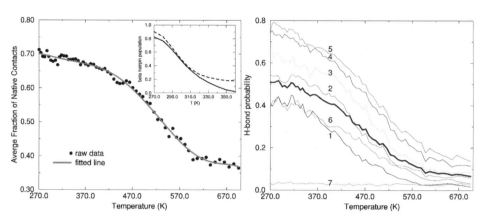

Fig. 5. **(A)** The average fraction of native contacts (population of β-hairpin) as a function of temperature. The population of β sheet decreases with temperature monotonically, but decays too slowly with temperature. The experimental results (solid line) *(31)* and simulation results from Thirumalai et al. (dash line) *(56)* are shown in the inset. **(B)** Temperature dependence of the probability of forming individual native hydrogen bonds. The thick solid line shows the average probability overall all native hydrogen bonds. (Reproduced with permission from **ref. 23**.)

Figure 4 shows the free energy contour map (in units of kT) and representative structures at each local free energy basin. The free energy landscape reveals a few interesting features: (1) the overall free energy contour map shows an "L" shape, indicating that the folding mechanism is likely driven by the hydrophobic core collapse. If it were driven by a hydrogen bond zipping mechanism as proposed from experiment *(31,46)*, the shape would have been a more "diagonal" one in the two-dimensional (2D) graph. (2) There are four states at biological temperature: the native folded state (F), the unfolded state (U) and two intermediates, a "molten globule" state, which is similar to Pande and coworkers' state H 48,49 and a partially folded state (P). (3) The intermediate state H shows a compact hydrophobic core but with no β-strand hydrogen bonds. These results in general are consistent with what has been found by others *(31,48,50,51,56)*. However, there are also some significant differences. One important difference is in the folding mechanism. Eaton et al. *(31,46)* proposed a hydrogen bond zipping mechanism in which folding initiates at the turn and propagates toward the tails, so that the hydrophobic clusters, from which most of the stabilization derives, form relatively late during the reaction. Our analysis shows no evidence of this hydrogen bond zipping model. Pande et al. *(48)* and Garcia et al. *(51)* found similar results using the CHARMM and AMBER force fields, namely, the β-hairpin system first folds into a compact H state before it folds into the native structure.

The fluorescence quantum yield experiment determines the temperature dependence of the β-sheet population for this hairpin *(31)*. It is of interest to see if all-atom force field simulations can reproduce this temperature dependence. Thirumalai et al. *(56)* have used the average fraction of native contacts to estimate the β-hairpin population, and here we follow the same approach. **Figure 5A** shows the average fraction of native contacts vs temperature. For comparison, we also plotted the experimental data and the simulation results from Thirumalai et al. Thirumalai et al.'s simulation gives excellent agreement with experiment even though a simple off-lattice model was used. The experimental data show that the β-hairpin population decreases monotonically with temperature, with about 80% population at 270 K, 40% at 310 K, and about 0% greater than 360 K *(31)*. Our simulation also shows a monoatomic decrease with temperature, but the population decays much more slowly than the experiment. The β-hairpin population is about 71% at 270 K and 66% at 310 K, which is in reasonable agreement with the experiment; but the populations at higher temperatures are too high as compared to the experiment. Interestingly, Karplus et al. found a β-hairpin population of about 75% at 300 K, 68% at 335 K, and 55% at 395 K using CHARMM *(57)*, which is similar to our OPLSAA results. Thus, it seems that both force fields show slower decay in the β-hairpin population at higher temperatures than 310 K, even though the populations near biological temperature are reasonable. Given that most of the modern force fields are parameterized at room temperature, this should not be too surprising. The water models might have also contributed to the errors in our simulations *(58)*. The constant volume simulation used is also unrealistic at high temperatures. The high temperature simulations at constant volume imply high pressures, which could stabilize the folded β-hairpin if its volume change on unfolding is positive.

Another interesting point is to determine which of the native hydrogen bonds play a significant role in the folding process. Thus, we calculated the probability of each individual hydrogen bond over a wide range of temperatures. All seven native backbone hydrogen bonds are included, and they are numbered from the tail to the turn (same as in Karplus et al. *[50]* and Garcia et al. *[51]*), i.e., the one closest to the tail is numbered as 1, and the one closest to the turn is numbered as 7. **Figure 5B** shows the probability of each hydrogen bond and their average as a function of temperature from 270 to 690 K. The average hydrogen bond probability decreases steadily with temperature, indicating a steady decrease of β-sheet content, as previously shown. The gradual decrease of the hydrogen bond probability with temperature also indicates that there is no cooperative transition in the hydrogen bonds with temperature. This also suggests that the hairpin formation is a broad two-state transition *(31,56)*. Hydrogen bonds 4 and 5 (which are closer to the turn) are the most stable ones, whereas 1 and 2 are much easier to break when water molecules attack the hairpin. Overall, the variations of hydrogen bond probabilities with temperature agree very well with Karplus et al.'s results from the CHARMM force field *(50)*, and also agree reasonably well with Garcia et al.'s results from the AMBER force field except at the lower temperatures (below the biological temperature 310 K) where Garcia et al. found α-helical structures and thus smaller β-sheet hydrogen bond probabilities *(51)*. The average probability of the β sheet hydrogen bonds in our model is 49% at 282 K, which is very close to the 54% found by Karplus et al. This is in reasonable agreement with the estimate of 42% from NMR experiment based on chemical shift data, and 40% from Garcia et al.'s simulation using the AMBER force field. Moreover, both our results and Karplus et al.'s results show that hydrogen bonds 7 and 6 have lower probabilities compared with other hydrogen bonds, because these hydrogen bonds are near the turn and it is difficult for them to achieve an ideal N-H...O angle owing to geometrical constraints. This differs from Thirumalai et al.'s results where hydrogen bonds 7 and 6 have the highest probabilities of being formed, which as a result favors the hydrogen bond zipping mechanism *(56)*.

Other useful properties, such as the α-helical content, NOE pair distances, and folding transition temperatures, can also be obtained from the REMD simulations. Furthermore, large-scale REMD simulations can be used to provide necessary data for the force field and solvation model assessment. We and other researchers have compared various protein force fields and their combination with solvation models in terms of the folding free energy landscape, lowest free energy structure, α-helical content, and temperature dependence *(59–61)*. We have also compared the explicit solvent model vs the implicit solvent models, including both the Generalized Born (GB) and Poisson–Boltzmann (PB) models *(59,62)*. These large-scale simulations have provided extensive data and insights for the possible fix of the problems and further improvement of solvation models and protein force fields. The future work will include applications to the folding of larger and more complex protein systems, such as membrane proteins and multidomain proteins, and also to other related fields such as the protein–protein interactions and other derived replica exchange methods (*see* **Note 6**).

4. Notes

1. Either a temperature swap or a configuration (coordinates and velocities) swap can be used in REMD for replica exchange (*see* **Eq. 9**). To reduce the communication cost in the parallel computing, the temperature swap is recommended. With the temperature swap, a post-analysis is required to obtain the ensemble averages at each individual temperature following the one-to-one temperature replica trace shown in **Eq. 3**. Of course, even with the configuration swap, the communication cost is usually minimal compared to the computational cost, because the swaps happen every hundreds or thousands of MD steps.

2. Another way to utilize molecular dynamics in REM is through HMC *(64)*. HMC also uses molecular dynamics as the underlying sampling engine. In this method, one starts with a configuration and samples the momenta of the particles from a Maxwell distribution. Molecular dynamics is then used to move the whole system for a certain number of steps, and finally one accepts or rejects the move by a Metropolis criterion based on exp(-H/kT) where H is the Hamiltonian of the whole system. Because the final configuration will be accepted or rejected based on a Metropolis criterion, a larger time step is often used in HMC. In general, HMC is more efficient for smaller systems, such as protein in a continuum solvent *(59)*, whereas the velocity rescaling approach is more appropriate for the explicit solvent simulations. Either way, one advantage of REMD is that the new advances in MD algorithms can be readily incorporated into replica exchange scheme.

3. The current description of REMD is based on the canonical ensemble, or constant volume and constant temperature (NVT) simulations. For grand canonical ensembles, or constant pressure and constant temperature (NPT) simulations, the similar approach can be used, except that the Hamiltonian of the system in **Eq. 15** becomes $H = KE + U(q) + PV$, where P and V are the pressure and the volume of the system, respectively. The PV term will be included in the Metropolis criterion **Eq. 15** as well.

4. For the REMD method to work efficiently, the acceptance ratio between the neighboring replicas should be roughly a constant. This can be easily achieved when the heat capacity of the system does not change much with temperature, because the optimal temperature sequence will be distributed exponentially *(60,63)*. However, when the heat capacity changes significantly with temperature or when the system is at or near the folding transition, a more careful procedure and longer trial runs need to be used to obtain the optimal temperature sequence.

5. The major advantage of REMD over other generalized ensemble methods, such as multicanonical algorithm *(12)* and simulated tempering *(15,16)*, lies in the fact that the weight factor is *a priori* known (*see* **Eq. 5**), whereas in the latter algorithms the determination of the weight factors can be nontrivial for complex systems and very time consuming.

6. Other related REMs have also been proposed, such as the replicas in the potential energy space for free energy perturbation *(65)*, replicas based on the Tsallis statistics (q-jump) *(66)*, and hydrophobic-aided replica exchange *(67)*. Because of the page limitation, we are not going to describe these implementations in detail, and interested readers can refer to the literatures for details.

Acknowledgments

The author would like to thank Angel Garcia for sending graphs for the α-helix simulations, and Yawen Bai for many helpful guidelines. Also, the author would like to thank Bruce Berne, Jed Pitera, William Swope, Vijay Pande, and Yuko Okmotto for numerous discussions and comments.

References

1. Brooks, C. L., Onuchic, J. N., and Wales, D. J. (2001) Taking a walk on a landscape. *Science* **293**, 612–613.
2. Dobson, C. M., Sali, A., and Karplus, M. (1998) Protein folding: a perspective from theory and experiment. *Angrew Chem. Int. Edit. Engl.* **37**, 868–893.
3. Brooks, C. L., Gruebele, M., Onuchic, J. N., and Wolynes, P. G. (1998) Chemical physics of protein folding. *Proc. Natl. Acad. Sci. USA* **95**, 11,037–11,038.
4. Zhou, Y. and Karplus, M. (1999) Interpreting the folding kinetics of helical proteins. *Nature* **401**, 400–403.
5. Zhou, R., Huang, X., Margulius, C. J., and Berne, B. J. (2004) Hydrophobic collapse in multidomain protein folding. *Science* **305**, 1605–1609.
6. Daggett, V. and Levitt, M. (1993) Realistic simulations of native-protein dynamics in solution and beyond. *Annu. Rev. Biophys. Biomol. Struct.* **22**, 353–380.
7. Duan, Y. and Kollman, P. A. (1998) Pathways to a protein folding intermediate observed in a 1-microsecond simulation in aqueous solution. *Science* **282**, 740–744.
8. Frantz, D. D., Freeman, D. L., and Doll, J. D. (1990) Reducing quasi-ergodic behavior in Monte Carlo simulations by J-Walking. Applications to atomic clusters. *J. Chem. Phys.* **93**, 2769–2784.
9. Freeman, D. L., Frantz, D. D., and Doll, J. D. (1992) Extending j walking to quantum systems: applications to atomic clusters. *J. Chem. Phys.* **97**, 5713.
10. Zhou, R. and Berne, B. J. (1997) Smart walking: a new method for boltzmann sampling of protein confromations. *J. Chem. Phys.* **107**, 9185–9196.
11. Andricioaei, I. and Straub, J. E. (1997) On Monte Carlo and molecular dynamics methods inspired by Tsallis statistics: methodology, optimization, and application to atomic clusters. *J. Chem. Phys.* **107**, 9117–9124.
12. Berg, B. A. and Neuhaus, T. (1991) Multicanonical algorithms for first order phase transitions. *Phys. Lett. B.* **267**, 249–253.
13. Hukushima, K. and Nemoto, K. (1996) Exchange monte carlo method and application to spin glass simulations. *J. Phys. Soc. Japan* **65**, 1604–1608.
14. Marinari, E., Parisi, G., and Ruiz-Lorenzo, J. J. (1998) Numerical simulations of spin glass systems. In: *Spin Glass and Random Fields*, (Young, A. P., ed.), World Scientific, Singapore, pp. 59.
15. Lyubarsev, A. P., Martsinovski, A. A., Shevkunov, S. V., and Vorontsov-Velyaminov, P. N. (1992) New approach to monte-carlo calculation of the free-energy-method of expanded ensembles. *J. Chem. Phys.* **96**, 1776–1793.
16. Marinari, E. and Parisi, G. (1992) Simulated tempering: a new Monte Carlo scheme. *Europhys. Lett.* **19**, 451–458.
17. Stolovitzky, G. and Berne, B. J. (2000) Catalytic tempering: a method for sampling rough energy landscapes by monte carlo *Proc. Natl. Acad. Sci. USA* **97**, 11,164–11,169.
18. Piela, L., Kostrowicki, J., and Scheraga, H. A. (1989) The multipleninima problem in the conformational analysis of molecules. Deformation of the protein energy hypersurface by the diffusion equation method. *J. Phys. Chem.* **93**, 3339–3346.

19. Kostrowicki J. A. and Scheraga, H. A. (1992) Application of the diffusion equiation method for global optimization to oligopeptides. *J. Phys. Chem.* **96,** 7442–7449.

20. Berne, B. J. and Straub, J. E. (1997) Novel methods of sampling phase space in the simulation of biological systems. *Curr. Opin. Struct. Biol.* **7,** 181–189.

21. Sugita, Y. and Okamoto, Y. (2000) Replica-exchange multicanonical algorithm and multi-canonical replica-exchange method for simulating systems with rough energy landscape. *Chem. Phys. Lett.* **329,** 261–270.

22. Garcia, A. E. and Sanbonmatsu, K. Y. (2002) Alpha-helical stabilization by side chain shielding of backbone hydrogen bonds. *Proc. Nat. Acad. Sci. USA* **99,** 2782–2787.

23. Zhou, R., Berne, B. J., and Germain, R. (2001) The free energy landscape for beta-hairpin folding in explicit water. *Proc. Natl. Acad. Sci. USA* **98,** 14,931–14,936.

24. Rhee, Y. M. and Pande, V. S. (2003) Multiplexed-replica exchange molecular dynamics method for protein folding simulation. *Biophys. J.* **84,** 775–786.

25. Ohkubo, Y. Z. and Brooks, C. L. (2003) Exploring flory's isolated-pair hypothesis: Statistical mechanics of helixcoil transitions in polyalanine and the c-peptide from rnase a. *Proc. Natl. Acad. Sci. USA* **100,** 13,916–13,921.

26. Zhou, R. (2003) Trp-cage: folding free energy landscape in explicit water. *Proc. Natl. Acad. Sci. USA* **100,** 13,280–13,285.

27. Nymeyer, H. and Garcia, A. E. (2003) Interfacial folding of a membrane peptide: replica exchange simulations of walp in a dppc bilayer. *Biophys. J.* **84,** 381A.

28. Zhou, R. (2003) Trp-cage: folding free energy landscape in explicit water. *Proc. Natl. Acad. Sci. USA* **100,** 13,280–13,285.

29. Im, W., Feig, M., and Brooks, C. L. (2003) An implicit membrane generalized born theory for the study of structure, stability, and interactions of membrane proteins. *Biophys. J.* **85,** 2900–2918.

30. Kokubo, H. and Okamoto, Y. (2004) Self-assembly of transmembrane helices of bacteriorhodopsin by a replica-exchange monte carlo simulation. *Chem. Phys. Lett.* **392,** 168–175.

31. Munoz, V., Thompson, P. A., Hofrichter, J., and Eaton, W. A. (1997) Folding dynamics and mechanism of β-hairpin formation. *Nature* **390,** 196–199.

32. Zhou, R. (2004) Sampling protein folding free energy landscape: coupling replica exchange method with p3me/respa algorithm. *J. Mol. Graph Model.* **22,** 451–463.

33. Williams, S., Causgrove, T. P., Gilmanshin, R., et al. (1996) Fast events in protein folding: Helix melting and formation in a small peptide. *Biochemistry* **35,** 691–697.

34. Lockhart, D. J. and Kim, P. S. (1993) Electrostatic screening of charge and dipole interactions with the helix backbone. *Science* **260,** 198–202.

35. Thompson, P. A., Eaton, W. A., and Hofrichter, J. (1997) Laser temperature jump study of the helix<==>coil kinetics of an alanine peptide interpreted with a 'kinetic zipper' model. *Biochemistry* **36,** 9200–9210.

36. Lednev, I. K., Karnoup, A. S., Sparrow, M. C., and Asher, S. A. (2001) Transient UV Raman spectroscopy finds no crossing barrier between the peptide alpha-helix and fully random coil conformation. *J. Am. Chem. Soc.* **123,** 2388–2392.

37. Kitchen, D. B., Hirata, F., Westbrook, J. D., Levy, R. M., Kofke, D., and Yarmush, M. (1990) Conserving energy during molecular dynamics simulations of water, proteins and proteins in water. *J. Comp. Chem.* **11,** 1169–1180.

38. Sayle, R. A. and Milner-White, E. J. (1995) Rasmol: biomolecular graphics for all. *Trends Biochem. Sci.* **20,** 374–376.

39. Jorgensen, W. L., Maxwell, D., and Tirado-Rives, J. (1996) Development and testing of the opls all-atom force field on conformational energetics and properties of organic liquids. *J. Am. Chem. Soc.* **118,** 11,225–11,236.

40. Hockney, R. W. and Eastwood, J. W. (1989) *Computer Simulation Using Particles.* Adam Hilger, Bristol-New York, NY.

41. Tuckerman, M., Berne, B. J., and Martyna, G. J. (1992) Reversible multiple time scale molecular dynamics. *J. Chem. Phys.* **97,** 1990–2001.

42. Zhou, R. and Berne, B. J. (1995) A new molecular dynamics method combining the reference system propagator algorithm with a fast multipole method for simulating proteins and other complex systems. *J. Chem. Phys.* **103,** 9444–9459.

43. Zhou, R., Harder, E., Xu, H., and Berne, B. J. (2001) Efficient multiple time step method for use with ewald and particle mesh ewald for large biomolecular systems. *J. Chem. Phys.* **115,** 2348–2358.

44. Vila, J. A., Ripoll, D. R., and Scheraga, H. A. (2000) Physical reasons for the unusual alpha-helix stabilization afforded by charged or neutral polar residues in alanine-rich peptides. *Proc. Natl. Acad. Sci. USA* **97,** 13,075–13,079.

45. Sundaralingam, M. and Sekharudu, Y. (1989) Water-inserted alpha-helical segments implicate reverse turns as folding intermediates. *Science* **244,** 1333–1337.

46. Munoz, V., Henry, E. R., Hofrichter, J., and Eaton, W. A. (1998) A statistical mechanical model for β-hairpin kinetics. *Proc. Natl. Acad. Sci. USA* **95,** 5872–5879.

47. Blanco, F. J., Rivas, G., and Serrano, L. (1994) A short linear peptide that folds in a native stable β-hairpin in aqueous solution. *Nature Struc. Bio.* **1,** 584–590.

48. Pande, V. S. and Rokhsar, D. S. (1999) Molecular dynamics simulations of unfolding and refolding of a β-hairpin fragment of protein g. *Proc. Natl. Acad. Sci. USA* **96,** 9062–9067.

49. Zagrovic, B., Sorin, E. J., and Pande, V. S. (2001) β-hairpin folding simulation in atomistic detail using an implicit solvent model. *J. Mol. Biol.* **313,** 151–169.

50. Dinner, A. R., Lazaridis, T., and Karplus, M. (1999) Understanding β-hairpin formation. *Proc. Natl. Acad. Sci. USA* **96,** 9068–9073.

51. Garcia, A. E. and Sanbonmatsu, K. Y. (2001) Exploring the energy landscape of a β hairpin in explicit solvent. *Proteins* **42,** 345–354.

52. Roccatano, D., Amadei, A., Nola, A. D., and Berendsen, H. J. (1999) A molecular dynamics study of the 41-56 β-hairpin from b1 domain of protein g. *Protein Sci.* **10,** 2130–2143.

53. Kolinski, A., Ilkowski, B., and Skolnick, J. (1999) Dynamics and thermodynamics of β-hairpin assembly: insights from various simulation techniques. *Biophys. J.* **77,** 2942–2952.

54. Ma, B. and Nussinov, R. (2000) Molecular dynamics simulations of a β-hairpin fragment of protein G: balance between side-chain and backbone forces. *J. Mol. Bio.* **296,** 1091–1104.

55. Ferrenberg, A. M. and Swendsen, R. H. (1989) Optimized Monte Carlo data analysis. *Phys. Rev. Lett.* **63,** 1195–1198.

56. Klimov, D. K. and Thirumalai, D. (2000) Mechanisms and kinetics of beta-hairpin formation. *Proc. Natl. Acad. Sci. USA* **97,** 2544–2549.

57. Dinner, A. R. (1999) Monte carlo simulations of protein folding. PhD Thesis, Harvard University, Cambridge, MA.

58. Walser, P., Mark, A. E., and van Gunsteren, W. F. (2000) On the temperature and pressure dependence of a range of properties of a type of water model commonly used in high-temperature protein unfolding simulations. *Biophys. J.* **78,** 2752–2760.

59. Zhou, R. and Berne, B. J. (2002) Can a continuum solvent model reproduce the free energy landscape of a β-hairpin folding in water? *Proc. Natl. Acad. Sci. USA* **99,** 12,777–12,782.

60. Zhou, R. (2003) Free energy landscape of protein folding in water: explicit vs. implicit solvent. *Proteins* **53,** 148–161.

61. Yoda, T., Sugita, Y., and Okamoto, Y. (2000) Comparisons of force fields for proteins by generalized-ensemble simulations. *J. Chem. Phys.* **113,** 6042–6051.

62. Zhou, R., Krilov, G., and Berne, B. J. (2004) Comment on "can a continuum solvent model reproduce the free energy landscape of a beta-hairpin folding in water?." *J. Phys. Chem. B.* **108,** 7528–7530.

63. Sugita, Y. and Okamoto, Y. (1999) Replica-exchange molecular dynamics method for protein folding. *Chem. Phys. Lett.* **314,** 141–151.

64. Duane, S., Kennedy, A. D., Pendleton, B. J., and Roweth, D. (1987) Hybrid Monte Carlo. *Phys. Lett. B* **195,** 216–222.

65. Sugita, Y., Kitao, A., and Okamoto, Y. (2000) Multidimensional replica-exchange method for free-energy calculations. *Chem. Phys. Lett.* **329,** 261–270.

66. Whitfield, T. W., Bu, L., and Straub, J. E. (2002) Generalized parallel sampling. *Physica A* **305,** 157–171.

67. Liu, P., Huang, X., Zhou, R., and Berne, B. J. (2006) Hydrophobic aided replica exchange method. *J. Chem. Phys.* **110,** in press.

13

Estimation of Folding Probabilities and Φ Values From Molecular Dynamics Simulations of Reversible Peptide Folding

Francesco Rao, Giovanni Settanni, and Amedeo Caflisch

Summary

Molecular dynamics simulations with an implicit model of the solvent have allowed to investigate the reversible folding of structured peptides.

For a 20-residue antiparallel β-sheet peptide, the simulation results have revealed multiple folding pathways. Moreover, the conformational heterogeneity of the denatured state has been shown to originate from high enthalpy, high entropy basins with fluctuating non-native secondary structure, as well as low enthalpy, low entropy traps. An efficient and simple approach to estimate folding probabilities from molecular dynamics simulations has allowed to isolate conformations in the transition state ensemble and to evaluate Φ values, i.e., the effects of mutations on the folding kinetics and thermodynamic stability. These molecular dynamics studies have provided evidence that, if interpreted by neglecting the non-native interactions, Φ values overestimate the amount of native-like structure in the transition state.

Key Words: Protein folding; energy landscape; transition state ensemble; denatured state ensemble; implicit solvent molecular dynamics.

1. Introduction

Energy landscape theory provides a framework for the description of the kinetics and thermodynamics of condensed phases. In the past years, it has been extensively applied to the analysis of protein folding *(1–5)*. Although proteins are essential macromolecules for life and are responsible for most cellular functions, the process by which proteins reach their functional structure are not fully understood *(6)*. Within the energy landscape framework, protein folding is envisioned to proceed along a moderately rough funnel-shaped effective energy surface *(2,7)*. The overall shape of the landscape arises from a strong energetic driving force to

From: *Methods in Molecular Biology, vol. 350: Protein Folding Protocols*
Edited by: Y. Bai and R. Nussinov © Humana Press Inc., Totowa, NJ

the native global minimum. This energetic bias is necessary to overcome the conformational search problem associated with finding the native state of the protein within a biologically reasonable time frame* *(2,8)*. The roughness of the surface is determined by local energy minima arising from the many competing interactions that are possible between the residues. Energetic traps are sequence-related and arise when non-native but stabilizing contacts form as the chain folds. The number and depth of such energetic traps influence both the thermodynamic and kinetic aspects of folding.

Experimental data *(9)* indicate that folding for many small proteins is a first-order transition in which the polypeptide chain passes from a free energy basin associated with low order and mainly stabilized by entropy to a free energy basin characterized by a highly ordered dominant conformation of the chain and mainly stabilized by favorable intraprotein interactions. The conformations populating the barrier dividing the two main free energy basins constitute the transition state ensemble (TSE). Understanding the characteristics of the TSE will allow the identification of the events that determine the folding rate and more in general the folding process itself. For this reason many studies have tried to characterize the TSE of proteins. Experimental data on the TSE of proteins have been mainly obtained by a widely diffused technique known as Φ-value analysis *(10)*. This technique consists of measuring the change of the height of the free energy barrier relative to the change in stability upon a single-point mutation. The denatured state is taken as reference. In this way it is possible to estimate the amount of native structure in the TSE around the mutated residue. This mainly energetic information however does not provide the atomic resolution that one would like to reach and the interpretation of the experiments is not always straightforward, as will be explained next.

Molecular dynamics (MD) is a very useful simulation approach to study the flexibility of proteins at atomic level of detail *(11,12)*. Since the first MD simulation of a protein in vacuo published in 1977 *(13)*, much progress has been made to increase the accuracy of the models and reliability of the simulations. Moreover, computer performances have evolved dramatically. However, even for a small protein it is not yet feasible to simulate reversible folding with a high-resolution approach, e.g., MD simulations with an all-atom transferable model. In this chapter we will show that despite their limitations, computer simulations are an important tool for the investigation of the energy landscapes governing protein folding.

The characterization of the TSE of protein folding has attracted the attention of many theoretical and computational studies *(14–19)*. By definition,

---

*In contrast with the astronomical amount of time needed by a random search in the configuration space of the protein (Levinthal's paradox).

TSE conformations have a 50% probability of reaching the folded state before unfolding (p_{fold}). Because p_{fold} is computationally very expensive, often the TSE of proteins has been identified on the basis of the projections of the phase space of the protein onto one or two order parameters, i.e., by selecting structures belonging to poorly populated regions of projected free energy landscapes in between the highly populated folded and unfolded state. In what follows we will show the possible problems related to this approach and will present a technique that has been developed to estimate the folding probability of the structures sampled along equilibrium folding-unfolding MD simulations. Such technique has been used to characterize the TSE of folding of the structured peptide Beta3s, a designed 20-residue, three-stranded antiparallel β-sheet *(20,21)*. We will also discuss how Φ values have been measured *in silico* for this peptide and how their structural interpretation matches the TSE obtained using p_{fold}.

2. Methods

2.1. Molecular Dynamics Simulations

All simulations and part of the analysis of the trajectories were performed with the program CHARMM *(11)*. Beta3s was modeled by explicitly considering all heavy atoms and the hydrogen atoms bound to nitrogen or oxygen atoms (PARAM19 force field *[11]*). A mean field approximation based on the solvent accessible surface was used to describe the main effects of the aqueous solvent on the solute *(22)*.

2.2. Clusterization

The 500,000 conformations saved along the 10-μs simulation time of Beta3s *(23)* were clustered by the leader algorithm *(24)*. Briefly, the first structure defines the first cluster and each subsequent structure is compared with the set of clusters found so far until the first similar structure is found. If the structural deviation (*see* below) from the first conformation of all of the known clusters exceeds a given threshold, a new cluster is defined. The leader algorithm is very fast even when analyzing large sets of structures like in the present work. The results presented here were obtained with a structural comparison based on the Distance Root Mean Square (DRMS) deviation considering all distances involving C_{α} and/or C_{β} atoms and a cutoff of 1.2 Å. This yielded 78,183 clusters for Beta3s. The DRMS and root mean square deviation of atomic coordinates (upon optimal superposition) have been shown to be highly correlated *(16)*. The DRMS cutoff of 1.2 Å was chosen on the basis of the distribution of the pairwise DRMS values in a subsample of the wild-type trajectories. The distribution shows two peaks that originate from intra- and intercluster distances. The cutoff is located at the minimum between the first and second DRMS peak.

The main findings of this chapter are valid also for clusterization based on secondary structure similarity *(19,25)*.

2.3. Definition of TSE

Each cluster i contains $n_f(i)$ snapshots committed to fold out of its total number of snapshots $N(i)$. A cluster j belonging to TSE by definition has an asymptotic $P_f = 0.5$, i.e., if we could extend our simulations so that $N(j) \to \infty$ then $n_f(j)/N(j) \to 0.5$. This means that, if the commitment of each snapshot is considered as an independent binary variable (i.e., 1 or 0 , for a snapshot committed to fold or not, respectively), then the number $n_f(j)$ of snapshots with commitment 1 in a cluster belonging to TSE will follow a binomial distribution with probability $p = 0.5$:

$$P_{N(j)}(n_f(j)) = \binom{N(j)}{n_f(j)} p^{n_f(j)}(1-p)^{(N(j)-n_f(j))} = \binom{N(j)}{n_f(j)} 0.5^{N(j)} \qquad (1)$$

Thus, it can be tested if cluster X belongs to TSE by checking that $n_f(X)$ is compatible with a binomial distribution with $p = 0.5$, i.e., $n_f(X)$ has to belong to a likelihood range of values centered around $N(X)/2$. This is done by verifying that the probability to have a hypothetical number n of fold-committed snapshots outside of the range from $n > m(X)$ to $N(X)-n_f(X)$ (that is twice the probability to have $n > m(X) = max(n_f(X), N(X) - n_f(X))$) is larger than a given likelihood confidence threshold λ (e.g., $\lambda = 0.2$ to allow for clusters with three snapshots to belong to the TSE if $n_f = 1$ or $n_f = 2$ because, in both cases, $2 \cdot 1/8 > 0.2$). In mathematical terms:

$$X \in \text{TSE} \Leftrightarrow \sum_{i=m(X)+1}^{N(X)} 2 \cdot P_{N(X)}(i) > \lambda \qquad (2)$$

In practice, the latter condition allows TSE clusters with few snapshots to have a larger spread of P_f^C (*see* below for P_f^C definition) around 0.5 than large TSE clusters, because in the approximation of the actual P_f (*see* **Subheading 4.2.**) by P_f^C the error is larger for smaller cluster size.

3. Projection of the Free Energy Landscape on Order Parameters

A common way to investigate and display the free energy landscape is to study it as a function of one or more *order parameters*, i.e., suitably chosen macroscopic quantities that distinguish the different states of the protein. For example, it is common in the study of protein folding to use the fraction of native contacts Q *(21,26)*. Q is a good *order parameter* in the sense that it distinguishes the unfolded from the folded state: unfolded conformations typically have small Q, while by definition Q is close to 1 in the native state. The free energy of a protein as a function of Q can be written as:

$$F(Q) = U(Q) - TS(Q) \tag{3}$$

where $F(Q)$, $U(Q)$ and $S(Q)$ are the average free energy, potential energy, and configurational entropy, respectively for the configurations with Q native contacts.

Free energy projections on order parameters have been used to analyze many aspects of protein folding. Stable *states* are associated with local free energy minima of the projected landscape. The depth of the minima is considered proportional to the stability of the states associated to them and the barriers between different minima indicate activation energies between states. In many cases, this approach reveals a surprisingly *simple* two-state picture for protein folding (**Fig. 1**, bottom).

Order parameters are also used as reaction coordinates to monitor the dynamics of the protein *(14)*. However, using free energy projections for the study of the kinetics of protein folding requires knowledge of a good reaction coordinate, which is not easily accessible and/or identifiable *(27)*. Given the complexity of protein folding and the large number of degrees of freedom involved, few simple reaction coordinates would be desirable for its description, even though they might miss essential aspects of the process *(19,26,28)*. Good reaction coordinates for studying the kinetics of protein folding should satisfy two assumptions:

1. The order parameter(s) should allow to distinguish the various states of the system.
2. Within a minimum of the projected free energy landscape, conformations should interconvert rapidly.

A first consequence of assumption 1 is that every value of the order parameter (or the combination of different order parameters) identifies only one state of the system. Assumption 2, stated in a different way, says that all the conformations in a state are kinetically homogeneous. In many cases at least one of these assumptions is not true. For example, if Q is used as order parameter, the conformations with half of the native contacts formed do not generally take similar times to reach the native state, as has been shown for the three-stranded antiparallel β-sheet peptide, Beta3s (**Fig. 1**; *[29]*). In fact, several order parameters are based on a comparison with a reference structure like the native state (i.e., rmsd, Q, and so on).

Structures that have a native-like values of the order parameters (i.e., high Q, small rmsd, and so on) satisfy a large set of tight constraints on the coordinates of their atoms (i.e., high Q means that a large number of distances between pairs of atoms has to be smaller that a certain tight threshold). As soon as the value of the order parameter becomes less native-like, the number of these constraints decreases or the threshold becomes loose (depending on the order parameter).

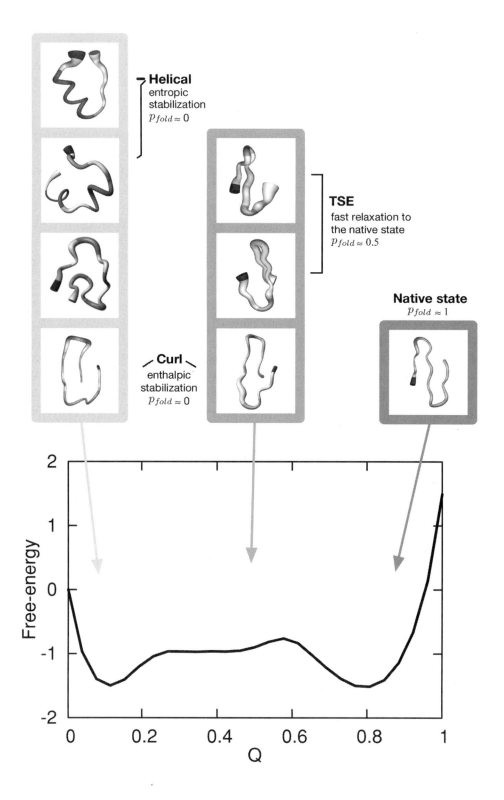

Helical
entropic
stabilization
$p_{fold} \approx 0$

TSE
fast relaxation to
the native state
$p_{fold} \approx 0.5$

Native state
$p_{fold} \approx 1$

Curl
enthalpic
stabilization
$p_{fold} \approx 0$

This means that a larger and more diverse region of the phase space of the protein projects into the same value of the order parameter. In other words, structures having the same non-native-like order parameter have non-homogeneous structural properties. In **Fig. 1**, some representative conformations of Beta3s with ≈10, ≈50, and ≈80% of the native contacts are shown, from left to right, respectively *(19)*. A hydrogen bond (HB) is defined as native if the distance between the hydrogen and oxygen atoms is lower than 2.5 Å for more than two-thirds of the conformations belonging to the most populated cluster *(21)*. A side chain contact (SC) is defined as native if the distance between the center of mass of the two residues averaged over the most populated cluster is smaller than 6.5 Å. Q identifies uniquely one state only when almost all the native contacts are formed, i.e., the native state. For $Q < 70$–80% many heterogenous conformations can have the same number of native contacts *(19)*.

Most of the time, these conformations are structurally and kinetically heterogeneous (e.g., the $Q \approx 0.5$ conformations with $p_{fold} \approx 0$ and $p_{fold} \approx 0.5$ in **Fig. 1**). In other words, although Q can discriminate between fully folded and fully denatured structures, it does not help in distinguishing structures with properties intermediate between the native and denatured state. Folding times t_{fold} for conformations with half or less of the native contacts formed (central column in **Fig. 1**) can differ as much as two orders of magnitude. Indeed, structures with Q as large as 0.7 may have $t_{fold} \approx 10^2$ ns and, vice versa, Q as low as 0.3 may correspond to structures with $t_{fold} \approx 10^0$ ns.

Of course it can be objected that, in order to optimally describe the thermodynamics and the kinetics of a peptide or a protein, suitable combinations of order parameters can always be found *(30)*. Even if this possibility exists, it is either very difficult to find and/or very specific for the system under study.

4. The Folding Probability

In the last section, it has been shown that the analysis of the kinetics of a peptide or a protein through near-equilibrium free energy projections can be misleading. Even the energetic barrier between the native and denatured state cannot be reliably estimated from such projections. However, projections are necessary to describe an otherwise very complex system like the one consisting of 10^2–10^5 atoms of a protein. To overcome this problem one has to find a

---

Fig. 1. (*Opposite page*)   Free energy projections on order parameters. In the case of Beta3s *(21)*, the fraction of native contacts does not necessarily identify structurally and kinetically homogeneous conformations. In the first, second, and third column, conformations with ≈10, ≈50, and ≈80% of native contacts Q are shown, respectively. The projected free energy shows no evidence of the structurally and kinetically heterogeneity of the denatured state of Beta3s (*see* **Subheading 3.**).

projection specifically suited for the wanted features to extract from the simulation, i.e., in the present case, the kinetics of protein folding. The folding probability p_{fold} of a protein conformation saved along a Monte Carlo or MD trajectory is the probability to fold before unfolding *(14)*. This order parameter defines the kinetics of protein folding because it allows the distinction of structures belonging to the native free energy basin ($p_{fold} = 1$), the unfolded free energy basin ($p_{fold} = 0$), and the free energy barrier ($p_{fold} \approx 0.5$). In principle, it also allows the detection of pathway intermediates in the form of large populations of structures with $0 << p_{fold} << 1$. In other words, it represents the kinetic distance of a structure from the folded state. As in the case of other order parameters, conformations with the same $p_{fold} << 1$ may be structurally different; however, they will have the same kinetic distance from the native state and, in particular, if $p_{fold} \approx 0.5$ they will be unequivocally members of the TSE.

The measure of p_{fold} consists of starting a large number of trajectories from putative TSE structures with varying initial distribution of velocities and counting the number of those that fold within a "commitment" time which has to be chosen much longer than the shortest time scales of conformational fluctuations and much shorter than the average folding time *(16)*. The concept of p_{fold} calculation originates from a method for determining transmission coefficients, starting from a known transition state *(31)* and the identification of simpler transition states in protein dynamics (e.g., tyrosine ring flips) *(32)*. The approach has been used to identify the otherwise very elusive folding TSE by atomistic Monte Carlo off-lattice simulations of small proteins with a Gō potential *(16,18)*, as well as implicit solvent MD *(15,19)* and Monte Carlo *(17)* simulations with a physico-chemical-based potential. The number of trial simulations needed for the reliable evaluation of p_{fold} makes the estimation of the folding probability computationally very expensive. For this reason, we have recently proposed a method to estimate folding probabilities for *all* structures visited in an equilibrium folding-unfolding trajectory without any additional simulation *(25)*. This method has been applied to the Beta3s peptide and to a large set of its mutants *(29)*, as will be shown in **Subheading 4.2**.

4.1. Folding Probability of a Single MD Snapshot

For the computation of p_{fold}, a criterion (λ) is needed to determine when the system reaches the folded state. Given a clusterization of the structures, a natural choice for λ is the visit of the most populated cluster, which for structured peptides and proteins is not degenerate (other criteria are also possible, e.g., fraction of native contacts Q larger than a given threshold). Given λ and a commitment time (τ_{commit}), the folding probability $p_{fold}(i)$ of an MD snapshot i is computed as *(14,16)*:

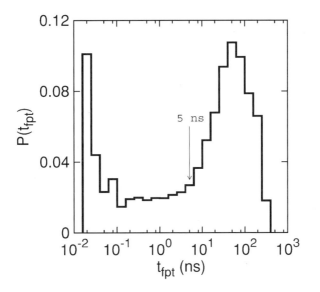

Fig. 2. Probability distribution for the first passage time (fpt) to the most populated cluster (*folded state*) of the DRMS 1.2 Å clusterization of Beta3s.

$$p_{fold}(i) = \frac{n_f(i)}{n_t(i)} \qquad (4)$$

where $n_f(i)$ and $n_t(i)$ are the number of trials started from snapshot i, which reach within a time τ_{commit} the folded state and the total number of trials, respectively.

Every simulation started from snapshot i can be considered as a Bernoulli trial of a random variable θ with value 1 (folding within τ_{commit}) or 0 (no folding within τ_{commit}). The variable θ has average and variance on the average of the form:

$$\langle \theta \rangle = p_{fold} = \frac{1}{n_t} \sum_{i=1}^{n_t} \theta_i$$

$$\sigma_{\langle\theta\rangle}^2 = \frac{1}{n_t} p_{fold}(1 - p_{fold}) \qquad (5)$$

where n_t is the total number of trials and the accuracy on the p_{fold} value increases with n_t.

In **Fig. 2** the distribution of the first passage time (fpt) to the folded state of Beta3s is shown. The double peak shape of the distribution provides evidence for the different time scales between intrabasin and interbasin transitions. A value of 5 ns is chosen for τ_{commit} because events with smaller time scales correspond to the diffusion within the native free energy basin, while events with larger time scales are transitions from other basins to the native one, i.e., folding/unfolding events *(23)*.

4.2. Folding Probability of a Cluster of Similar Conformations

Conformations that are structurally similar have been shown to have the same kinetic behavior *(25)*, hence they have similar values of p_{fold}. (Note that the opposite is not necessarily true as already mentioned and as more extensively explained in the next section for the TSE and the denatured state.) Snapshots saved along a trajectory are first grouped in structurally similar clusters. Then, the τ_{commit}-segment of MD trajectory following each snapshot is analyzed to check if the folding condition λ is met (i.e, the snapshot "folds"). For each cluster, the ratio between the snapshots, which lead to folding and the total number of snapshots in the cluster is defined as the cluster-p_{fold} (P_f^C; throughout the text uppercase P and lowercase p refer to folding probability for clusters and individual snapshots, respectively). This value is an approximation of the p_{fold} of any single structure in the cluster which is valid if the cluster consists of structurally similar conformations. In other words, the occurrence of the folding event for the snapshots of a given cluster can be considered as a Bernoulli trial of a random variable θ. The average of θ and variance on the average for the set of snapshots belonging to a given cluster α can be written as:

$$P_f^C[\alpha] = \langle\theta\rangle = \frac{1}{W}\sum_{i=1}^{W}\theta_i, \quad i \in \alpha$$

$$\sigma_{\langle\theta\rangle}^2 = \frac{1}{W}P_f^C(1 - P_f^C)$$

(6)

where W is the number of snapshots in cluster α. P_f^C is the average folding probability over a set of structurally homogeneous conformations. Using the clustering and the folding criterion λ introduced previously, values of P_f^C for the 78,183 clusters of Beta3s can be computed by **Eq. 6**, i.e., the number of conformations of the cluster that fold within 5 ns divided by the total number of conformations belonging to the cluster.

$$P_f[\alpha] = \frac{1}{W}\sum_{i=1}^{W}p_{fold}(i), \quad i \in \alpha$$

(7)

which is measured by starting several simulations from each snapshot i in the cluster α with W snapshots, is well approximated by P_f^C whose evaluation is straightforward.

To compare the values of P_f^C with those obtained from the standard approach *(14)*, folding probabilities P_f were computed for the structures of 37 clusters by starting several 5-ns MD runs from each structure and counting those that fold

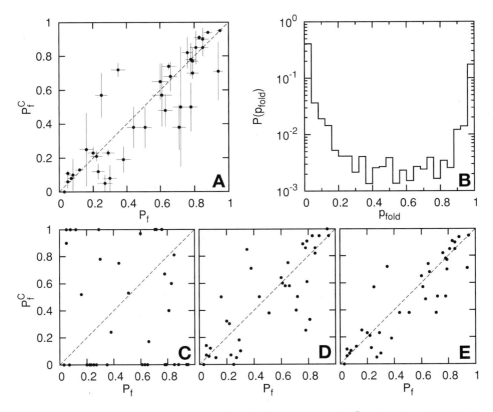

Fig. 3. Cluster folding probability P_f^C. (A) Scatter plot of P_f^C vs P_f. The DRMS 1.2 Å clusterization and the folding criterion λ (reaching the most populated cluster within τ_{commit} = 5 ns) were used. (B) Probability distribution of the p_{fold} value for the 500,000 snapshots saved along the 10-μs MD trajectory of Beta3s. The folding probability for snapshot i is computed as $p_{fold}(i) = P_f^C[\alpha]$ for $i \in \alpha$. (C–E) Scatter plot of P_f^C vs P_f for 1.0, 5.0, and 10 μs of simulation time, respectively.

(Eqs. 4 and 7). The 37 clusters chosen among the 78,183 include both high- and low-populated clusters with P_f^C values evenly distributed in the range between 0 and 1. In the case of large clusters, a subset of snapshots is considered for the computation of P_f. In those cases W is replaced in Eq. 7 by $W_{sample} < W$ that is the number of snapshots involved in the calculation. Namely, for the 37 clusters previously mentioned, a correlation of 0.89 between P_f^C and P_f is found with a slope of 0.86 (*see* Fig. 3A), indicating that the procedure is able to estimate folding probabilities for clusters on the folding–transition barrier ($P_f \approx 0.5$) as well as in the folding ($P_f \approx 1.0$) or unfolding ($P_f \approx 0.0$) regions. The error bars for P_f^C in Fig. 3A are derived from the definition of variance given in Eq. 6. In the same spirit of Eq. 6 the folding probability P_f and its variance are written as:

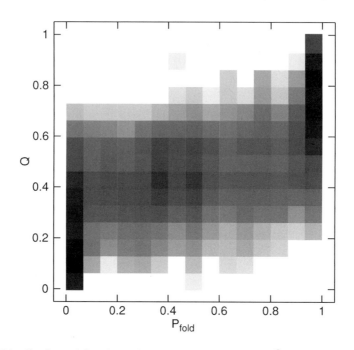

Fig. 4. Distribution of fraction of native contacts Q and P_f^C in the wild-type Beta3s simulations. The gray scale from black to white corresponds to high and low density, respectively. Although structures with very large Q ($Q > 0.8$) or very low Q ($Q < 0.2$) have P_f^C close to 1 or 0, respectively, conformations with intermediate values of Q span all the allowed spectrum of P_f^C values.

$$P_f = \langle \theta \rangle = \frac{1}{N} \sum_{i=1}^{N} \theta_i$$

$$\sigma_{\langle \theta \rangle}^2 = \frac{1}{N} P_f (1 - P_f)$$

(8)

where $N = \sum n_t$ is the total number of runs and θ is equal to 1 or 0, if the run folded or unfolded, respectively. Note that the same number of runs n_t has been used for every snapshot of a cluster. The large vertical error bars in **Fig. 3A** correspond to clusters with less than 10 snapshots. The largest deviations between P_f and P_f^C are around the 0.5 region. This is owing to the limited number of crossings of the folding barrier observed in the MD simulation (**Fig. 3B**, around 70 events of folding *[23]*). Improvements in the accuracy for the estimation of P_f are achieved as the number of folding events, i.e., the simulation time, increases (**Fig. 3C–E**).

The validity of P_f^C as an approximation of P_f, is robust with respect to the choice of the clusterization. Similar results can be obtained also with different flavors of conformation space partitioning, as long as they group together structurally homogeneous conformations, e.g., clusterization based on root mean square deviation of atomic coordinates (RMSD) or secondary structure strings *(25)*. The latter are appropriate for structured peptides but not for proteins with irregular secondary structure because of string degeneracy. Note that partitions based on order parameters (like native contacts) are usually unsatisfactory and not robust. This is mainly owing to the fact that clusters defined in this way are characterized by large structural heterogeneities *(19)*.

Interestingly, there is no correspondence between the number of native contacts formed and p_{fold} (**Fig. 4**). In other words, it would have been impossible to simply use the order parameter Q to extract TSE conformations. This result shows again that the indiscriminate use of free energy projections on order parameters can be misleading and kinetic properties cannot, in general, be inferred from the thermodynamic analysis.

5. The Transition State Ensemble Defined Using the Folding Probability

The folding probability of structure i is estimated as $p_{fold}(i) = P_f^C[\alpha]$ for $i \in \alpha$. This approximation allows one to plot the pairwise RMSD distribution of Beta3s structures with $p_{fold} > 0.51$ (native state), $0.49 < p_{fold} < 0.51$ (TSE), and $p_{fold} < 0.49$ (denatured state) (**Fig. 5A**). For the native state, the distribution is peaked around low values of RMSD (≈ 1.5 Å) indicating that structures with $p_{fold} > 0.51$ are structurally similar and belong to a nondegenerate state. The statistical weight of this group of structures is 49.4% and corresponds to the expected statistics for the native state because the simulations are performed at the melting temperature. In the case of TSE, the distribution is broad because of the coexistence of heterogeneous structures. This scenario is compatible with the presence of multiple folding pathways. Beta3s folding was already shown to involve two main average pathways depending on the sequence of formation of the two hairpins *(19,21)*. Here, a *naive* approach based on the number of native contacts *(21,25)* is used to structurally characterize the folding barrier. TSE structures with number of native contacts of the first hairpin greater than the ones of the second hairpin are called type I conformations (**Fig. 5B**), otherwise they are called type II (**Fig. 5C**). In both cases the transition state is characterized by the presence of one of the two native hairpins formed while the rest of the peptide is mainly unstructured. These findings are also in agreement with the complex network analysis of Beta3s reported recently *(19)*. Finally, the denatured state shows a broad pairwise RMSD distribution around even larger values of RMSD (≈ 5.5 Å), indicating the presence of highly heterogeneous conformations.

6. Φ-Value Analysis by MD Simulations

The p_{fold} values of all conformations saved along a reversible folding MD trajectory can be used to isolate the TSE. However, the information on TSE derived

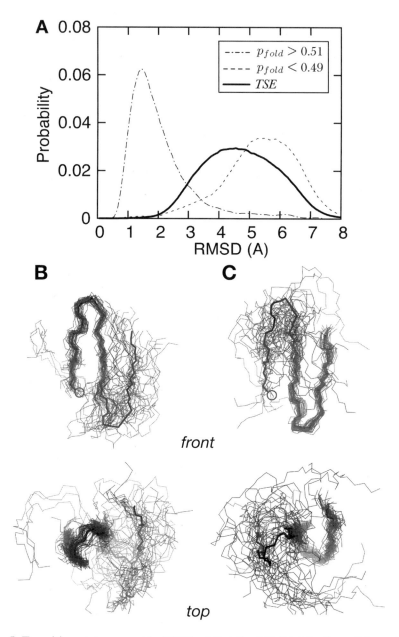

Fig. 5. Transition state ensemble (TSE) of Beta3s. (**A**) Distribution of the values of pairwise RMSD for structures with $p_{fold} > 0.51$ (native state), $0.49 < p_{fold} < 0.51$ (TSE), and $p_{fold} < 0.49$ (denatured state). (**B**) Type I and (**C**) type II transition states (thin lines). Structures are superimposed on residues 2–11 and 10–19 with an average pairwise RMSD of 0.81 and 0.82 Å for type I and type II, respectively. For comparison, the native state is shown as a thick line with a circle to label the N-terminus.

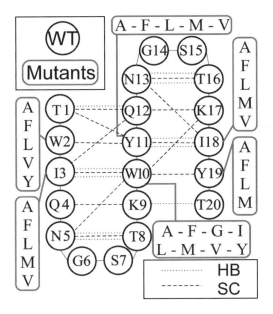

Fig. 6. Schematic representation of the Beta3s peptide, where the wild-type sequence and the mutants are indicated. The backbone hydrogen bonds (dotted lines) and side chain contacts (dashed lines) common to most of the peptides are reported. HB, hydrogen bond; SC, side chain contacts; WT, wild-type.

from protein folding experiments is represented by the Φ values. As we will see in more detail, the Φ value of a residue is the change in the activation free energy of folding relative to the change in stability of the protein on mutation of the residue. Φ values have been usually interpreted as the fraction of native contacts formed at TSE by the mutated residue. This interpretation, however, does not allow one to consider non-native interactions that may form at TSE and is not able to explain anomalous Φ values (i.e., those out of the 0 to 1 range). Thus, we have extensively tested the standard interpretation of Φ values by evaluating them from the folding and unfolding rates measured in equilibrium MD simulations of wild-type Beta3s and a large number of single-point mutants.

Thirty-two single-point mutations of the hydrophobic and aromatic side chains W2, I3, W10, Y11, I18, and Y19 were investigated (**Fig. 6**). The six sites of mutation are distributed along the sequence of the peptide, two for each strand. Between four and eight mutations have been studied for each site. Six of the 32 mutations are nondisruptive (I3A, I3V, Y11F, I18A, I18V, and Y19F), six mutations are conservative but change the steric properties of the side chain (I3M, Y11L, Y11M, I18M, Y19L, and Y19M), and the remaining 20 mutations are radical but acceptable because, in most of the cases, they do not significantly change the TSE of the peptide. This is probably because of the fact that the side chains of Beta3s are not fully buried in a densely packed hydrophobic core as is the case in larger proteins *(33)*. Ten MD runs of 2 μs each (total of 20 μs for each mutant) with different initial velocities were performed with the Berendsen thermostat at 330 K, which is close to the melting temperature of wild-type Beta3s *(34)*. A time step of 2 fs was used and the coordinates were saved every 20 ps for a total of 10^6 conformations for each mutant. During the 20-μs simulation time between 57 and 120 folding events were observed for every mutant (**Table 1**), thus providing sufficient statistical sampling for the kinetic analysis. The small statistical error is supported by the small difference in the native population measured for each individual mutant on two disjoint equal-size subsets of the trajectories (5% on average, the largest being 13%).

The native structure of the wild-type, i.e., the three-stranded anti-parallel β-sheet with turns at G6-S7 and G14-S15, is also the most populated in all the mutants, as shown by the cluster analysis of the trajectories (**Table 1**). The only exception is Y11V, which has a more distorted native state and has not been considered for further analysis. Moreover, there is no predominant structure in the denatured state for any of the mutants. The number of folding and unfolding events observed along the trajectories ranges from 57 to 120 and 64 to 127, respectively (**Table 1**). Interestingly, the values of the stability change upon mutation, calculated with **Eq. 2**, and show that all mutants are less stable than wild-type Beta3s except for W10F and I3V, which are essentially as stable as Beta3s. This result is not unexpected because Beta3s is a designed peptide whose sequence was carefully optimized for its fold *(20)*.

As in the kinetic experiments used to measure experimental Φ values, free energy changes with respect to wild-type are computed from the folding and unfolding rates. The fraction of native contacts Q has been computed along the trajectories of all peptides. A folding (unfolding) event occurs when, along the trajectory, Q first reaches values larger than 0.85 (lower than 0.15) immediately after a previous unfolding (folding) event *(21)*. All the trajectories are started from the folded state, thus, the first event is always an unfolding. The average time separation between a folding (unfolding) event and the previous unfolding (folding) event, is the folding (unfolding) time τ_f (τ_u). The folding and unfolding rates are $k_f = 1/\tau_f$ and $k_u = 1/\tau_u$, respectively. Setting the free energy of the denatured state as reference:

$$\Delta\Delta G_{TS-D}^{kin} = RT \log \left(\frac{k_f^{WT}}{k_f^{mut}} \right) \tag{9}$$

$$\Delta\Delta G_{N-D}^{kin} = RT \log \left(\frac{k_f^{WT}}{k_f^{mut}} \cdot \frac{k_u^{mut}}{k_u^{WT}} \right) \tag{10}$$

Table 1
Stability, Folding/Unfolding Rates, and Φ Values of the Mutants

Mutation [a]	W_{highQ} [b] (%)	Nat.Cont. [c]	W_{lowQ} [d] (%)	τ_f [e] (ns)	N_f [f]	τ_u [g] (ns)	N_u [h]	$\Delta\Delta G_{N-D}^{kin}$ [i] (kcal/mol)	$\Delta\Delta G_{TS-D}^{kin}$ [i] (kcal/mol)	Φ [i,j]
WT	21.4	19.3 ± 1.7	2.9	70 ± 10	92	67 ± 6	94			
W2A	26.5	18.1 ± 2.3	3.5	107 ± 14	108	63 ± 6	114	-0.32 ± 0.15	-0.28 ± 0.13	0.87 ± 0.57
W2F	33.5	18.8 ± 2.2	3.4	106 ± 14	97	82 ± 8	103	-0.14 ± 0.16	-0.27 ± 0.13	-
W2L	24.9	18.2 ± 2.2	6.3	109 ± 16	101	63 ± 5	111	-0.34 ± 0.16	-0.30 ± 0.14	0.87 ± 0.57
W2V	23.6	18.3 ± 2.3	4.4	124 ± 17	95	62 ± 6	102	-0.43 ± 0.16	-0.38 ± 0.13	0.89 ± 0.45
W2Y	21.9	18.5 ± 2.4	6.4	129 ± 21	93	65 ± 6	98	-0.43 ± 0.16	-0.41 ± 0.14	0.95 ± 0.49
I3A	19.9	18.7 ± 2.2	3.9	137 + 18	92	64 ± 5	101	-0.48 ± 0.15	-0.44 ± 0.13	0.93 ± 0.40
I3F	33.0	18.8 ± 2.1	3.3	121 ± 22	83	93 ± 8	91	-0.15 ± 0.17	-0.36 ± 0.15	-
I3L	28.5	18.5 ± 2.4	3.9	119 ± 19	94	72 ± 7	101	-0.31 ± 0.17	-0.35 ± 0.14	1.1 ± 0.77
I3M	30.2	18.9 ± 2.2	5.4	108 ± 19	94	81 ± 9	102	-0.16 ± 0.17	-0.29 ± 0.15	-
I3V	37.2	18.6 ± 2.1	5.2	124 ± 18	75	109 ± 10	83	-0.06 ± 0.16	-0.38 ± 0.14	-
W10A	31.8	19.5 ± 2.1	5.0	161 ± 21	74	95 ± 10	79	-0.32 ± 0.16	-0.55 ± 0.13	1.7 ± 0.93
W10F	41.3	18.7 ± 2.2	3.8	77 ± 9	120	78 ± 6	127	0.04 ± 0.14	-0.06 ± 0.12	-
W10G	12.7	19.3 ± 2.2	3.1	212 ± 32	60	68 ± 9	69	**-0.72**± 0.17	-0.73 ± 0.14	**1.0 ± 0.31**
W10I	30.8	18.3 ± 2.1	6.0	129 ± 17	77	88 ± 9	83	-0.23 ± 0.16	-0.40 ± 0.13	-
W10L	20.8	18.8 ± 2.2	4.2	166 ± 22	81	58 ± 5	87	**-0.67**± 0.16	-0.57 ± 0.13	**0.86 ± 0.28**
W10M	18.4	19.0 ± 2.2	6.6	155 ± 21	82	52 ± 5	91	**-0.68**± 0.16	-0.52 ± 0.13	**0.76 ± 0.26**
W10V	17.2	17.8 ± 2.5	6.7	259 ± 40	57	65 ± 11	64	**-0.88**± 0.19	-0.86 ± 0.14	**0.98 ± 0.26**
W10Y	26.2	19.0 ± 2.1	3.5	118 ± 15	94	77 ± 7	98	-0.26 ± 0.15	-0.35 ± 0.13	-
Y11A	5.7	18.1 ± 2.0	2.3	249 ± 38	64	30 ± 3	71	**-1.37**± 0.17	-0.84 ± 0.14	**0.61 ± 0.13**
Y11F	33.1	19.1 ± 2.2	4.4	138 ± 20	73	112 ± 12	79	-0.11 ± 0.16	-0.45 ± 0.14	-
Y11L	14.8	18.6 ± 2.1	4.8	169 ± 23	76	54 ± 6	83	**-0.72**± 0.16	-0.58 ± 0.13	**0.81 ± 0.26**
Y11M	11.3	18.0 ± 2.2	3.5	152 ± 24	95	35 ± 3	105	**-0.94**± 0.16	-0.51 ± 0.14	**0.54 ± 0.18**
Y11V	5.7	17.0 ± 2.7	7.4							
I18A	12.3	18.5 ± 2.3	2.4	168 ± 22	80	53 ± 6	88	**-0.73**± 0.16	-0.58 ± 0.13	**0.79 ± 0.25**
I18F	21.3	19.0 ± 2.0	3.2	159 ± 23	74	72 ± 8	83	-0.50 ± 0.17	-0.54 ± 0.14	1.1 ± 0.46
I18L	22.2	19.0 ± 2.2	4.4	145 ± 19	73	94 ± 9	81	-0.26 ± 0.16	-0.48 ± 0.13	-
I18M	28.9	18.8 ± 2.2	4.8	97 ± 15	99	77 ± 6	106	-0.13 ± 0.16	-0.22 ± 0.14	-
I18V	29.6	18.8 ± 2.3	3.2	124 ± 20	87	86 ± 9	93	-0.22 ± 0.17	-0.38 ± 0.14	-
Y19A	20.7	18.6 ± 2.4	7.4	123 ± 18	90	84 ± 8	95	-0.23 ± 0.16	-0.37 ± 0.14	-
Y19F	29.2	18.4 ± 2.2	3.8	130 ± 18	92	71 ± 7	98	-0.37 ± 0.16	-0.41 ± 0.13	1.1 ± 0.59
Y19L	30.0	18.3 ± 2.2	3.2	117 ± 17	83	88 ± 8	89	-0.17 ± 0.16	-0.34 ± 0.13	-
Y19M	17.5	18.5 ± 2.3	6.2	155 ± 26	68	97 ± 10	76	-0.28 ± 0.17	-0.52 ± 0.15	-

[a]Mutants in italics are radical but acceptable and mutations in Roman are conservative.
[b]Statistical weight of the three most populated clusters with $Q \geq 16/24$.
[c]Average number of contacts in the three most populated clusters with $Q \geq 16/24$.
[d]Statistical weight of the three most populated with $Q < 16/24$.
[e]Average folding time.
[f]Number of folding events.
[g]Average unfolding time.
[h]Number of unfolding events.
[i]The standard deviations have been obtained by propagation of the error on τ_f and τ_u.
[j]Dashes indicate "unreliable" Φ values due to $|\Delta\Delta G_{N-D}^{kin}| < 0.3$ kcal/mol. Boldface emphasize the "reliable" Φ values and the corresponding large stability changes (33). The multipoint Φ values are 0.77, 0.60, 0.79, 0.46, 0.72, and 1.23 for W2, I3, W10, Y11, I18, and Y19, respectively.

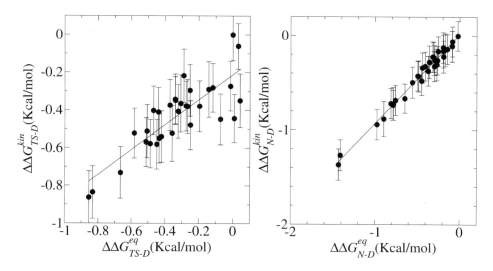

Fig. 7. Comparison between free energy changes calculated with the kinetic and P_f^C data. The correlation coefficient is 0.83 and 0.99 for $\Delta\Delta G_{TS-D}$ and $\Delta\Delta G_{N-D}$, respectively.

The Φ value is $\Phi = \Delta\Delta G_{TS-D}^{kin}/\Delta\Delta G_{N-D}^{kin}$. Values of $\Delta\Delta G_{TS-D}^{kin}$ and $\Delta\Delta G_{N-D}^{kin}$ from multiple mutations at the same site can be displayed on a single plot. The slope of the corresponding regression line is called the multipoint Φ value *(33,35)*.

Clusters are assigned to the native state, the TSE and the denatured state assemble according to their P_f^C. Their statistical weights are W_N, W_{TS} and W_D, respectively; these values can be used to evaluate relative free energies by a different equation with respect to the kinetically evaluated $\Delta\Delta G^{kin}$. In the canonical ensemble $\Delta G_{TS-D}^{eq} = -RT \log(W_{TS}/W_D)$ and $\Delta G_{N-D}^{eq} = -RT \log(W_N/W_D)$. As shown in **Fig. 7**, an excellent match is observed between the $\Delta\Delta G_{N-D}^{kin}$ and $\Delta\Delta G_{N-D}^{eq}$ values (correlation coefficient of 0.99) and a good correlation between $\Delta\Delta G_{TS-D}^{kin}$ and $\Delta\Delta G_{TS-D}^{eq}$ (correlation coefficient of 0.83). The agreement represents a consistency check for the parameters used to define folding and unfolding events. That activation free energy differences computed with the two sets of data show larger discrepancies than changes in stability, is owing to the difficulty in sampling the TSE.

6.1. Accuracy of Two-Point and Multipoint Φ Values

Figure 8 shows the Φ values extracted from the simulations as a function of the change in free energy of folding upon mutation (*see also* **Table 1**). Because of the difficulties in the interpretation of Φ values, as many mutants as possible have been considered and the resulting Φ values divided into classes of "reliable", "tolerable", and "unreliable" according to the size of the induced stability change

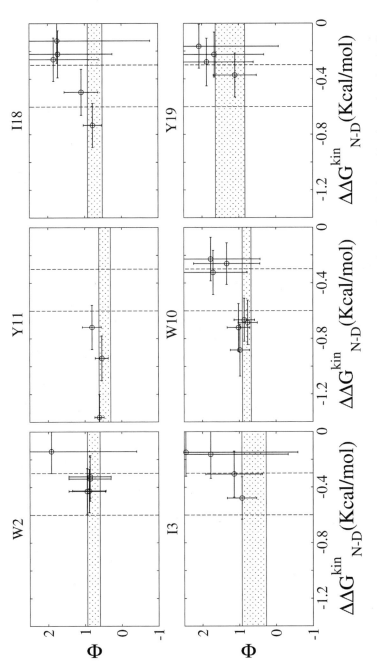

Fig. 8. Φ values as a function of change in the native state stability upon mutation. (The shadowed horizontal region indicates one standard deviation around the multipoint Φ value. The Φ values span a wide range and become "anomalous" for $|\Delta\Delta G_{N-D}^{kin}|$ smaller than about 0.3 kcal/mol. The Φ values corresponding to mutations with $|\Delta\Delta G_{N-D}^{kin}| > 0.3$ are mainly in the "normal" range, i.e., between 0 and 1, and are in agreement with the multipoint Φ value. Vertical dashed lines are drawn at $\Delta\Delta G_{N-D}^{kin} = -0.3$ kcal/mol and $\Delta\Delta G_{N-D}^{kin} = -0.6$ kcal/mol. The Φ value of mutations I3V, W10F, and Y11F are located outside of the plot boundaries.) The graphs are ordered according to the antiparallel β-sheet topology of Beta3s with vertical orientation of the three strands, and the N- and C-terminus on the top-left and bottom-right, respectively.

243

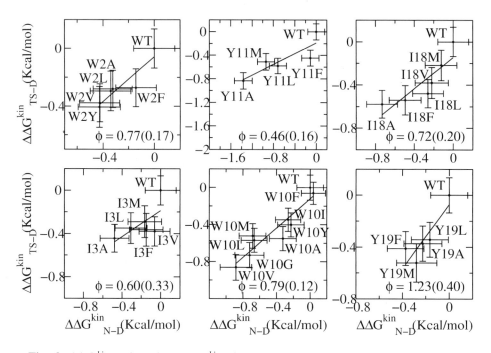

Fig. 9. $\Delta\Delta G^{kin}_{TS-D}$ plotted vs $\Delta\Delta G^{kin}_{N-D}$ for all the mutants grouped according to the mutation site along the structure of Beta3s. The optimal regression line (including the wild-type data point) is plotted and its slope, i.e., the multipoint Φ value, is reported in the lower-right corner of each graph with the standard deviation derived from the fit in parentheses. The correlation coefficient is 0.91, 0.67, 0.93, 0.86, 0.87, and 0.88 for W2, I3, W10, Y11, I18, and Y19 mutants, respectively.

$|\Delta\Delta G^{kin}_{N-D}|$. The deviations from the 0 to 1 range are large for "unreliable" Φ values, i.e., for mutations with $|\Delta\Delta G^{kin}_{N-D}|<0.3$, in agreement with previous observations (35). Indeed, in the "unreliable" class the deviation can be observed for both radical mutations (e.g., I3F, W10A, Y19A) as well as for nondisruptive mutations (e.g., I3V, Y11F, and I18V). For "tolerable" Φ values, i.e., $0.3 \leq |\Delta\Delta G^{kin}_{N-D}| < 0.6$, the deviation from the 0–1 interval is less frequent but the relative error is large. The eight "reliable" Φ values ($|\Delta\Delta G^{kin}_{N-D}|\geq 0.6$) are all in the range 0 to 1 and have a small standard deviation. In a small-structured peptide, like Beta3s, most residues have a relatively large exposed surface area in the folded state so that conservative mutations generally induce small free energy changes. Indeed, among the six conservative mutations only I18A falls in the "reliable" class. For this reason more radical mutations have been also investigated.

The multipoint Φ of Beta3s as extracted from the simulations are reported in **Fig. 9**. The good linear relationship between $\Delta\Delta G^{kin}_{TS-D}$ and $\Delta\Delta G^{kin}_{N-D}$, observed in mutants of W2, W10, Y11, and Y19, supports the validity of the multipoint analysis for these residues and indicates a substantial similarity among the folding TSEs of those peptides. In mutants of I3 the linear correlation is less strong than the others and in I18 there is a change in the slope for $\Delta\Delta G^{kin}_{N-D} < -0.3$ kcal/mol. A possible explanation for the presence of a linear relationship in the multipoint plots is the partial flexibility of the native state of Beta3s *(19)*. Its partially exposed non-polar side chains, which have been mutated in this work, are involved in less specific interactions with the rest of the peptide than buried side chains in the hydrophobic core of larger proteins. Because of the partial flexibility, the mutations do not affect only specific interactions but produce an effect that is spread over the large available set of contacts and thus averaged over them. This averaging of the effects of mutations in the native state may translate into a simple linear dependence of the effects in the TS. In this context, deviations from linearity may indicate TSE shifts.

In multipoint plots different local probes of the same residue are forced in a single fit which can yield wrong estimates *(36)*. As an example, in the I → V → A → G mutation series the I → V measures interactions originating from tertiary structure contacts, the V → A a mixture of tertiary and secondary structure interactions, whereas the A → G reports almost exclusively on secondary structure formation *(36)*. In a framework *(37)* or diffusion–collision *(38)* mechanism of folding, the "tertiary" Φ values will most probably be lower than "secondary" Φ values, even for the same residue. In the case of Beta3s, where the formation of β-sheet backbone hydrogen bonds and long-range contacts between side chains are concomitant events (*see* **Fig. 4** in **ref. 21**), different mutations probe the formation of the same level of structure (i.e., the β-sheet) with no distinction between secondary and tertiary components. This supports the validity of the multipoint analysis for Beta3s, which we do not want to generalize to proteins with more complex folds.

Given the peculiarities of Beta3s, i.e., concomitant formation of secondary and tertiary structure and partial flexibility of its folded state, multipoint Φ values may add information on the accuracy of the two-point Φ values. Indeed, "reliable" and "tolerable" Φ values fall mostly within a standard deviation from the corresponding multipoint Φ value (**Fig. 8**), whereas "unreliable" Φ values show large deviations. Five of the six multipoint Φ values of Beta3s are larger than 0.5. For diffuse TSEs of proteins of about 100 residues, Φ values around 0.2–0.3 have been measured experimentally *(39,40)*. The high Φ values of Beta3s are probably owing to the small size of the peptide. Because of its small size a large part of the native interactions of the hydrophobic residues is already present in the rate-limiting step.

6.2. Structural Interpretation of Φ Values

In each snapshot a van der Waals contact is defined when the distance between two heavy atoms is smaller than 6 Å. $p_N(i)$ and $p_{TS}(i)$ measure the fraction of native

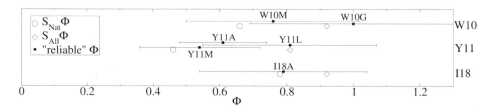

Fig. 10. Comparison between "reliable" two-point Φ values (filled squares) of mutants with TSE similar to wild-type, and the structure of wild-type TSE as measured by $S_x\Phi$ values (open symbols). The structural Φ values are the ratio between the number of contacts formed in TSE and native state. $S_{Nat}\Phi$ takes into account only native contacts, whereas $S_{All}\Phi$ includes native and non-native contacts. The two-point Φ values tend to overestimate the degree of nativeness of the TSE (measured by $S_{Nat}\Phi$) because of the presence of specific non-native interactions.

and TSE structures, respectively, in which the contact i is formed. If $p_N(i) > 0.66$ the contact i belongs to the set of the native contacts (NC). The structural Φ value:

$$S_{Nat}\Phi(R) = \frac{1}{M_{NC(R)}} \frac{\sum\limits_{i\in NC(R)} p_{TS}(i)}{\sum\limits_{i\in NC(R)} p_N(i)}, \qquad (11)$$

where $M_{NC}(R)$ is the number of native contacts of residue R, represents an estimate of the degree of nativeness of residue R at the TSE. This measure has been used in the past to give a structural interpretation to experimental Φ values (41–43). An estimate of the relevance of non-native interactions at the TSE is obtained by extending the sum to all possible contacts (AC), including contacts not present in the NC set:

$$S_{All}\Phi(R) = \frac{1}{M_{AC(R)}} \frac{\sum\limits_{i\in AC(R)} p_{TS}(i)}{\sum\limits_{i\in AC(R)} p_N(i)} \qquad (12)$$

Both $S_{Nat}\Phi$ and $S_{All}\Phi$ profiles of Beta3s provide a detailed picture of its TSE. It is useful to compare them with the "reliable" Φ values, i.e., those derived from mutations that do not significantly change the TSE of the peptide (e.g., W10M, W10G, Y11A, Y11L, Y11M, and I18A). Such comparison allows for the assessment of the standard interpretation of the Φ as the ratio between contacts formed at TSE and native state (Fig. 10). The comparison reveals that, within their error, the two-point Φ values are in agreement with both $S_x\Phi$. However, the former tend to overestimate the degree of native structure present at the TSE (i.e., "reliable" Φ > $S_{Nat}\Phi$) because specific non-native interactions are formed at the TSE *(29)*. More generally speaking, the presence of specific non-native contacts, distinguishing the TSE conformations from other structures having the same native interactions but different non-native interactions, makes the standard interpretation of Φ values not completely appropriate. Namely, neglecting non-native interactions may prevent a complete understanding of the factors that are responsible for protein folding.

7. Conclusions

Despite its very simple native topology, the 20-residue structured peptide Beta3s has been shown, using MD simulations with implicit solvent, to have multiple folding pathways *(21)* and a very heterogeneous denatured state consisting of both high enthalpy, high entropy basins, and low enthalpy, low entropy traps *(19)*. Furthermore, folding-unfolding equilibrium simulations of Beta3s and several single-point mutants have been used to evaluate folding probabilities of Beta3s conformations *(25)* and Φ values of several of its residues *(29)*, respectively. The latter, calculated from folding and unfolding rates measured from the MD trajectories, are reliable if the stability loss upon mutation is larger than about 0.6 kcal/mol, in agreement with experimental observations. Another interesting simulation result is that Φ values tend to overestimate the nativeness of the TSE, when interpreted neglecting the non-native interactions. The next challenge is to generalize the simulation results obtained with Beta3s to other structured peptides and small proteins.

References

1. Abkevich, V., Gutin, A., and Shakhnovich, E. (1994) Free-energy landscape for protein-folding kinetics: intermediates, traps, and multiple pathways in theory and lattice model simulations. *J. Chem. Phys.* **101,** 6052–6062.
2. Dill, K. and Chan, H. (1997) From Levinthal to pathways to funnels. *Nat. Struct. Biol.* **4,** 10–19.
3. Frauenfelder, H., Sligar, S., and Wolynes, P. (1991) The energy landscapes and motions of proteins. *Science* **254,** 1598–1603.
4. Saven, J., Wang, J., and Wolynes, P. (1994) Kinetics of protein-folding: the dynamics of globally connected rough energy landscapes with biases. *J. Chem. Phys.* **101,** 11037–11043.
5. Wang, J., Onuchic, J., and Wolynes, P. G. (1996) Statistics of kinetic pathways on biased rough energy landscapes with applications to protein folding. *Phys. Rev. Lett.* **76,** 4861–4864.
6. Daggett, V. and Fersht, A. (2003) Is there a unifying mechanism for protein folding? *Trends Biochem. Sci.* **28,** 18–25.
7. Leopold, P. E., Montal, M., and Onuchic, J. N. (1992) Protein folding funnels: a kinetic approach to the sequence-structure relationship. *Proc. Natl. Acad. Sci. USA* **89,** 8721–8725.

8. Karplus, M. (1997) The Levinthal paradox: yesterday and today. *Fold. Des.* **2,** S69–S75.

9. Fersht, A. R. (1999) *Structure and Mechanism in Protein Science: Guide to Enzyme Catalysis and Protein Folding,* W. H. Freeman, New York, NY.

10. Fersht, A. R., Matouschek, A., and Serrano, L. (1992) The folding of an enzyme. 1. Theory of protein engineering analysis of stability and pathway of protein folding. *J. Mol. Biol.* **224,** 771–782.

11. Brooks, B., Bruccoleri, R., Olafson, B., States, D., Swaminathan, S., and Karplus, M. (1983) Charmm: a program for macromolecular energy, minimization, and dynamics calculations. *J. Comput. Chem.* **4,** 187–217.

12. Karplus, M. and Kuriyan, J. (2005) Chemical theory and computation special feature: molecular dynamics and protein function. *Proc. Natl. Acad. Sci. USA* **102,** 6679–6685.

13. McCammon, J. A., Gelin, B. R., and Karplus, M. (1977) Dynamics of folded proteins. *Nature* **267,** 585–590.

14. Du, R., Pande, V., Grosberg, A., Tanaka, T., and Shakhnovich, E. (1998) On the transition coordinate for protein folding. *J. Chem. Phys.* **108,** 334–350.

15. Gsponer, J. and Caflisch, A. (2002) Molecular dynamics simulations of protein folding from the transition state. *Proc. Natl. Acad. Sci. USA* **99,** 6719–6724.

16. Hubner, I. A., Shimada, J., and Shakhnovich, E. I. (2004) Commitment and nucleation in the protein G transition state. *J. Mol. Biol.* **336,** 745–761.

17. Lenz, P., Zagrovic, B., Shapiro, J., and Pande, V. S. (2004) Folding probabilities: a novel approach to folding transitions and the two-dimensional Ising-model. *J. Chem. Phys.* **120,** 6769–6778.

18. Li, L. and Shakhnovich, E. I. (2001) Constructing, verifying, and dissecting the folding transition state of chymotrypsin inhibitor 2 with all-atom simulations. *Proc. Natl. Acad. Sci. USA* **98,** 13014–13018.

19. Rao, F. and Caflisch, A. (2004) The protein folding network. *J. Mol. Biol.* **342,** 299–306.

20. De Alba, E., Santoro, J., Rico, M., and Jimenez, M. (1999) De novo design of a monomeric three-stranded antiparallel beta-sheet. *Protein Sci.* **8,** 854–865.

21. Ferrara, P. and Caflisch, A. (2000) Folding simulations of a three-stranded antiparallel beta-sheet peptide. *Proc. Natl. Acad. Sci. USA* **97,** 10780–10785.

22. Ferrara, P., Apostolakis, J., and Caflisch, A. (2002) Evaluation of a fast implicit solvent model for molecular dynamics simulations. *Proteins* **46,** 24–33.

23. Cavalli, A., Haberthur, U., Paci, E., and Caflisch, A. (2003) Fast protein folding on downhill energy landscape. *Protein Sci.* **12,** 1801–1803.

24. Hartigan, J. (1975) *Clustering Algorithms,* Wiley, New York, NY.

25. Rao, F., Settanni, G., Guarnera, E., and Caflisch, A. (2005) Estimation of protein folding probability from equilibrium simulations. *J. Chem. Phys.* **122,** 184901.

26. Chan, H. S. and Dill, K. A. (1998) Protein folding in the landscape perspective: Chevron plots and non-Arrhenius kinetics. *Proteins* **30,** 2–33.

27. Pande, V., Grosberg, A., Tanaka, T., and Rokhsar, D. (1998) Pathways for protein folding: is a new view needed? *Curr. Opin. Sltruct. Biol.* **8,** 68–79.

28. Krivov, S. and Karplus, M. (2004) Hidden complexity of free energy surfaces for peptide (protein) folding. *Proc. Natl. Acad. Sci. USA* **101,** 14766–14770.
29. Settanni, G., Rao, F., and Caflisch, A. (2005) Value analysis by molecular dynamics simulations of reversible folding. *Proc. Natl. Acad. Sci. USA* **102,** 628–633.
30. Best, R. and Hummer, G. (2005) Reaction coordinates and rates from transition paths. *Proc. Natl. Acad. Sci. USA* **102,** 6732–6737.
31. Chandler, D. (1978) Statistical mechanics of isomerization dynamics in liquids and the transition state approximation. *J. Chem. Phys.* **68,** 2959–2970.
32. Northrup, S. H., Pear, M. R., Lee, C. Y., McCammon, J. A., and Karplus, M. (1982) Dynamical theory of activated processes in globular proteins. *Proc. Natl. Acad. Sci. USA* **79,** 4035–4039.
33. Fersht, A. R. and Sato, S. (2004) Value analysis and the nature of protein-folding transition states. *Proc. Natl. Acad. Sci. USA* **101,** 7976–7981.
34. Cavalli, A., Ferrara, P., and Caflisch, A. (2002) Weak temperature dependence of the free energy surface and folding pathways of structured peptides. *Proteins* **47,** 305–314.
35. Sanchez, I. E. and Kiefhaber, T. (2003) Origin of unusual values in protein folding: evidence against specific nucleation sites. *J. Mol. Biol.* **334,** 1077–1085.
36. Fersht, A. R. (2004) Relationship of Leffler (Bronsted) values and protein folding values to positions of transition-state structures on reaction coordinates. *Proc. Natl. Acad. Sci. USA* **101,** 14338–14342.
37. Baldwin, R. L. and Rose, G. D. (1999) Is protein folding hierarchic? II. Folding intermediates and transition states. *Trends Biochem. Sci.* **24,** 77–83.
38. Karplus, M. and Weaver, D. L. (1976) Protein folding dynamics. *Nature* **260,** 404–406.
39. Daggett, V., Li, A. J., Itzhaki, L. S., Otzen, D. E., and Fersht, A. R. (1996) Structure of the transition state for folding of a protein derived from experiment and simulation. *J. Mol. Biol.* **257,** 430–440.
40. Itzhaki, L. S., Otzen, D. E., and Fersht, A. R. (1995) The structure of the transition-state for folding of chymotrypsin inhibitor-2 analyzed by protein engineering methods: evidence for a nucleation-condensation mechanism for protein-folding. *J. Mol. Biol.* **254,** 260–288.
41. Li, A. J. and Daggett, V. (1996) Identification and characterization of the unfolding transition state of chymotrypsin inhibitor 2 by molecular dynamics simulations. *J. Mol. Biol.* **257,** 412–429.
42. Settanni, G., Gsponer, J., and Caflisch, A. (2004) Formation of the folding nucleus of an SH3 domain investigated by loosely coupled molecular dynamics simulations. *Biophys. J.* **86,** 1691–1701.
43. Vendruscolo, M., Paci, E., Dobson, C. M., and Karplus, M. (2001) Three key residues form a critical contact network in a protein folding transition state. *Nature* **409,** 641–645.

14

Packing Regularities in Biological Structures Relate to Their Dynamics

Robert L. Jernigan and Andrzej Kloczkowski

Summary

The high packing density inside proteins leads to certain geometric regularities and also is one of the most important contributors to the high extent of cooperativity manifested by proteins in their cohesive domain motions. The orientations between neighboring nonbonded residues in proteins substantially follow the similar geometric regularities, regardless of whether the residues are on the surface or buried, a direct result of hydrophobicity forces. These orientations are relatively fixed and correspond closely to small deformations from those of the face-centered cubic lattice, which is the way in which identical spheres pack at the highest density. Packing density also is related to the extent of conservation of residues, and we show this relationship for residue packing densities by averaging over a large sample or residue packings. There are three regimes: (1) over a broad range of packing densities the relationship between sequence entropy and inverse packing density is nearly linear, (2) over a limited range of low packing densities the sequence entropy is nearly constant, and (3) at extremely low packing densities the sequence entropy is highly variable. These packing results provide important justification for the simple elastic network models that have been shown for a large number of proteins to represent protein dynamics so successfully, even when the models are extremely coarse grained. Elastic network models for polymeric chains are simple and could be combined with these protein elastic networks to represent partially denatured parts of proteins. Finally, we show results of applications of the elastic network model to study the functional motions of the ribosome, based on its known structure. These results indicate expected correlations among its components for the step-wise processing steps in protein synthesis, and suggest ways to use these elastic network models to develop more detailed mechanisms, an important possibility because most experiments yield only static structures.

Key Words: Elastic network models; protein packing; protein dynamics; conformational transition; Gaussian network model; anisotropic network model; harmonic analysis.

From: *Methods in Molecular Biology, vol. 350: Protein Folding Protocols*
Edited by: Y. Bai and R. Nussinov © Humana Press Inc., Totowa, NJ

1. Introduction

Globular proteins are compact and usually quite densely packed *(1)*, even to the extent that their interior have sometimes been viewed as being solid-like *(2,3)*; however, there are still numerous voids and cavities in protein interiors *(4)*. The importance of tight packing is widely acknowledged and is thought to be important for protein stability *(5,6)*, for nucleation of protein folding *(7–9)*, and for the design of novel proteins *(10)*. In conjunction with nucleation, it has previously been postulated that the conservation of amino acid residues through evolution may include essential tightly packed sites *(7–9,11)*. Much regularity exists in the geometries within proteins. For example, it is well known that proteins have strong biases for certain backbone and side chain torsion angles *(12)*. High packing density also imposes some regularity on nonbonded geometries; for example, there are some biases for the orientations of side chains *(13–15)*. This preference for side chain orientations could imply some relationship between sequence conservation and structure, which has been poorly understood *(16,17)*. The aim of our chapter is to show how these regularities in the packing inside proteins can provide motivation for the recent advancements in studies of protein dynamics with the elastic network (EN) models to explain large conformational changes in proteins and assemblies of proteins and nucleic acids, such as the ribosome. The study of these conformational changes is extraordinarily important because they relate closely to the functions and mechanisms of biological macromolecules.

In the next section, we will discuss the problem of packing regularities in proteins. In the following section, EN theory will be described. Because this theory was first developed for polymers, we will briefly discuss some polymer network cases for which analytical solutions are known. We will mention the applicability of these solutions for analysis of loops in proteins. We will present the formalism of EN theories for proteins: the Gaussian network model (GNM) and the anisotropic network model (ANM) of proteins presented here are based on a harmonic approximation. We will consider hierarchical levels of coarse graining and the mixed coarse graining (MCG) where the functionally important parts of proteins are modeled with higher accuracy than the rest. In the final section, we will present some results obtained for the functional dynamics of the ribosome.

2. Packing Regularities in Proteins

The standard coarse-grained assumption of one geometric point for each residue is equivalent to an assumption of equal-sized residues. This implies that residues pack similarly as spheres. This packing regularity is evident in the angular directions, but not in the spatial distances because different types of amino acids have different sizes. If the packing was of equal sized spheres, then residues would pack as in **Fig. 1**.

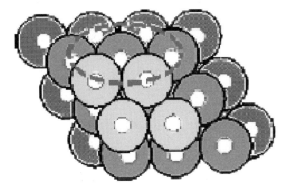

Fig. 1. Packing of uniform-sized spheres in their highest density form. Although amino acids do not pack exactly in this face-centered cubic lattice way, they do pack with similar angular preferences, but do not preserve the fixed distances between the centers of nearest neighbors.

Their angular distributions are preserved, however, and actual residues pack with the angles shown in **Fig. 2** (**Table 1**).

Residue points are taken as the C^β atoms and close residues within a radius of 6.8 Å are considered. Clusters obtained from a set of nonhomologous protein structures were rotated to find an optimal superimposition. The central residue defines the origin of a spherical reference system with the rigid body-rotated directional vectors from different clusters intersecting the surface of a unit sphere. The most populous regions on the surface of this sphere define the most probable coordination directions, characterized by two spherical angles, θ-polar and ϕ-azimuthal. When there is less than a full set of neighbors, sites are gradually filled as the coordination number increases by clusters of nearby points. It is important to note that the filling is not sparse or staggered, but instead, the relatively close sites are filled first, thus, maintaining an approximately constant density excluding the solvent space. The optimal protein architecture can be viewed as a distorted, incomplete face-centered cubic packing. This approximate cubic closest-packed geometry appears as a generic characteristic of the residue packing architecture for proteins. Note that face-centered cubic packing is the densest way to pack uniform spheres in space.

Although these local orientation clusters are present in proteins, the ability to sustain substitutions of amino acids depends on other aspects of protein structure, such as the available space, its ability to accommodate side chains of a specific shape, and the change in stabilizing interactions upon substitution. Thus, it is difficult to make sense of individual substitutions. However, if we include a sufficiently large sample and average results, it becomes clearer as will be seen next.

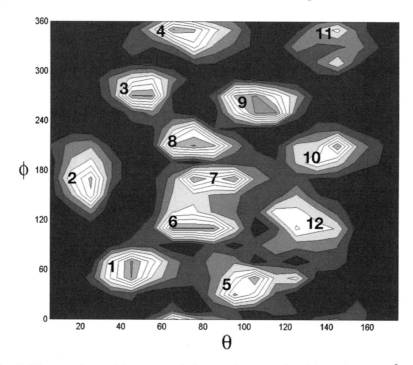

Fig. 2. The angular positions occupied around a central residue within 6.5 Å having 12 or more neighbors. The numbers index the angular density maxima identified in **Table 1** *(13–15)*.

2.1. Sequence Alignments and Structure

Employing sequence alignments in conjunction with molecular modeling has proven to provide significant improvements for protein structure prediction *(18,19)*. One key assumption in homology-based modeling is that conserved regions share structural similarities, but the structural basis of this connection has not been clearly established. Although there have been many individual demonstrations of connections between sequence conservation and structural properties *(20)*, there are no definitive studies on this subject. Establishing direct connections between sequences and structural features have proven difficult; hence the limited number of successes at protein design and the limited comprehension of mutagenesis. Recent applications that incorporate sequence variability into structure predictions have often lead to enhanced results *(21)*, indicating that empirical measures of sequence variability are useful by themselves, even if their full implications are not well understood.

Table 1
Coordination States of Surface-to-Core Central Residues[a]

Coordination number		Coordination states (degrees)											
		1	2	3	4	5	6	7	8	9	10	11	12
3 < m < 4	θ	40	45			95	90						
	φ	30	170			50	110						
(3 < m < 14)	θ	40	35	45	95	105	55	90					
	φ	10	200	285	350	50	115	180					
m ≥ 10	θ	45	45	45	95	105	60	100	85	105	140		
	φ	40	180	280	360	60	100	140	240	300	220		
m ≥ 12	θ	45	25	50	70	100	75	80	75	105	140	145	130
	φ	60	170	280	340	40	120	160	220	260	200	330	120
fcc lattice	θ	35	35	35	90	90	90	90	90	90	145	145	145
	φ	30	150	270	360	60	120	180	240	300	210	330	90

[a]Coordination states of surface-to-core central residues for groups of amino acids having different numbers of neighboring residues. m. Residues with neighbor coordination numbers of 3–4 are principally surface residues, all residues are included in $3 < m < 14$, core residues are represented by $m \geq 10$, and residues having $m \geq 12$ include the most densely packed residues (13–15).

2.2. Sequence Variability

Entropy is defined as the sum over the physical states i of a system, $-\sum_i p_i \ln p_i$. Developing such a measure for the probabilities of individual residues at each place in a structure is straightforward, and these types of computed Shannon entropies for protein sequences have proven useful and have been shown to correlate with entropies calculated from local physical parameters, including backbone geometry *(22)*. A deeper exploration of the connection between sequence entropy and protein structure is useful for developing a better understanding of protein stability and function.

Although extending the investigations of packing in proteins for greater detail could prove to be informative, we have chosen to investigate the coarse-grained packing of the points representing neighboring amino acids similarly to what was developed in the previous section. The results we will see are quite general, even if not so immediately useful for protein structure predictions, and we have established a strong relationship between packing density and sequence conservation. Specifically we generate a large set of aligned protein sequences over a diverse sample of 130 nonredundant protein sequences, all having structures in the PDB, and calculate sequence entropies for each individual residue position in the alignment. These are then compared with a corresponding local flexibility as defined by the C^α packing density calculated from the corresponding structures. Similar comparisons can also be made between residue hydrophobicity and packing density.

Sequence alignments were generated using BLASTP *(23)* searching through GenBank available from the National Center of Biotechnology Information (NCBI, 2002). Alignments are not included if bit scores fall below 100 and they are required to be at a level greater than or equal to 40% of the best score. Exploratory calculations showed that 40% of the BLASTP bit score was a reasonable threshold for calculations of sequence entropy and its relationship with density. Also a minimal set of sequences is required, and a maximum number of 100 alignments are typically included. This yields a set of 7143 aligned protein sequences. The average number of alignments per query is 55. The frequency distribution of the BLASTP bit scores for all 130 sets of alignments is consistent with the right-skewed (i.e., positive skew) distribution for a randomized set of BLAST scores *(24)*. Here the mean, median, and the overall range of BLASTP bit scores for all 7143 alignments were 408, 354, and 100–1793. Sequence entropy S_k at amino acid position k was calculated from:

$$S_k = -\sum_{j=1,20} p_{jk} \ln p_{jk} + \sum_{j=1,20} p_j \ln p_j \qquad (1)$$

where p_{jk} is the probability at some amino acid sequence position k derived from the frequency f_{jk} for an amino acid type j at sequence position k for all

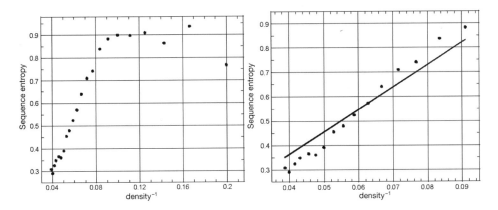

Fig. 3. Relationship between sequence variability, expressed as sequence entropy and inverse residue packing density. These results are shown for a set of 113 proteins. Note that there are two different regions seen in the figure on the left, a nearly linear region for high packing densities and a more constant region for low packing densities. On the right the data for a broad range of the higher packing densities have been fit with a straight line *(27)*.

aligned residues and p_j is the probability of amino acid type j in all alignments. Gaps were ignored. The second term corresponds to the random case, which has been subtracted *(25)*. C^α packing densities are calculated from the atomic coordinates. The radius for inclusion in the packing calculation was 9 Å. Here, we discuss the extent to which the inverse of this local packing density, as a measure of local flexibility *(26)*, is correlated with sequence variability. Calculated sequence entropies for individual proteins when compared with the inverse C^α packing density do not agree with one another. For example, three individual proteins gave slopes of the best-fit lines of 13.0, 6.1, and 4.3. This appears to be because the sample sizes within individual proteins are insufficiently large. In our sample over the entire set of proteins there were a total of 41,543 residues. The results shown in **Fig. 3** *(27)* were first averaged for each protein and then averaged over the set of proteins within each interval of density.

There are two major regions corresponding to high and low densities visible in the correlation plots of sequence entropy vs inverse packing density. The first region for higher packing densities has a steep slope, and corresponds to densities of 12–25 C^α atoms (inverse densities from 0.040–0.083). There is the hint of a slight sigmoidal shape to the curve. The second region on the right of the figure includes a significant number of residues but appears to be nearly constant in sequence entropy; this region involves lower packing densities ranging from 6 to 11. It is quite logical that in the low packing density regime, changes in sequence entropy should be unaffected by their packing. The surprise is that there is constant sequence variability for the low-residue packing density regime.

2.3. Flexibility, Hydrophobicity, and Sequence Entropy

Previously, a strong correlation was reported between computed displacements using ENs, which directly reflects residue packing *(20,28,29)* and the measured hydrogen exchange (HX). Regions of high packing density naturally resist HX, because of both stability and inaccessibility. Here, we have gone further to relate the calculated inverse C^α packing density from X-ray structures to sequence variability. Strong linear correlations are observed between sequence entropy and the inverse packing density, except for the lowest range of density. This provides a quantitative relationship between these two quantities, and can provide an important structural measure for determining likely nondisruptive sites for mutagenesis.

There is also a strong correlation between calculated sequence entropy and hydrophobicity. Clearly, correlations between the sequence variabilities as reflected in the sequence entropies and the corresponding hydrophobicities are consistent with the average behavior for residues within a given packing density range. Still, this observed correlation between average sequence entropy and hydrophobicity is noteworthy, but both density and hydrophobicity are reflecting fundamental properties related to the extent of burial. The critical importance of hydrophobicity for folding of model protein chains *(30,31)* is well known. This is consistent with the fact that key hydrophobic residues can be described as buried or tightly packed *(7–9)*. Packing and the resulting interactions associated with hydrophobicity were reported by Dima and Thirumalai *(32)* to be manifested in pairs and triplets of contacts. Our calculation of residue packing density represents a coarse-grained counting of such contacts. It is possible to calculate more detailed measures than those used here, by using full atomic representations. Such calculations would depend upon the residue's environment in more detailed ways than is given by the present simple residue density. Progress in this direction could assist with protein design, a closely related problem *(10,22,33–38)*. Further efforts are clearly required to achieve such a goal; however, the present results point the way toward such a goal. Here, packing at the residue level for coarse-grained structures has been shown to exhibit an extremely strong connection to sequence conservation. Why is the averaging here necessary? One possible explanation is that the large number of combinatorial ways in which a group of residues' atoms can be packed together requires averaging over large numbers of occurrences in order to obtain a meaningful single coarse-grained value.

3. EN Models

Originally, EN theory was developed for networks of polymer chains. James and Guth *(39–45)* developed in the forties a theory of a "phantom network" whose behavior is dictated only by the connectivity of the network chains if the effect of the excluded volume of polymer chains is neglected. The chains are phantom-like; i.e., they are able to pass readily through one another. They assumed that the network is composed of cross-linked Gaussian chains. They assumed that there are two types of network junctions. Junctions at the surface of the rubber are fixed and deform affinely with the macroscopic strain, whereas the junctions inside the network are free to fluctuate about their mean positions. The configurational partition function Z_N of the network is the product of the configurational partition functions of the individual Gaussian chains *(46)*.

$$Z_N = C \prod_{i<j} \exp[-\frac{1}{2} \sum_i \sum_j \gamma_{ij}(R_i - R_j)^2] \tag{2}$$

where R_i is the position of junction i, C is a normalization constant, and the matrix elements γ_{ij} are defined as:

$$\gamma_{ij} = \begin{cases} \dfrac{3}{2 < r_{ij}^2 > 0} & \text{if i and j are connected by a chain} \\ 0 & \text{if i and j are not connected} \end{cases} \tag{3}$$

where 0 is the mean square end-to-end vector of the chain between junctions i and j in the undeformed state. The diagonal elements are defined so that the summation of all matrix elements in a given row (or column) is zero.

$$\gamma_{ii} = -\sum_j \gamma_{ij} \tag{4}$$

This means that all statistical properties of the network depend on the connectivity matrix Γ defined in **Eqs. 3–4**. One can easily calculate fluctuations of junctions in the phantom network model, and correlations between these fluctuations for the ideal infinite network with the topology of a tree (with network junctions having functionality ϕ, the number of chains connected at each junction). An example of such a tree-like network is shown in **Fig. 4**.

These quantities are related to the matrix elements of the inverse matrix Γ^{-1}. The mean square fluctuations of the end-to-end vector $<(\Delta r)^2>$ of polymer chains depend on the functionality ϕ of the network as:

$$<(\Delta r)^2> = \frac{2}{\phi} <r^2> 0 \tag{5}$$

It can be shown that correlations are related to Γ_{ij}^{-1}

$$<(\Delta R_i \cdot \Delta R_j)> = \frac{3}{2} \Gamma_{ij}^{-1} \tag{6}$$

The problem of fluctuations of junctions and chains in random phantom networks has been studied in detail by Kloczkowski, Mark, and Erman *(47,48)*.

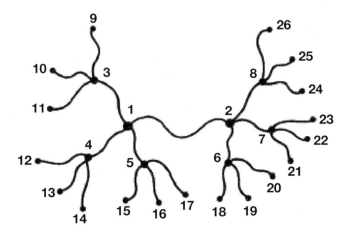

Fig. 4. The first three tiers of a unimodal symmetrically grown tree-like network with functionality $\phi = 4$. When this is applied to denatured proteins, these network connections will include both sequence (chain) connections and nonbonded interactions.

In the simplest case, when junctions i and j are directly connected by a single chain the solution of the problem converges to the following simple formula:

$$
\begin{bmatrix}
<(\Delta R_i)^2> & <(\Delta R_i \cdot \Delta R_j)> \\
<(\Delta R_i \cdot \Delta R_j)> & <(\Delta R_j)^2>
\end{bmatrix}
= <r^2>_0
\begin{bmatrix}
\dfrac{\phi - 1}{\phi(\phi - 2)} & \dfrac{1}{\phi(\phi - 2)} \\[2ex]
\dfrac{1}{\phi(\phi - 2)} & \dfrac{\phi - 1}{\phi(\phi - 2)}
\end{bmatrix}
\tag{7}
$$

where ϕ is the functionality of the network. In the case of two junctions m and n separated by d other junctions along the path joining m and n the solution of the problem is:

$$
\begin{bmatrix}
<(\Delta R_m)^2> & <(\Delta R_m \cdot \Delta R_n)> \\
<(\Delta R_m \cdot \Delta R_n)> & <(\Delta R_n)^2>
\end{bmatrix}
= <r^2>_0
\begin{bmatrix}
\dfrac{\phi - 1}{\phi(\phi - 2)} & \dfrac{1}{\phi(\phi - 2)(\phi - 1)^d} \\[2ex]
\dfrac{1}{\phi(\phi - 2)(\phi - 1)^d} & \dfrac{\phi - 1}{\phi(\phi - 2)}
\end{bmatrix}
\tag{8}
$$

and the fluctuations of the distance r_{mn} are

$$
<(\Delta r_{mn})^2> = \frac{2}{\phi(\phi - 2)(d + 1)} \frac{(\phi - 1)^{d+1} - 1}{(\phi - 1)^d} <r^2_{mn}>_0
\tag{9}
$$

The problem was also solved for the general case of fluctuations of points along the chains in the network and correlations of fluctuations among such points. To deal with this case it was assumed that each chain between two

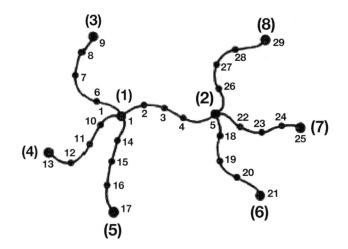

Fig. 5. Tetrafuctional network with additional bifunctional junctions that separate each chain into four subchains of equal length. Such segments can ideally represent loops on the surfaces of proteins.

functional junctions is composed of *n* Gaussian segments connected by bifunctional junctions to form a chain, as shown in **Fig. 5.**

Because of this the diagonal elements γ_{ii} of the connectivity matrix are ϕ if the index *i* corresponds to the ϕ functional junction and 2 for the bifunctional junction. The off-diagonal elements γ_{ij} are -1 if *i* and *j* are directly connected by a chain segment, and zero otherwise.

For the infinite number of tiers in the tree-like network the solution of the problem has the following form

$$
\begin{bmatrix} <(\Delta R_i)^2> & <(\Delta R_i \cdot \Delta R_j)> \\ <(\Delta R_i \cdot \Delta R_j)> & <(\Delta R_j)^2> \end{bmatrix} =
$$

$$
<r^2>_0 \begin{bmatrix} \dfrac{\phi-1}{\phi(\phi-2)} + \dfrac{\zeta(1-\zeta)(\phi-2)}{\phi} & \dfrac{1}{\phi(\phi-2)(\phi-1)^d} + \dfrac{[1+\zeta(\phi-2)][(\phi-1)-\theta(\phi-2)]}{\phi} \\ \dfrac{1}{\phi(\phi-2)(\phi-1)^d} + \dfrac{[1+\zeta(\phi-2)][(\phi-1)-\theta(\phi-2)]}{\phi} & \dfrac{\phi-1}{\phi(\phi-2)} + \dfrac{\theta(1-\theta)(\phi-2)}{\phi} \end{bmatrix}
\tag{10}
$$

where $\zeta = \dfrac{i-1}{n}$ and $\theta = \dfrac{j-1}{n}$ are the fractional distances of sites *i* and *j* from the nearest ϕ-functional junctions on their left side (as shown in **Fig. 6**) with $0 < \zeta, \theta < 1$, and d is the number of ϕ-functional junctions between sites *i* and *j*. Our preliminary results show that **Eq. 10** is applicable also to proteins loops.

3.1. EN Models of Proteins: GNM

The EN models developed for polymer networks have been successfully applied to proteins in many different applications by us (**49–70**) and many others

Fig. 6. Two points i and j of the network separated by $d = 3$ tetrafunctional junctions. The positions of points i and j are measured with respect to the nearest multifunctional junction from the left as fractions ζ, θ of the contour length of the chain between multifunctional junctions.

(71–96). The basic difference between elastomeric polymer networks and proteins is that a protein is a collapsed polymer with many nonbonded interactions, whereas for polymers these nonbonded interactions could be (as for the phantom chain models) completely neglected. The simplest EN model of proteins is the GNM. In the GNM it is assumed that fluctuations of residues about their mean positions are spherically symmetric. (The ANM discussed in **Subheading 3.2.** assumes that these fluctuations are represented by ellipsoids.) For the usual level of one point per amino acid, usually taken as the C^{α} atoms, all close points within a cutoff distance R_c are connected with identical deformable springs. Although this is a kind of uniform material model of proteins, it provides a representation of the mechanics of the intact structure, and does not have the level of detail of a potential energy function that would be needed for describing the folding process or how the structure was arrived at. It is assumed that the initial structure (usually a well-resolved crystal structure) is lowest in energy. The energy of each spring increases with deviations $\Delta\mathbf{R}$ away from the initial structure, with $\Delta\mathbf{R}$ following a Gaussian distribution. The total energy of the network at is given by

$$V_{tot} = (\gamma/2)\; tr\; [\Delta\mathbf{R}^{\mathrm{T}}\Gamma\Delta\mathbf{R}] \tag{11}$$

where Γ is the Kirchhoff (contact) matrix defined on the basis of the cutoff distance R_c and tr is the trace of the matrix. Then the average changes in position, given either as the correlation between the displacements of pairs of residues i and j or as the mean-square fluctuations for a single residue, i, are

$$<\Delta R_i^2> = <\Delta\mathbf{R}_i \cdot \Delta\mathbf{R}_i> = (1/Z_N)\int(\Delta\mathbf{R}_i \cdot \Delta\mathbf{R}_i)\,\exp\{-V_{tot}/kT\}\,d\{\Delta\mathbf{R}\}$$
$$= (3kT/2\gamma)[\Gamma^{-1}]_{ii} \tag{12}$$

The most direct validation of this approach has been made by comparing the computed fluctuation magnitudes with the Debye–Waller temperature factors usually reported by crystallographers as B-factors

$$B_i = 8\pi^2 <\Delta R_i^2>/3 \tag{13}$$

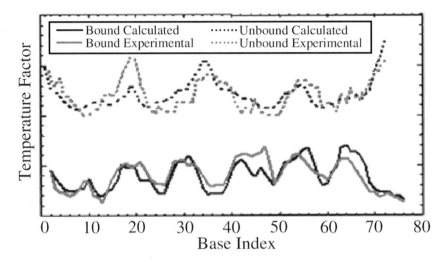

Fig. 7. Comparison of the computed (black) and experimental (gray) temperature factors, for tRNA^{Gln}, with and without its cognate. Notably the changes in nucleotide mobilities occurring upon binding of the tRNA to the protein are well reproduced by the Gaussian network model computations.

Usually excellent agreement is achieved by this approach when B-factors of high-resolution structures are compared. Another test was a comparison with H/D exchange data *(66)*, where the protection factors and/or free energy of exchange from experiments for a series of proteins were closely reproduced/ explained by the theoretical (GNM) results. Other tests (performed after mode-decomposition of GNM dynamics, *see* **Subheading 3.2.**) include a comparison of the most constrained regions in the GNM with conserved residues and/or folding nuclei, comparison of hinge sites (minima in the slowest mode fluctuation curve) and cross-correlations between domain motions, with those revealed in experimental studies.

The applicability of the GNM to oligonucleotides, or protein–RNA complexes is supported by our studies. **Figure 7** shows how the GNM methodology closely reproduces the experimental B-factors measured for tRNA in isolated form, and when bound to Gln synthetase.

Another study that lends support to the use of the GNM is the application to HIV-1 reverse transcriptase (RT), where RTs in different forms (bound to DNA, or to inhibitors) were analyzed to infer the mechanism of action, consistent with experimental data *(60,65,97)*.

3.2. EN Models of Proteins: ANM

The GNM provides information on the *sizes* of fluctuations, $<\Delta R_i^2>$, and on their cross-correlations, $<\Delta \mathbf{R}_i \cdot \Delta \mathbf{R}_j>$. The ANM is the extension that includes fluctuation vector *directions (68)*. In the ANM, the potential is defined as a function of inter-residue distances R_{ij} as

$$V_2(\vec{r}_{ij}) = \frac{1}{2}\gamma(R_{ij} - R_{ij}^0)^2 H(R_c - R_{ij}) \tag{14}$$

where R_{ij}^0 is the equilibrium (native state) distance between residues i and j, and $H(x)$ is the Heaviside step function equal to 1 if $x > 0$, and zero otherwise. V can be rewritten as an expression similar in form to **Eq. 13**, where Γ is replaced by the $3N \times 3N$ Hessian matrix $\mathbf{H}_{ANM}$ of the second derivatives of the potential with respect to positions, similar to classical normal mode analysis (NMA) of equilibrium structures. H_{ANM} provides information on the individual components, ΔX_i, ΔY_i, and ΔZ_i of the deformation vectors $\Delta \mathbf{R}_i$, and their correlations, whereas GNM yields only the magnitudes of ΔR_i and the correlations between ΔR_i vectors, only.

The contributions of the individual EN modes to the observed motion are found by eigenvalue decomposition of $\mathbf{G}$(GNM) (or $\mathbf{H}$[ANM]). The individual eigenvectors represent the normal modes of motion. The decomposition is represented as $\Gamma^{-1} = \Sigma \lambda_k^{-1} [\mathbf{u}_k \mathbf{u}_k^T]$, where λ_k is the kth eigenvalue and $\mathbf{u}_k$ is the kth eigenvector. Note that the solution requires using singular value decomposition to remove the zero eigenvalue terms corresponding to rigid-body translations and rotations in order to obtain the remaining eigenvalues and eigenvectors. In general, the result is a small number of modes that contribute most, and a large number that have only small contributions *(62)*. This is consistent with the general viewpoint that proteins are highly cohesive materials, and that the collective motions, which correspond to domain-like motions, are the most related to function. The localized, high-frequency motions do not significantly contribute to the overall motions, but are perhaps related to stability *(20,29)*. We note that the use of harmonic potentials restrict the amplitude of motions, and truly large transitions are not observed with the present EN models *(91,92)*.

We have recently applied the ANM *(68)* to the ribosome (these results will be shown in detail in **Subheading 4.1.**). This and a recent study by another group *(98)*, which also employs the ANM methodology, provide firm evidence for the utility and applicability of the EN models to represent and predict the dynamics of not only proteins, but also other biological structures such as protein–RNA complexes and the ribosome.

3.3. Hierarchical Levels of Coarse Graining

The hypothesis is that the most cooperative (lowest frequency) modes of motion are insensitive to the details of the model and parameters, and can be satisfactorily captured by adopting low-resolution EN models. **Figure 8** gives the results from an application *(63)* to two large proteins of the hierarchical

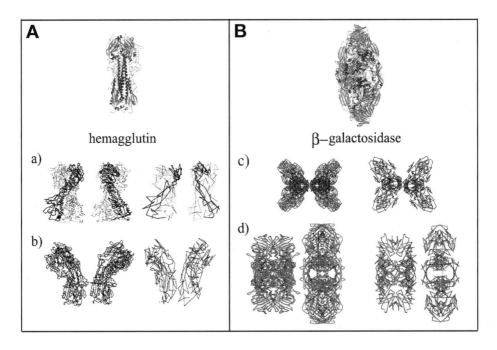

Fig. 8. Motions of (**A**) influenza virus hemagglutinin A (HA) and (**B**) β-galactosidase at different levels of detail. The top diagrams show the native structures, and the diagrams (**A–D**) show their "deformed" conformations predicted by the anisotropic network model. (**A**) and (**B**) show the ribbon diagram (left) and coarse-grained Gaussian network model representation (right) of deformed HA structures, induced by the slowest two collective modes of motion. Their counterparts for β-galactosidase are shown in **C** and **D**. The coarse-grained models are constructed using $m = 10$, i.e., groups of 10 consecutive residues are condensed into unified nodes of the elastic network model *(52,61,62)*.

coarse-graining scheme we have recently introduced *(62)*, which support this hypothesis.

The idea is to adopt a low-resolution EN representation, where each node stands for a group of m residues condensed into a single interaction site. Calculations performed for different levels of coarse-graining ($m = 1, 2, 10, 40$) have shown that the slowest modes of motion are preserved regardless of the level of coarse-graining. Part A compares the slowest mode (a global torsion) predicted for influenza virus hemagglutinin A (HA) using EN models of $m = 1$ (left) and 10 (right). One of the monomers is shown in black for visualizing the motion. The two symmetric structures shown in each case refer to the "deformed" conformations between which the original structure fluctuates, thus, disclosing the potential global changes in structure favored by the equilibrium topology of inter-residue contacts. Part B shows the second most probable mode, a cooperative bending of the trimer. The functional implications of these modes have been discussed in our earlier study *(52)*. Panel B shows similar results for β-galactosidase, confirming that the global motions of the system of N residues are almost identically reproduced by adopting a coarse-grained ($m = 10$, $N/10$ nodes) model.

The ANM **computation time** for a given structure scales with the third power of the number of EN nodes. The use of a model of $N/10$ sites would increase the computational speed by three orders of magnitude. Application of the hierarchical coarse graining to HA showed that the global dynamics are accurately captured even with an EN model of only $N/40$ nodes. The change in the coordinates of every m^{th} residue is computed in this approach and by repeating this approach m times, each time shifting the selected node residue by one residue along the sequence permits us to reconstruct the fluctuation profile of all N residues, with a net reduction in computation time by a factor of m^2.

3.4. Mixed Coarse Graining

A MCG approach *(52)* has also been introduced for EN models, where a protein's native conformation is represented by different regions having high and low resolution. The aim here is to capture the dynamics of the interesting parts in structures at higher resolution and retain the remainder of the structure at lower resolution. As a result, the total number of nodes in a system can be kept sufficiently low for computational tractability.

4. Modeling Ribosome Functional Dynamics

The X-ray crystal structure of the 30S subunit from the *Thermus thermophilus* was reported by Wimberly et al. *(99)*. The crystal structure of the entire assembly of the 70S ribosome from *T. thermophilus* has been reported by Yusupov et al. *(100)*. We performed ANM analysis on the structure of the 30S subunit reported from Wimberly et al. (PDB code 1J5E), the 50S subunit from Yusupov et al. (PDB code 1GIY) and the 70S ribosome structure reported by Yusupov et al. *(100)* (PDB structures 1gix and 1giy). The 30S subunit of Wimberly et al. *(99)* contains full coordinates for the rRNAs, whereas the structure of the 70S ribosome by Yusupov et al. *(100)* contains only the P atoms of the rRNAs and the C^α atoms of the proteins, except for the three tRNAs and mRNA. The cutoff distance for interaction is taken as 15 Å. We used one interaction site on the P atom per nucleotide of the rRNAs and tRNAs and one interaction site on the C^α atom of the proteins. The cutoff distance between the $C^\alpha - C^\alpha$ atoms is still 15Å, but the cutoff distance between P–P and P–C^α atoms is increased to 24 Å. The dimensions of the ANM contact matrices (equivalent to NMA Hessian) are $16,266 \times 16,266$ for the 30S subunit, and $29,238 \times 29,238$ for the 70S. We utilized the software BLZPACK

CP

L1 stalk

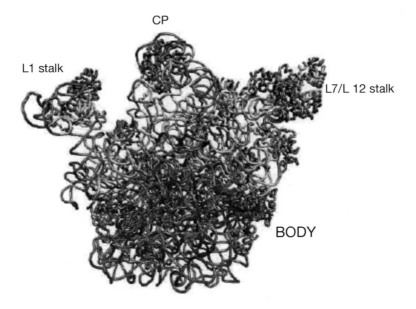

L7/L 12 stalk

BODY

Fig 9. Structure of the 50S subunit black- and white-coded according to the deforma-tion energy averaged over the 10 slowest (dominant) modes. The interfacial regions between these domains and the body is highly flexible, indicated by the high deforma-tion energies found by anisotropic network model, and shown by the lighter portions above. This flexibility ensures the functional mobilities of the L1 stalk, CP, and L7/L12 domains *(49)*.

of Marques and Sanejouand *(88)*. This yields a specific number of eigenvalues and eigenvectors for this matrix and is quite rapid.

4.1. ANM Results for the 50S Subunit and the 70S Assembly

Flexibility in the 50S subunit predicted by the ANM is illustrated in **Fig. 9**. The figure displays the 50S subunit from *T. thermophilus* (PDB code: 1GIY), viewed from the 30S interface side. Residues are black- and white-coded according to the deformation energy averaged over the 10 slowest modes. The landmarks on the structures are: center protuberance (CP), L1 stalk, L7/L12 stalk, and the body. There are four dynamic domains: L1 stalk, L7/L12 stalk, CP, and the body. The two stalks and the CP are found by GNM/ANM (not shown) to be highly mobile, and their relative motions oppose each other against the body for the first few slowest modes. For example, in mode 1, only the L1 stalk is mobile and it moves toward the CP. In mode 2, we see the L1 and L7/L12 stalks move out of the plane of the page, and the CP moves in the oppo-site direction. Motions of the 70S assembly in the slowest modes can be viewed in http://ribosome. bb.iastate.edu/70SnKmode/. The deformable residues are found to be located mostly at the interface between the two subunits, an indication that the two subunits form separate dynamic domains relative to each other. A recent study compared the differences between the atomic positions from low-resolution cryo-EM density maps of the 70S ribosome in two different functional states *(98,101,102)* and reported that they were related by a ratchet-like motion, which is the type of motion we observe in several of the slowest modes of motion. In an additional experimental observation from cryo-EM, the 30S subunit was observed to rotate counter-clockwise (viewed from the 30S solvent side) when EF-G binds, reducing the opening between the CP and L1 stalk and bifurcating the L7/L12 stalk. This motion was most similar to our computed mode 3. In this mode, when the two stalks pull toward the CP, the 30S subunit rotates counter-clockwise viewed from the 30S solvent side. The L7/L12 stalk also undergoes a large conformational change that may be linked to an observed bifurcation. The study of Tama et al. *(98)* also reported that their mode 3 resembled the experimentally observed ratchet-like motion. However, we have seen a much more significant rearrangement in the deformation of the L7/L12 stalk in this mode than they report. Another difference occurs for mode 1, where they have only the motion of the L1 stalk, whereas our mode 1 includes a counter-rotation of the two subunits and the motions of the two stalks.

Analysis of the motions of the structural subunits and their relative motions within the 70S ribosome give further information. **Table 2** gives the orientation correlations between the motions of the ribosome structural units averaged over the 100 slowest modes.

The motions of the 30S and 50S subunits are almost perfectly anticorrelated (i.e., coupled but rotation in opposite directions).

This could arise from both the counter-rotation and from the opening-closing motions. Their motions are weakly correlated with the tRNAs, because the tRNAs and the mRNA move linearly and in a direction tangential to the rotational motions of the 30S and 50S subunits. The 30S and mRNA are positively correlated, whereas the 50S and mRNA are anticorrelated. This indicates that the mRNA moves in the same direction as the 30S subunit, and in the opposite direction to the 50S subunit. In addition, the A-tRNA and the P-tRNA are strongly and positively correlated with each other, but they are less correlated with the E-tRNA. A strong correlation between the A-tRNA and the P-tRNA indicates that the A-tRNA and the P-tRNA move together in the same direction. This may indicate that the translocation of the A-tRNA to the P site and the translocation of the P-tRNA to the E site are likely to occur simultaneously. However, we see that the E-tRNA motion is not so strongly correlated with either the A-tRNA or P-tRNA, but nonetheless more strongly correlated with its neighboring P-site tRNA than with the A-site tRNA *(49)*.

Table 2
Correlations Between the Motions of Ribosome Structural Units (49)[a]

	50S	A-tRNA	P-tRNA	E-tRNA	mRNA
30S	−0.99	−0.06	−0.10	−0.07	0.50
50S		−0.01	0.03	−0.01	−0.55
A-tRNA			0.77	0.17	0.42
P-tRNA				0.31	0.39
E-tRNA					0.19

[a]From the anisotropic network model analysis of the PDB files:1GIX + 1GIY.

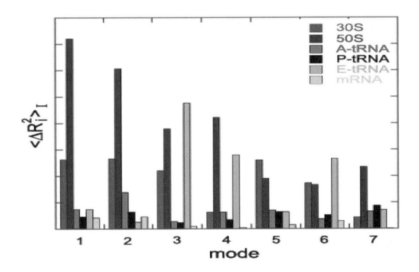

Fig. 10. Amplitude of motions induced by the first seven slowest modes on 70S assembly subunits, averaged over residues forming these subunits. The 50S shows especially large displacements in the slowest two modes. The E-site tRNA exhibits large displacements in modes 3, 4, and 6. The order of the bars for each mode (left to right) is that given in the key (top to bottom) (49).

Figure 10 shows the amplitudes of motions in the slowest modes, revealing the high mobility of the 50S and 30S subunits and the E-site tRNA during the collective motions of the 70S assembly.

Analysis of the deformation energies within each structural unit reveals that the large mobility of the E-tRNA is mainly owing to intersubunit movements that relate to the exit of the E-tRNA from the assembly. Further analysis of the tRNA motions has revealed that the rigid body directions of their motions correspond closely to the directional vectors connecting the center of the tRNA to the center of its neighboring tRNA site *(49)*. We anticipate learning further details of the mechanism of the ribosome by further analysis and application of the MCG approach.

5. Conclusion

We have discussed the effects of residue packing density on protein structure and how these observations provide some justification for the EN models. The basic approaches used in these models originates in polymer theory where it has been shown possible to represent a variety of systems having different connectivities. These approaches outlined could, for example, be combined with the EN protein models to represent the disordered parts of proteins.

The EN models for proteins are robust and do not depend on the details of protein structure; consequently they can be applied to extremely large systems, which can be possible only by approximating structures with relatively highly coarse-grained models. We have shown the results of EN simulations on the ribosome, which compare favorably with the electron micrograph images of the ratchet motions in the ribosome. The correlated motions seen overall correspond closely to what would be expected for the functional processing steps of the ribosome.

Acknowledgement

We acknowledge the financial support from the National Institutes of Health grant R01GM072014-01.

References

1. Richards, F. M. (1974) The interpretation of protein structures: total volume, group volume and packing density. *J. Mol. Biol.* **82,** 1–14.
2. Hermans, J. and Scheraga, H. A. (1961) Structural studies of ribonuclease. IV. Abnormal ionizable groups. *J. Am. Chem. Soc.* **83,** 3283–3292.
3. Richards, F. M. (1997) Protein stability: still an unsolved problem. *Cell. Mol. Life Sci.* **53,** 790–802.
4. Liang, J. and Dill, K. A. (2001) Are proteins well-packed? *Biophys. J.* **81,** 751–766.
5. Eriksson, A. E., Baase, W. A., Zhang, X. J., et al. (1992) Response of a protein structure to cavity-creating mutations and its relation to the hydrophobic effect. *Science* **255,** 178–183.
6. Privalov, P. L. (1996) Intermediate states in protein folding. *J. Mol. Biol.* **258,** 707–725.
7. Ptitsyn, O. B. (1998) Protein folding and protein evolution: common folding nucleus in different subfamilies of c-type cytochromes? *J. Mol. Biol.* **278,** 655–666.
8. Ptitsyn, O. B. and Ting, K. L. (1999) Non-functional conserved residues in globins and their possible role as a folding nucleus. *J. Mol. Biol.* **291,** 671–682.

9. Ting, K. L. and Jernigan, R. L. (2002) Identifying a folding nucleus for the Lysozyme/alpha-lactalbumin family from sequence conservation clusters. *J. Mol. Evol.* **54,** 425–436.

10. Dahiyat, B. I. and Mayo, S. L. (1997) De novo protein design: fully automated sequence selection. *Science* **278,** 82–87.

11. Mirny, L., Abkevich, V. L., and Shakhnovich, E. I. (1998) How evolution makes proteins fold quickly. *Proc. Natl. Acad. Sci. USA* **95,** 4976–4981.

12. Solis, A. D. and Rackovsky, S. (2002) Optimally informative backbone structural propensities in proteins. *Proteins* **48,** 463–486.

13. Bagci, Z., Kloczkowski, A., Jernigan, R. L., and Bahar, I. (2003) The origin and extent of coarse-grained regularities in protein internal packing. *Proteins* **53,** 56–67.

14. Bagci, Z., Jernigan, R. L., and Bahar, I. (2002) Residue packing in proteins: uniform distribution on a coarse-grained scale. *J. Chem. Phys.* **116,** 2269–2276.

15. Bagci, Z., Jernigan, R. L., and Bahar, I. (2002) Residue coordination in proteins conforms to the closest packing of spheres. *Polymer* **43,** 451–459.

16. Jones, D. T. (2000) Protein structure prediction in the postgenomic era. *Curr. Opin. Struct. Biol.* **10,** 371–379.

17. Baker, D. and Sali, A. (2001) Protein structure prediction and structural genomics. *Science* **294,** 93–96.

18. Bryant, S. H. and Lawrence, C. E. (1993) An empirical energy function for threading protein sequence through the folding motif. *Proteins* **16,** 92–112.

19. Marti-Renom, M. A., Stuart, A. C., Fiser, A., Sanchez, R., Melo, F., and Sali, A. (2000) Comparative protein structure modeling of genes and genomes. *Annu. Rev. Biophys. Biomol. Struct.* **29,** 291–325.

20. Demirel, M. C., Atilgan, A. R., Jernigan, R. L., Erman, B., and Bahar, I. (1998) Identification of kinetically hot residues in proteins. *Protein Sci.* **7,** 2522–2532.

21. Kloczkowski, A., Ting, K. L., Jernigan, R. L., and Garnier, J. (2002) Combining the GOR V algorithm with evolutionary information for protein secondary structure prediction from amino acid sequence. *Proteins* **49,** 154–166.

22. Koehl, P. and Levitt, M. (2002) Protein topology and stability define the space of allowed sequences. *Proc. Natl. Acad. Sci. USA* **99,** 1280–1285.

23. Altschul, S. F., Madden, T. L., Scaffer, A. A., et al. (1997) Gapped BLAST and PSI-BLAST: a new generation of protein database search programs. *Nucleic Acids Res.* **25,** 3389–3402.

24. Altschul, S. F., Boguski, M. S., Gish, W., and Wooten, J. C. (1994) Issues in searching molecular sequence databases. *Nat. Genet.* **6,** 119–129.

25. Gerstein, M. and Altman, R. B. (1995) Average core structures and variability measures for protein families: Application to the immunoblobulins. *J. Mol. Biol.* **251,** 161–175.

26. Haliloglu, T., Bahar, I., and Erman, B. (1997) Gaussian dynamics of folded proteins. *Phys. Rev. Lett.* **79,** 3090–3093.

27. Liao, H., Yeh, W., Chiang, D., Jernigan, R. L., and Lustig, B. (2005) Protein sequence entropy is closely related to packing density and hydrophobicity. *Protein Eng. Des. Sel.* **18,** 59–64.

28. Bahar, I. (1999) Dynamics of proteins and biomolecular complexes. Inferring functional motions from structure. *Rev. Chem. Eng.* **15,** 319–347.
29. Bahar, I., Atilgan, A. R., Demirel, M. C., and Erman, B. (1998) Vibrational dynamics of folded proteins: Significance of slow and fast motions in relation to function and stability. *Phys. Rev. Lett.* **80,** 2733–2736.
30. Hinds, D. A. and Levitt, M. (1994) Exploring conformational space with a simple lattice model for protein structure. *J. Mol. Biol* **243,** 668–682.
31. Dill, K. A., Bromberg, S., Yue, K., et al. (1995) Principles of protein folding: a perspective from simple exact models. *Protein Sci.* **4,** 561–602.
32. Dima, R. I. and Thirmalai, D. (2004) Asymmetry in the shapes of folded and denatured states of proteins. *J. Phys. Chem. B* **108,** 6564–6570.
33. Li, H., Tang, C., and Wingreen, N. S. (1998) Are protein folds atypical? *Proc. Natl. Acad. Sci. USA* **95,** 4987–4990.
34. Buchler, N. E. and Goldstein, R. A. (1999) Effect of alphabet size and foldability requirements on protein structure designability. *Proteins* **34,** 113–124.
35. Shih, C. T., Su, Z. Y., Gwan, J. F., Hao, B. L., Hsieh, C. H., and Lee, H. C. (2000) Mean-field HP model, designability and alpha-helices in protein structures. *Phys. Rev. Lett.* **84,** 386–389.
36. Tiana, G., Broglia, R. A., and Provasi, D. (2001) Designability of lattice model heteropolymers. *Phys. Rev. E. Stat. Nonlin. Soft. Matter Phys.* **64,** 011904.
37. Larson, S. M., England, J. L., Desjarlais, J. R., and Pande, V. S. (2002) Thoroughly sampling sequence space: large-scale protein design of structural ensembles. *Protein Sci.* **11,** 2804–2813.
38. England, J. L., Shakhnovich, B. E., and Shakhnovich, E. I. (2003) Natural selection of more designable folds: a mechanism for thermophilic adaptation. *Proc. Natl. Acad. Sci. USA* **100,** 8727–8731.
39. James, H. M. and Guth, E. (1941) Elastic and thermoelastic properties of rubber like materials. *Ind. Eng. Chem.* **33,** 624–629.
40. James, H. M. and Guth, E. (1942) Theory of rubber elasticity for development of synthetic rubbers. *Ind. Eng. Chem.* **34,** 1365–1367.
41. James, H. M. and Guth, E. (1943) Theory of the elastic properties of rubber. *J. Chem. Phys.* **11,** 455–481.
42. James, H. M. and Guth, E. (1944) Theory of the elasticity of rubber. *J. Appl. Phys.* **15,** 294–303.
43. James, H. M. and Guth, E. (1947) Theory of the increase in rigidity of rubber during cure. *J. Chem. Phys.* **15,** 669–683.
44. James, H. M. and Guth, E. (1949) Simple presentation of network theory of rubber, with a discussion of other theories. *J. Polym. Sci.* **4,** 153–182.
45. James, H. M. and Guth, E. (1953) Statistical thermodynamics of rubber elasticity. *J. Chem. Phys.* **21,** 1039.
46. Flory, P. J. (1976) Statistical thermodynamics of random networks. *Proc. Roy. Soc. London* **A351,** 351–380.
47. Kloczkowski, A., Mark, J. E., and Erman, B. (1989) Chain dimensions and fluctuations in random elastomeric networks I. Phantom Gaussian networks in the undeformed state. *Macromolecules* **22,** 1423–1432.

48. Kloczkowski, A., Mark, J. E., and Erman, B. (1992) The James-Guth Model in modern theories of neutron scattering from polymer networks. *Comput. Polym. Sci.* **2**, 8–31.
49. Wang, Y., Rader, A. J., Bahar, I., and Jernigan, R. L. (2004) Global ribosome motions revealed with elastic network model. *J. Struct. Biol.* **147**, 302–314.
50. Sen, T. Z., Kloczkowski, A., Jernigan, R. L., et al. (2004) Predicting binding sites of hydrolase-inhibitor complexes by combining several methods. *BMC Bioinformatics* **5**, 205.
51. Navizet, I., Lavery, R., and Jernigan, R. L. (2004) Myosin flexibility: structural domains and collective vibrations. *Proteins* **54**, 384–393.
52. Kurkcuoglu, O., Jernigan, R. L., and Doruker, P. (2004) Mixed levels of coarse-graining of large proteins using elastic network model succeeds in extracting the slowest motions. *Polymer* **45**, 649–657.
53. Kundu, S. and Jernigan, R. L. (2004) Molecular mechanism of domain swapping in proteins: an analysis of slower motions. *Biophys. J.* **86**, 3846–3854.
54. Kim, M. K., Jernigan, R. L., and Chirikjian, G. S. (2003) An elastic network model of HK97 capsid maturation. *J. Struct. Biol.* **143**, 107–117.
55. Kim, M. K., Chirikjian, G. S., and Jernigan, R. L. (2002) Elastic models of conformational transitions in macromolecules. *J. Mol. Graph. Model.* **21**, 151–160.
56. Kim, M. K., Jernigan, R. L., and Chirikjian, G. S. (2002) Efficient generation of feasible pathways for protein conformational transitions. *Biophys. J.* **83**, 1620–1630.
57. Keskin, O., Jernigan, R. L., and Bahar, I. (2000) Proteins with similar architecture exhibit similar large-scale dynamic behavior. *Biophys. J.* **78**, 2093–2106.
58. Keskin, O., Durell, S. R., Bahar, I., Jernigan, R. L., and Covell, D. G. (2002) Relating molecular flexibility to function: a case study of tubulin. *Biophys. J.* **83**, 663–680.
59. Keskin, O., Bahar, I., Flatow, D., Covell, D. G., and Jernigan, R. L. (2002) Molecular mechanisms of chaperonin GroEL-GroES function. *Biochemistry* **41**, 491–501.
60. Jernigan, R. L., Bahar, I., Covell, D. G., Atilgan, A. R., Erman, B., and Flatow, D. T. (2000) Relating the structure of HIV-1 reverse transcriptase to its processing steps. *J. Biomol. Struct. Dyn.* **11**, 49–55.
61. Doruker, P. and Jernigan, R. L. (2003) Functional motions can be extracted from on-lattice construction of protein structures. *Proteins* **53**, 174–181.
62. Doruker, P., Jernigan, R. L., and Bahar, I. (2002) Dynamics of large proteins through hierarchical levels of coarse-grained structures. *J. Comput. Chem.* **23**, 119–127.
63. Doruker, P., Jernigan, R. L., Navizet, I., and Hernandez, R. (2002) Important fluctuation dynamics of large protein structures are preserved upon coarse-grained renormalization. *Int. J. Quant. Chem.* **90**, 822–837.
64. Sen, T. Z., Feng, Y., Garcia, J., Kloczkowski, A., and Jernigan, R. L. (2006) The extent of cooperativity of protein motions observed with Elastic Network Model is similar for atomic and coarser-grained models. *J. Chem. Theory Comput.*, in press.
65. Bahar, I., Erman, B., Jernigan, R. L., Atilgan, A. R., and Covell, D. G. (1999) Collective motions in HIV-1 reverse transcriptase: examination of flexibility and enzyme function. *J. Mol. Biol.* **285**, 1023–1037.

66. Bahar, I., Wallqvist, A., Covell, D. G., and Jernigan, R. L. (1998) Correlation between native-state hydrogen exchange and cooperative residue fluctuations from a simple model. *Biochemistry* **37,** 1067–1075.

67. Bahar, I. and Jernigan, R. L. (1999) Cooperative fluctuations and subunit communication in tryptophan synthase. *Biochemistry* **38,** 3478–3490.

68. Atilgan, A. R., Durell, S. R., Jernigan, R. L., Demirel, M. C., Keskin, O., and Bahar, I. (2001) Anisotropy of fluctuation dynamics of proteins with an elastic network model. *Biophys. J.* **80,** 505–515.

69. Kim, M. K., Jernigan, R. L., and Chirikjian, G. S. (2005) Rigid-cluster models of conformational transitions in macromolecular machines and assemblies. *Biophys. J* **89,** 43–55.

70. Bahar, I. and Jernigan, R. L. (1998) Vibrational dynamics of transfer RNAs: comparison of the free and synthetase-bound forms. *J. Mol. Biol.* **281,** 871–884.

71. Zheng, W. and Doniach, S. (2003) A comparative study of motor-protein motions by using a simple elastic-network model. *Proc. Natl. Acad. Sci. USA* **100,** 13,253–13,258.

72. Zheng, W. and Brooks, B. (2005) Identification of dynamical correlations within the myosin motor domain by the normal mode analysis of an elastic network model. *J. Mol. Biol.* **346,** 745–759.

73. Van Wynsberghe, A., Li, G., and Cui, Q. (2004) Normal-mode analysis suggests protein flexibility modulation throughout RNA polymerase's functional cycle. *Biochemistry* **43,** 13,083–13,096.

74. Tirion, M. M. (1996) Large amplitude elastic motions in proteins from a single-parameter, atomic analysis. *Phys. Rev. Lett.* **77,** 1905–1908.

75. Temiz, N. A., Meirovitch, E., and Bahar, I. (2004) Escherichia coli adenylate kinase dynamics: comparison of elastic network model modes with mode-coupling (15)N-NMR relaxation data. *Proteins* **57,** 468–480.

76. Tama, F. and Brooks, C. L., III (2005) Diversity and identity of mechanical properties of icosahedral viral capsids studied with elastic network normal mode analysis. *J. Mol. Biol.* **345,** 299–314.

77. Tama, F., Wriggers, W., and Brooks, C. L., III (2002) Exploring global distortions of biological macromolecules and assemblies from low-resolution structural information and elastic network theory. *J. Mol. Biol* **321,** 297–305.

78. Tama, F., Miyashita, O., and Brooks, C. L., III (2004) Flexible multi-scale fitting of atomic structures into low-resolution electron density maps with elastic network normal mode analysis. *J. Mol. Biol.* **337,** 985–999.

79. Micheletti, C., Lattanzi, G., and Maritan, A. (2002) Elastic properties of proteins: Insight on the folding process and evolutionary selection of native structures. *J. Mol. Biol.* **321,** 909–921.

80. Li, G. and Cui, Q. (2004) Analysis of functional motions in Brownian molecular machines with an efficient block normal mode approach: Myosin-II and Ca2+ -ATPase. *Biophys. J.* **86,** 743–763.

81. Kong, Y., Ming, D., Wu, Y., Stoops, J. K., Zhou, Z. H., and Ma, J. (2003) Conformational flexibility of pyruvate dehydrogenase complexes: a computational analysis by quantized elastic deformational model. *J. Mol. Biol.* **330,** 129–135.

82. Schuyler, A. D. and Chirijkian, G. S. (2005) Efficient determination of low-frequency normal modes of large protein structures by cluster-NMA. *J. Mol. Graph. Model* **24,** 46–58.

83. Kong, Y. F., Ma, J. P., Karplus, M., and Lipscomb, W. N. (2006) The allosteric mechanism of yeast chorismate mutase. A dynamic analysis. *J. Mol. Biol.* **356,** 237–247.

84. Delarue, M. and Sanejouand, Y. H. (2002) Simplified normal mode analysis of conformational transitions in DNA-dependent polymerases: the elastic network model. *J. Mol. Biol.* **320,** 1011–1024.

85. ben Avraham, D. and Tirion, M. M. (1995) Dynamic and elastic properties of F-actin: a normal-modes analysis. *Biophys. J.* **68,** 1231–1245.

86. Valadie, H., Lacapcre, J. J., Sanejouand, Y. H., and Etchebest, C. (2003) Dynamical properties of the MscL of Escherichia coli: a normal mode analysis. *J. Mol. Biol.* **332,** 657–674.

87. Sanejouand, Y. H. (1996) Normal-mode analysis suggests important flexibility between the two N-terminal domains of CD4 and supports the hypothesis of a conformational change in CD4 upon HIV binding. *Protein Eng* **9,** 671–677.

88. Marques, O. and Sanejouand, Y. H. (1995) Hinge-bending motion in citrate synthase arising from normal mode calculations. *Proteins* **23,** 557–560.

89. Reuter, N., Hinsen, K., and Lacapere, J. J. (2003) The nature of the low-frequency normal modes of the E1Ca form of the SERCA1 Ca^{2+}-ATPase. *Ann. N. Y. Acad. Sci.* **986,** 344–346.

90. Hinsen, K., Reuter, N., Navaza, J., Stokes, D. L., and Lacapere, J. J. (2005) Normal mode-based fitting of atomic structure into electron density maps: application to sarcoplasmic reticulum Ca-ATPase. *Biophys. J.* **88,** 818–827.

91. Hinsen, K., Thomas, A., and Field, M. J. (1999) Analysis of domain motions in large proteins. *Proteins* **34,** 369–382.

92. Hinsen, K. (1998) Analysis of domain motions by approximate normal mode calculations. *Proteins* **33,** 417–429.

93. Hinsen, K., Thomas, A., and Field, M. J. (1999) A simplified force field for describing vibrational protein dynamics over the whole frequency range. *J. Chem. Phys.* **111,** 10,766–10,769.

94. Beuron, F., Flynn, T. C., Ma, J., Kondo, H., Zhang, X., and Freemont, P. S. (2003) Motions and negative cooperativity between p97 domains revealed by cryo-electron microscopy and quantised elastic deformational model. *J. Mol. Biol.* **327,** 619–629.

95. Ming, D., Kong, Y., Lambert, M. A., Huang, Z., and Ma, J. (2002) How to describe protein motion without amino acid sequence and atomic coordinates. *Proc. Natl. Acad. Sci. USA* **99,** 8620–8625.

96. Ming, D., Kong, Y., Wakil, S. J., Brink, J., and Ma, J. (2002) Domain movements in human fatty acid synthase by quantized elastic deformational model. *Proc. Natl. Acad. Sci. USA* **99,** 7895–7899.

97. Temiz, N. A. and Bahar, I. (2002) Inhibitor binding alters the directions of domain motions in HIV-1 reverse transcriptase. *Proteins* **49,** 61–70.

98. Tama, F., Valle, M., Frank, J., and Brooks, C. L., III (2003) Dynamic reorganization of the functionally active ribosome explored by normal mode analysis and cryo-electron microscopy. *Proc. Natl. Acad. Sci. USA* **100,** 9319–9323.

99. Wimberly, B. T., Brodersen, D. E., Clemons, W. M., Jr., et al. (2000) Structure of the 30S ribosomal subunit. *Nature* **407,** 327–339.

100. Yusupov, M. M., Yusupova, G. Z., Baucom, A., et al. (2001) Crystal structure of the ribosome at 5.5 A resolution. *Science* **292,** 883–896.

101. Agrawal, R. K., Spahn, C. M., Penczek, P., Grassucci, R. A., Nierhaus, K. H., and Frank, J. (2000) Visualization of tRNA movements on the Escherichia coli 70S ribosome during the elongation cycle. *J. Cell Biol.* **150,** 447–460.

102. Frank, J. and Agrawal, R. K. (2000) A ratchet-like inter-subunit reorganization of the ribosome during translocation. *Nature* **406,** 318–322.

15

Intermediates and Transition States in Protein Folding

D. Thirumalai and Dmitri K. Klimov

Summary

The complex role played by intermediates is dissected using experimental data on apomyo-globin (apoMb), simple theoretical concepts, and simulations of kinetics of simple minimal off-lattice models. The folding of moderate-to-large-sized proteins often occurs through passage of an ensemble of intermediates. In the case of apoMb there is dominant kinetic intermediate **I** that also occurs at equilibrium. The cooperativity of transition of **U**↔**I** (**U** represents the ensemble of unfolded states) in apoMb at pH 4.0 is determined not only by the sequence but also by the anion concentration. Point mutations can substantially alter the cooperativity of formation of **I**. Another class of intermediates arise owing to bottlenecks in the rugged energy landscape that arises from topological frustration. As a result of the rough energy landscape, folding is predicted to follow the kinetic partitioning mechanism (KPM). According to KPM a fraction of molecules reaches the native state rapidly, while the remaining fraction is kinetically trapped in intermediates. The folding of lysozyme at pH 5.5 follows KPM. Our perspective also shows that the fraction of fast folding trajectories can be altered by changing pH, for example. These observations are clearly illustrated in simple off-lattice models of proteins. The simulations show that equilibrium inter-mediates occur "on-pathway" and have substantial probability to be revisited after the native state is reached, while kinetic intermediates are almost *never sampled* after native state is reached. In addition, kinetic intermediates are higher in free energy than equilibrium intermediates.

We also discuss the consequences of multiple routes and intermediates on the transition state ensemble (TSE) in folding. Whenever multiple routes to the native state dominate, Φ-values can be larger than unity or less than zero. There appears to be a relationship between the diversity of structures in the denatured state ensemble and the extent to which the TSE is plastic. Simulations of β-hairpins are used to illustrate these ideas.

1. Introduction

In the last decade there has been a great impetus to understand how proteins, that apparently fold in a "two-state" manner, reach the unique native state from a myriad of fluctuating unfolded conformation (*1–6*). Spurred in part by theory, fast folding experiments, simulations, and protein engineering (*7*) a detailed

From: *Methods in Molecular Biology, vol. 350: Protein Folding Protocols*
Edited by: Y. Bai and R. Nussinov © Humana Press Inc., Totowa, NJ

characterization of folding kinetics of small proteins has emerged *(6)*. There are general themes that are repeatedly observed in the folding of small single domain proteins: (1) in most cases polypeptide chain collapse and acquisition of the native state nearly coincide. For this class of proteins (the much studied CI2 is the best example) there is not a significant difference in the time scale for the formation of secondary and tertiary structures *(8)*, and (2) nucleation–collapse (or condensation) mechanism tidily rationalizes the way many simple proteins fold *(9,10)*. In general, there are multiple folding nuclei (MFN) *(11)* and the set of structures that correspond to the MFN constitutes the transition state ensemble (TSE) *(12)*. According to the MFN model certain tertiary contacts, that are determined by native topology, form with relatively high (say, >0.5) probability. Even residues that are perceived to be the key for folding seldom form all the contacts with unit probability in the TSE *(13,14)*. Thus, there is an inherent plasticity associated with the TSE. (3) The nature of the folding nuclei, that describes the TSE structures, can vary considerably depending on the precise native structure. The TSE can be highly polarized (a part of the polypeptide chain is structured in the TSE, whereas the rest of the chain is mobile and unstructured) as is the case in the SH3 domain family *(15,16)* or can be diffuse as in CI2 *(17)*. Altering the folding conditions can change the very nature of the TSE itself. (4) A key finding has been that native topology plays a dominant role in determining the folding rates *(18)*. It has been more recently appreciated that other factors, most notably the length of the protein, also play an important role *(19–21)*. (5) Single molecule experiments on a few proteins are beginning to show that the folding pathways, at least at the early stages, are highly heterogeneous *(22)*. The clear implication is that owing to variations in the structures of the denatured ensemble there is no "average" folding polypeptide chain.

Although many outstanding questions pertaining to folding of small single domain proteins remain unanswered, a general framework exists for describing the generic folding characteristics. In contrast, there are only few cases of proteins that fold via one or more intermediates, for which detailed characterization of intermediates and transition states have been made *(23)*. The role of intermediates has always been controversial *(24)*. They were postulated to be important, in part because their existence could rationalize the rapid folding of proteins. In particular, the existence of intermediates was thought to resolve the Levinthal paradox. Experimental characterization of intermediates is difficult, because they are short lived. Despite the difficulty in interpreting folding of proteins with intermediates, it is clear that for moderate-to-large-sized proteins they do readily form. As focus in protein folding research moves away from small single domain proteins it will be increasingly important to consider folding of proteins with multiple pathways and multiple intermediates.

In this chapter we have used experimental data and simulations of simple off-lattice-bead models to present tentative views on the effects of intermediates on protein folding. One of the best characterized systems that has a well-defined intermediate is apomyoglobin (apoMb) *(25)*. Pioneering NMR and hydrogen/deuterium exchange experiments have been used to show that a specific intermediate is populated both in equilibrium and under refolding kinetics *(26)*. In this case the intermediate is native-like and hence may be considered "on-pathway." The stability and cooperativity of formation of the intermediate is highly sensitive to mutations and nature of anions (*see* **Subheading 2.**). It is not clear whether this situation is generic, because simulations and recent fast folding experiments suggest that (at least, in the early stages) non-native intermediates can appear in the refolding of cytochrome c (Cyt c) and apoMb *(27,28)*. An example in which folding occurs by the kinetic partitioning mechanism (KPM) *(9,29)* is lysozyme, which folds by two parallel and vastly different pathways. According to the KPM, a fraction of molecules fold rapidly (reminiscent of two-state folders) without populating any discernible intermediates. The remaining fraction gets trapped in a long-lived ensemble of intermediates. The transition from the intermediates to the native state occurs on a much longer time scale. Using these case studies and results from off-lattice models we try to clarify the role of kinetic and equilibrium intermediates in protein folding.

2. Equilibrium Intermediates in apoMb

2.1. Cooperativity of Folding of pH 4.0 Intermediate of apoMb Depends on Mutations and the Nature of Anions

One of the best characterized systems is the refolding of apoMb, whose native state has eight helices labeled A through H (**Fig. 1**). At pH 4.0 apoMb forms a stable intermediate. NMR hydrogen exchange data *(25)* show that the pH 4.0 intermediate of apoMb, referred to as **I**, contains three helices A, G, and H. **I** is also a rapidly formed transient intermediate in the refolding kinetics of the wild-type apoMb *(26)*. Because **I** is unusually stable and is also involved in the folding kinetics, it is a good candidate for examining the properties of intermediates in protein folding. These characteristics of **I** prompted Luo et al. to investigate cooperativity of its formation by making glycine and proline mutations in the A and G helices *(30)*.

By using a combination of far-ultraviolet circular dichroism and Trp fluorescence (FL) for the wild-type (WT) and mutants Q8G, E109G, and E109P Luo et al. made a number of interesting observations. They are: (1) the midpoint C_m, of urea-induced unfolding has the largest value for the WT and decreases for all other mutants. The mutant E109P has the smallest value of C_m in solutions containing either 20 mM Na_2SO_4 or 50 mM $NaClO_4$. (2) The WT and the mutants

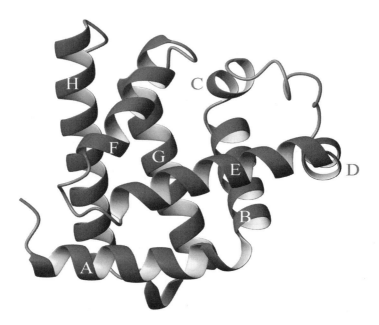

Fig. 1. Native structure of myoglobin (PDB access code 1MBC). The structure of apomyoglobin, which does not include heme, is believed to be very similar to that of myoglobin. The eight helices are labeled A through H. In the pH 4.0 intermediate of apomyoglobin helices A, G, and H are formed.

(except perhaps E109P) show two-state transitions with clearly identifiable values of C_m. In the presence of the stabilizing 50 mM ClO$_4$ anions even E109P exhibits nearly a cooperative two-state transition. The two-state transitions were established by the superposition of data from CD and Trp fluorescence.

Based on these observations and other assumptions, Luo et al. arrived at the major conclusion that cooperativity is intimately linked to the stability (relative to the unfolded state) of **I** itself. The stability of the folding reaction **U**↔**I** is ΔG_{H_2O}, which is the free energy difference between **U** and **I** states at zero urea concentration. Usually ΔG_{H_2O} is obtained by extrapolating the free energy at nonzero values of [urea] to 0 M concentration. This, of course, requires knowing the m values and introduces an additional parameter that needs to be measured or inferred from suitable experiments. If the m values for WT and all the mutants are nearly identical, then C_m, an easily measurable quantity, would give the most direct estimate of the stability itself *(31)* $\Delta G_{H_2O} \approx mC_m$. Luo et al. further argued that if m values of mutants are very different from WT, then deviations from the two-state behavior should be observed. However, because all the sequences exhibited two-state transitions, they assumed that the m values are nearly equal. These arguments were used to take the magnitude of C_m itself to be a measure of stability. Consequently, they concluded that the transition $U \leftrightarrow I$ is most cooperative for the WT and is the least for E109P. Because they assumed that C_m is proportional to ΔG_{H_2O}, the order of stability is WT > Q8G > E109G > E109P independent on the concentration of anions. Thus, cooperativity and stability are linked.

2.2. Data Analysis Using Cooperativity Measure

We have reanalyzed the data obtained by Luo et al. *(30)* using a simple measure of cooperativity *(32)*. We were prompted to undertake this exercise, because the assumption that C_m alone (without the precise determination that m values for all sequences are identical) determines the extent of cooperativity may not be valid. Relying on C_m alone ignores the slope change (given by m) and this can give incorrect assessment of cooperativity. Our measure of cooperativity, Ω_c, relies on the characteristics of the denaturant concentration dependence of the fraction of folded molecules around C_m. Consider

$$\Omega_c = \frac{C_m^2}{\Delta C} max \left| \frac{df}{dC} \right| \tag{1}$$

where f is the fraction of folded molecules, ΔC is the full width at half-maximum of df/dC, and $max[df/dC]$ is evaluated at $C = C_m$. From the measurements, such as the ones given in **Fig. 3** of **ref. *30*,** Ω_c can be easily calculated. In the experiments of Luo et al. f is the fraction of the intermediate **I**, so $f \equiv f_I$. Notice that the computation of Ω_c entails the dependence of f around C_m only. Thus, accurate fits are needed only around C_m and errors from baseline drifts do not affect Ω_c. For an infinitely sharp transition $\Omega_c \rightarrow \infty$.

The magnitude of Ω_c gives a good indication of whether the transition is two state or not. If $\Omega_c < 1$, then the transition is noncooperative and is not two state. We should emphasize that the inference about the nature of transitions based on Ω_c values should be made with caution. It is certainly the case that, if Ω_c is relatively large ($\gtrsim 5$), then the transition is cooperative, and can be treated using two-state model *(32)*. For our purposes here we will assume following Luo et al. that $U \leftrightarrow I$ is a two-state transition. This assumption can be tested by computing Ω_c.

For two-state systems one can represent the fraction of folded state f as *(31)*,

$$f = \frac{1}{1 + \exp\left(-\dfrac{\Delta G}{RT} \right)} \tag{2}$$

where $\Delta G = \Delta G_{H_2O} - mC$, C is the concentration of denaturant, and $\Delta G_{H_2O} = G_U - G_I$.

It can be shown that, if f obeys **Eq. 2**, then Ω_c in **Eq. 1** is given by *(32)*.

$$\Omega_c = \frac{1}{8}\left(\frac{\Delta G_{H_2O}}{RT}\right)^2 \frac{1}{ln(3+2\sqrt{2})} \cdot \tag{3}$$

Notice that **Eq. 3** only contains ΔG_{H_2O}, which, of course, is the direct measure of stability. Thus, **Eq. 3** shows that the cooperativity (as measured by Ω_c) and stability are directly linked. Our objective is to compute Ω_c according to **Eq. 3** using the experimental data of Luo et al. We will also determine the order of cooperativity of the WT and its mutants for different anions.

We have used the Trp FL experimental data *(30)* to compute Ω_c for the WT and the three mutants at 20 mM Na$_2$SO$_4$ and 50 mM NaClO$_4$. The dependence of f on C is obtained using the Santoro–Bolen (SB) six parameter fit to the relative Trp FL *(33)*. **Figure 2** shows the fitted denaturation curves computed using the SB method for two anions superimposed on experimental data. Since the free energy of stability $\Delta G_{H_2O}/RT$ is one of the parameters to be adjusted in the SB method, the calculation of Ω_c using **Eq. 3** is straightforward. We also calculated Ω_c directly from the data in **Fig. 3** of **ref.** *30* (which shows f as a function of C) for the WT and the three mutants at both anion salt concentrations. The results are summarized in **Table 1**. There are few conclusions that can be drawn from these results:

1. As suggested by Luo et al., cooperativity and stability for two-state transitions are directly related. In **Table 1** the largest values of Ω_c (i.e., most cooperative) are obtained for the most stable mutants. This is the case at both anion concentrations. **Equation 3** also suggests a new way to estimate $\Delta G_{H_2O}/RT$ directly from the unfolding data without the knowledge of the baselines.

2. Our analysis shows that the WT is not the most stable. The order of stability depends on the concentration and nature of anions. In 20 mM Na$_2$SO$_4$ the mutant Q8G is the most stable, whereas in 50 mM NaClO$_4$ E109G is the most stable. The stability order for SO$_4^{2-}$ anions is Q8G > E109G ≈ WT > E109P and for ClO$_4^-$ anions it is E109G > Q8G > WT > E109P. In both cases, E109P is the least stable and exhibits at best a weak cooperative transition. This is reflected in the very small values of Ω_c. Because Ω_c is less than or close to unity, we conclude that the transition in E109P is too broad to be classified as two state. Stability, and hence the degree of cooperativity, in the **U↔I** transition of apoMb and its mutants depends on a salt concentration.

The dependence of cooperativity on sequence and external conditions arises because of the variations in m values. **Table 1** lists the m values computed by the SB method. It is seen that the m values are not identical and this shows that the degree of exposure to solvent depends not only on the sequence, but also on the anion conditions. We also calculate the midpoint of denaturation curves for

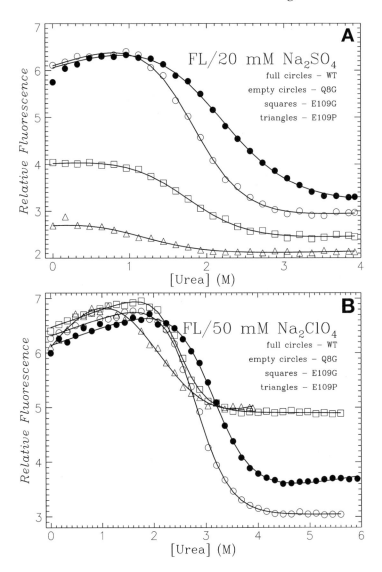

Fig. 2. Relative Trp fluorescence for the urea-induced unfolding of the pH 4.0 intermediate at T = 4°C for the wild-type (WT) and three mutants. The symbols are experimental points determined in **ref. 30**. The solid lines are the Santoro–Bolen (SB) fits to the data. Note that the measure of cooperativity Ω_c may be directly computed, when ΔG_{H_2O} is known. The adequate fit to the data by the SB method suggests that the unfolding transitions are two state. Note that the first two data points for WT near zero [urea] in **A** are not taken into account. (**A**) 20 m*M* Na_2SO_4 concentration; (**B**) 50 m*M* $NaClO_4$ concentration.

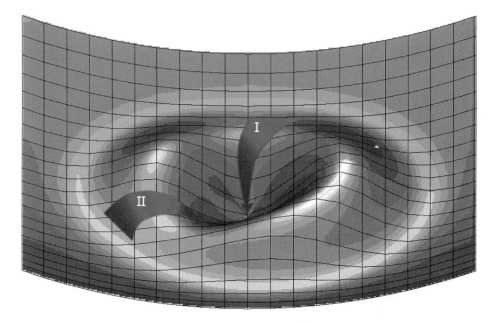

Fig. 3. A conceptual sketch of free energy landscape, which results in a kinetic partitioning mechanism. Two (fast, track I, and slow, track II) folding pathways are indicated by arrows.

Table 1
Thermodynamic Parameters of the Formation of pH 4.0 Intermediate I of WT Sequence and Three Mutants of apoMb at T = 4°C and pH = 4.2

	FL, 20 mM Na$_2$SO$_4$				FL, 50 mM NaClO$_4$			
Protein	ΔG_{H_2O}/RT	m/RT, M^{-1}	C_m, [urea]^a	$\Omega_c{}^b$	ΔG_{H_2O}/RT	m/RT, M^{-1}	C_m, [urea]^a	$\Omega_c{}^b$
WT	4.44	2.13	2.08	1.40 (1.71)	8.29	2.66	3.12	4.88
Q8G	5.33	3.00	1.77	2.02 (2.06)	8.49	3.05	2.78	5.11
E109G	4.37	2.54	1.72	1.36 (1.46)	9.64	3.92	2.46	6.59
E109P	2.33	2.30	1.01	0.38 (0.53)	2.93	1.57	1.86	0.61

[a]C_m values calculated as $\Delta G_{H_2O}/m$.
[b]Ω_c values computed using the SB fit and **Eq. 3**. Numbers in parenthesis give the values of Ω_c computed directly by fitting the data near C_m.

two-state systems as $C_m = \Delta G_{H_2O}/m$. The midpoint C_m may be identified with the location of the maximum in the derivative df/dC. These values of C_m coincide with those estimated from the condition $f(C_m) = 0.5$. We find that the calculated C_m values are consistent with those given in **Table 1** of Luo et al. Alternatively, we can use the approach proposed in remark (1) above to obtain stabilities and m values. We find that these ΔG_{H_2O} and m values are in a close agreement with those calculated from the SB analysis. **Figure 2** shows that the two-state assumption embedded in SB method provides an excellent fit to the data, especially to WT, Q8G, and E109G. Thus, we conclude that the folding transition $\mathbf{U} \leftrightarrow \mathbf{I}$ in apoMb at pH 4.0 is a cooperative two-state transition (excluding perhaps E109P).

2.3. Test of Robustness of the Conclusions

In the previous section we used the SB method to analyze the urea dependence of the relative FL *(33)*. This procedure accounts for the drifts in the baselines of the FL data at small and large values of the denaturant concentration. In order to assess the robustness of the conclusions we use a slightly different way to analyze the data.

The fraction of folded molecules f (in our case $\mathbf{N} = \mathbf{I}$) is inferred from the measured FL data as

$$f = \frac{F - F_U}{F_N - F_U} \tag{4}$$

where F is the FL at a given concentration (C) of the denaturant, F_U and F_N are the values of FL for the denatured and folded states, respectively. Following SB method we define

$$F_U = F_{int,U} + \alpha_U [C] \tag{5}$$

and

$$F_N = F_{int,N} + \alpha_N [C] \tag{6}$$

In the SB procedure **Eqs. 2, 4, 5**, and **6** express F in terms of six parameters ΔG_{H_2O}, m, $F_{int,U}$, $F_{int,N}$, α_U, and α_N. These are adjusted to give the best fit to the data. This data reduction was done for apoMb by Luo et al., and reanalyzed in the previous section. The six parameter fit yields a very robust fit of the data for two-state transitions, and hence is the standard method of choice.

We analyzed the FL data of Luo et al. differently in order to further affirm the general conclusions concerning the nature of cooperativity in the formation of pH 4.0 intermediate of apoMb and its mutants. The required parameters in **Eqs. 5** and **6**, namely, the intercepts and slopes, are computed directly from the FL data at small and large urea concentrations. We use a simple least square fit of the actual experimental data at small (large) urea concentration, which gives us $F_{int,N}$ and α_N ($F_{int,U}$ and α_U). Thus, these four parameters are no longer adjustable *(34)*. After fixing the four parameters, $\Delta G_{H_2O}/RT$ can be obtained by knowing f (or equivalently F) around the midpoint C_m alone as suggested in the previous section. The analysis of the FL data using our method of fixing the slopes and intercepts in **Eqs. 5** and **6** and the subsequent use of Ω_c to calculate $\Delta G_{H_2O}/RT$ and m/RT was carried out for the WT and the mutants Q8G, E109G, and E109P at 20 mM Na$_2$SO$_4$. The values of Ω_c are 1.71, 2.06, 1.46, and 0.53 for the WT, Q8G, E109G, and E109P, respectively. These values are given in **Table 1** in parenthesis. Because $\Omega_c \propto (\Delta G_{H_2O})^2$, we conclude, as in the previous section, that the order of stability in 20 mM Na$_2$SO$_4$ is Q8G > WT > E109G > E109P. Thus, both approaches (the SB method and the one described in this section) give the same ordering of cooperativity and, hence, stability for WT and the mutants.

A very closely related method was used to analyze thermal denaturation curves *(34)*. By manipulating the van't Hoff equation it was shown that the enthalpy change at the melting temperature T_m is related to the slope (df/dT) at T_m. To use that method, the parameters in baselines are determined manually as done here. This method was also examined by Allen et al. *(35)*. By analyzing the denaturation curves for C102T variant of yeast iso-1-ferricytochrome C *(36)*, they concluded that the values of the parameters near T_m are invariant to the uncertainty in the data near the baselines. In particular, even upon removal of the 20 data points in the low-temperature region, the values of the van't Hoff enthalpy and T_m are unchanged. Thus, hand fitting of even poorly defined baseline values (as done here) yields robust fits near C_m, which is all that is required to compute ΔG_{H_2O} and m.

We have shown using the dimensionless measure of cooperativity Ω_c, the SB analysis, and the data of Luo et al. for $U \leftrightarrow I$ transition in apoMb that the degree of cooperativity is directly related to stability. For two-state denaturant-induced transitions this result is very general as can be seen from **Eq. 3**. The analysis of the experimental data shows that the stability depends on the sequence and the anion conditions in a more subtle way than suggested *(30)*. This conclusion is robust and independent on the methods used to analyze the data (**Fig. 3** of Luo et al. or Trp FL).

Additional experiments *(37)* on the pH 4.0 folding intermediate of apoMb were done in the presence of 5% (volume) 2,2,2-trifluoroethanol (TFE) or 20 mM Na$_2$SO$_4$. It was showed that C_m values are roughly similar, when only one of the additives (5% TFE or 20 mM Na$_2$SO$_4$) is present *(37)*. When both 5% TFE and 20 mM Na$_2$SO$_4$ are added, C_m nearly doubles. However, ΔG_{H_2O} essentially remains unchanged indicating that the m value is halved compared to the case when only one solvent is added. This shows that correlating C_m with stability is not justified, unless independent measurements showing the constancy of m are made.

The analysis carried out here shows the difficulties in probing the nature of equilibrium intermediates in folding. Even for the well-studied case of apoMb, for which detailed NMR structural data are available at pH 4.0 *(25)*, there can be variations in the cooperativity of their formation. Thus, dissecting how the molecular interactions affect cooperativity by mutational analysis (such as the one carried out by Luo et al.) is difficult. It is necessary to obtain nuclear Overhauser effect constraints for mutants under differing anion concentrations to get full understanding of the interactions responsible for the cooperative folding of intermediates.

3. Topological Frustration and the Kinetic Partitioning Mechanism

Qualitative aspects of folding of sequences that form intermediates can be understood in terms of topological frustration (*38*). Natural protein sequences are about 55% hydrophobic and the linear density of hydrophobic residues is roughly constant along the sequence. As a result on any length scale *l* there is a propensity for certain residues to form tertiary contacts under folding conditions. The resulting structures, which arise owing to contacts between proximal residues, may be in conflict with the global native fold. The incompatibility of low energy structures on local and global scales has been termed as topological frustration, which is a natural consequence of the polymeric nature of proteins as well as the presence of competing heterogeneous interactions. Topological frustration in protein sequences makes the underlying energy landscape rugged, which implies that there are a multitude of local minima separated by barriers of varying heights. Most of these structures have relatively high free energies and are unstable owing to thermal fluctuations. However, some of these structures (whose number grows only as *lnN [39]*) may have low free energies and significant overlap with the native fold. The low energy misfolded conformations serve as kinetic traps that can impede the folding process (*40*).

The basic consequences of the complex free energy surface arising from topological frustration leads naturally to the KPM (*4,7,29*). To illustrate the key notions of KPM, consider an ensemble of denatured molecules under folding conditions (**Fig. 3**). This situation arises when the concentration of denaturant or temperature is decreased. A fraction of molecules Φ (the partition factor) would reach the native basin of attraction (NBA) rapidly without being trapped in the low energy minima (track I in **Fig. 3**). The remaining fraction would be captured by one or more of these minima and reach the NBA by activated transitions on a longer time scales (track II). The value of Φ depends on the sequence and is determined by $\sigma = (T_\theta - T_F)/T_\theta$, where T_θ and T_F are the collapse and folding transition temperatures (*4*). Therefore, the initial population of protein molecules partitions into fast and slow folders.

A detailed kinetic analysis for the remaining fraction of molecules $(1 - \Phi)$ showed that they reach the NBA by a three-stage multipathway mechanism (*7,9,40*). According to this model the polypeptide chain initially collapses to a compact structure. Such a collapse is nonspecific and occurs on a time scale of the order of 1 μs for a protein consisting of about 100 amino acids. The structures that are accessed on this time scale are determined by favorable local interactions. In the second stage, driven by energetic biases, the polypeptide chain diffusively searches for low energy structures. Because of the initial nonspecific collapse the polypeptide chain reaches one of low energy minima at the end of the second stage. The third stage involves activated transitions from one of such low energy native-like intermediate to the native basin of attraction. There are few pathways, which connect the native-like intermediates with the native state. This suggests that the transition states for the $(1 - \Phi)$ fraction of molecules, which occur late in the folding process, are sparse.

Detailed refolding experiments on hen-egg lysozyme verified the applicability of KPM *(41,42)*. The experiments by Kiefhaber *(41)* were the first to show that a fast track exists, in which lysozyme assembles very rapidly (on the order of tens of milliseconds). Interrupted refolding experiments demonstrated that about 15% ($\Phi \sim 0.15$) of folding molecules reach the folded state in about 50 ms, whereas the remaining fraction arrives to the native state in about 400 ms. A very detailed mapping of the parallel pathways in lysozyme has been made using a combination of the hydrogen exchange-labeling experiments and binding of a fluorescence-labeled inhibitor *(42)*. This study showed that about 25% ($\Phi = 0.25$ at pH 5.5 and 20°C) of the molecules fold by the fast track. The time scale for the assembly of the native state by the slow process exceeds 350 ms. Although the overall validity of KPM has been demonstrated in the refolding experiments on lysozyme, the nature of the transition states in the two pathways have not been clarified.

4. Illustrating the Role of Intermediates Using Minimal Off-Lattice Models of Proteins

The role of intermediates in the folding process can be illustrated using simple off-lattice models. We use the coarse-grained model of a β-hairpin forming sequence and Langevin dynamics simulations to illustrate the role of folding intermediates *(40)*. Veitshans et al. have shown that many of the aspects of rugged energy landscape expected in large proteins manifest themselves in structures that form β-hairpins *(40)*. With this in mind, we selected a sequence E, which folds to the native state by KPM *(40)*. Under weakly folding conditions (i.e., at the temperature $T \lesssim T_F$, the folding temperature) about a quarter of folding molecules become trapped in distinct low energy (intermediate) states before reaching the native conformation. To characterize the folding process 20 instantaneous structures were recorded for each 102 slow folding trajectories between the temperature quench and the first passage time to the native state $\tau_{F,m}$ (m is a trajectory index). These conformations were then energy minimized to reduce structural fluctuations. We have made a special effort to verify that changing the frequency of saving the instantaneous conformations does not qualitatively alter

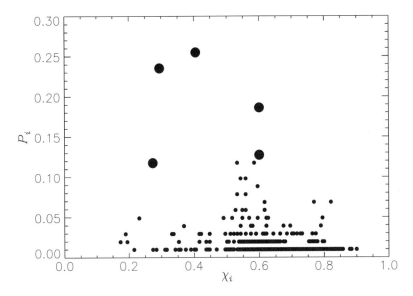

Fig. 4. The frequency of occurrence of intermediates P_i in slow folding trajectories as a function of their similarity to the native state (as measured by the overlap function χ_i [40]). Although a generic intermediate state rarely occurs in more than a few folding trajectories, there are some intermediates, which are sampled in more than 10% of folding pathways. Five most frequently occurring such states (marked by large circles) represent the equilibrium intermediates $\mathbf{I}_{E,i}$.

our findings. The resulting 2040 minimized structures comprise the dataset of folding intermediates. To compare these structures with the conformations visited after $\tau_{F,m}$ we have created the second dataset of 2420 equilibrium structures using the same methodology. The structures from both datasets were clustered into groups with the mutual structural overlap χ_{ij} (*40*) not exceeding 0.15 for any i and j members of each group. The intermediate states considered next are the representative structures from these structurally dissimilar groups. Our goal is to evaluate the characteristics of intermediates along the folding pathway and compare these with the structures sampled at equilibrium.

We first compute the frequency of appearance of various intermediate states along the folding pathways. **Figure 4** displays the fraction P_i of slow folding trajectories, which sampled at least once an intermediate i from the first dataset. Although the majority of intermediates does not occur frequently ($P_i < 0.1$), there are few which repeatedly reoccur. For further analysis we have selected five most frequently sampled intermediates $\mathbf{I}_{E,i}$ ($i = 1, ..., 5$) with $P_i \gtrsim 0.1$. These states occur approximately halfway to the native state as indicated by r_i (**Table 2**). Once the protein folds to these states, it becomes trapped for an appreciable fraction of total folding time $0.2 < P_i < 0.4$ (**Table 2**).

Table 2
The List of Equilibrium Intermediates

Intermediate	$E_{p,i}{}^{a}$	$x_i{}^{b}$	$P_i{}^{c}$	$r_i{}^{d}$	$P_i{}^{e}$	$R_{g,i}^{2}{}^{f}$
$\mathbf{I}_{E,1}$	−10.13	0.27	0.12	0.35	0.24	6.13
$\mathbf{I}_{E,2}$	−10.50	0.60	0.13	0.39	0.18	5.24
$\mathbf{I}_{E,3}$	−10.20	0.60	0.19	0.55	0.20	5.57
$\mathbf{I}_{E,4}$	−10.55	0.30	0.24	0.36	0.39	5.96
$\mathbf{I}_{E,5}$	−10.08	0.41	0.25	0.44	0.27	5.82

$^{a}E_{p,i}$ is the potential energy of an intermediate i in the units of $\varepsilon_h \approx 1.25$ kcal/mol.
$^{b}x_i$ is the overlap value for an intermediate i *(40)*.
$^{c}P_i$ is the fraction of slow folding trajectories, sampling an intermediate i.
$^{d}r_i$ is the ratio of the first passage times to i and to the native state.
$^{e}p_i = <\Delta\tau_{E(K),i,m}/\tau_{F,m}>$, where $\Delta\tau_{E(K),i,m}$ is the total lifetime of an intermediate i in a trajectory m, $\tau_{F,m}$ is the folding time for a trajectory m, and $< ... >$ implies the average over all slow folding trajectories.
$^{f}R_{g,i}^{2}$ is the radius of gyration squared for an intermediate i in the units of a. The native radius of gyration $R_{g,N}^{2} = 6.69a$ ($a = 3.8$ Å).

The crucial feature of the intermediates in **Table 2** is their accessibility after the native state is reached. **Figure 5** monitors folding process before and after the first passage time using the overlap function $\chi(t)$ *(40)*. The trajectory shown in the upper panel indicates that the protein rapidly reaches the intermediate $\mathbf{I}_{E,4}$ and remains trapped there for more than three-fourths of the folding time for a given trajectory before making a transition to the native conformation. Remarkably, the polypeptide chain repeatedly visits the *same* intermediate state $\mathbf{I}_{E,4}$ after folding. Qualitatively similar conclusions may be drawn from the analysis of other slow folding trajectories. For example, the one displayed in the lower panel of **Fig. 5** demonstrates that the polypeptide chain samples $\mathbf{I}_{E,4}$ state on the way to the native state, but makes excursions to the state $\mathbf{I}_{E,5}$ after reaching equilibrium. Furthermore, direct monitoring of occurrence of intermediate states $\mathbf{I}_{E,i}$ confirms that all five of those are sampled after folding is completed. For this reason, we classify the intermediates $\mathbf{I}_{E,i}$ ($i = 1, ..., 5$) as equilibrium intermediates. These are analogous to pH 4.0 intermediate in apoMb.

The examination of $\mathbf{I}_{E,i}$ structures shows that the equilibrium intermediates have low energy, relatively high native content (for $\mathbf{I}_{E,i}$ [$i = 1, 4, 5$] $\chi_i \sim 0.3 - 0.4$), and are as compact as the native structure (**Table 2**). The most common structural difference between the equilibrium intermediates and the native state is associated with non-native conformations of few dihedral angles. For example, in $\mathbf{I}_{E,4}$, which has the lowest potential energy E_p, the third dihedral angle in one of the β-strands is in the g^+ position instead of the native *trans* state.

It is also interesting to investigate the structural connectivity between $\mathbf{I}_{E,i}$ states during folding. The order of appearance of $\mathbf{I}_i$ states in slow folding

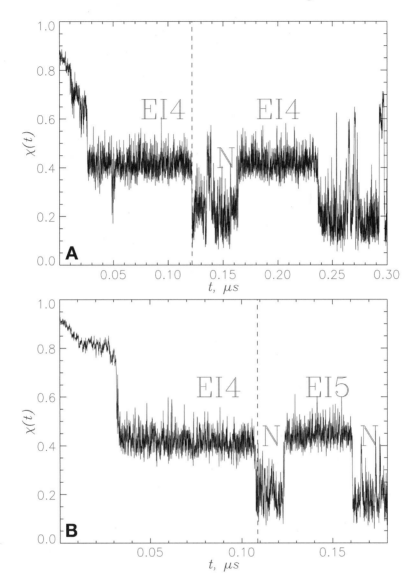

Fig. 5. Examples of folding trajectories, which sample the equilibrium intermediates. Folding trajectory in **A** rapidly folds to the intermediate state $\mathbf{I}_{E,4}$ (indicated as EI4 in the plot). After being trapped in $\mathbf{I}_{E,4}$ for most of the folding part of the trajectory, the protein makes a transition to the native state indicated by a dashed vertical line. The *same* intermediate is later revisited by a protein as a result of equilibrium thermal fluctuations. (**B**) Demonstrates that the protein samples different equilibrium intermediates after folding to **N**.

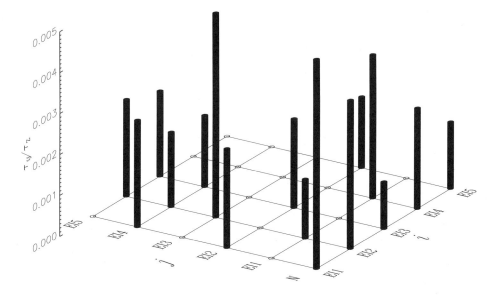

Fig. 6. The transition times τ_{ij} between various intermediate states $\mathbf{I}_{E,i}$ and $\mathbf{I}_{E,j}$ and the native state during folding. The transitions, which are not observed, have $\tau_{ij} = 0$. The plot shows that most of $\mathbf{I}_{E,i}$ are well interconnected. Therefore, as a rule protein passes through several $\mathbf{I}_{E,i}$ before reaching $\mathbf{N}$. τ_{ij} are scaled by the average slow folding time τ_F.

trajectories allows us to compute the average transition times τ_{ij} for any pair of $\mathbf{I}_{E,i}$ and $\mathbf{I}_{E,j}$ states. The values of τ_{ij} computed for five $\mathbf{I}_{E,i}$ and the native state are presented in **Fig. 6**, which suggests that most (although not all) equilibrium intermediates are well connected and transitions between them occur frequently. The analysis shows that as expected, all transition times τ_{ij} are considerably less than the lifetimes of $\mathbf{I}_{E,i}$ (**Table 2**). There are, however, several transitions $\mathbf{I}_{E,i} \to \mathbf{I}_{E,j}$, which are blocked. For example, no transitions from $\mathbf{I}_{E,1}$ to $\mathbf{I}_{E,3}$ or $\mathbf{I}_{E,5}$ are observed. Although native state is accessible from all $\mathbf{I}_{E,i}$, the transitions to $\mathbf{N}$ occur far more frequently from $\mathbf{I}_{E,i}$ with $i = 3$, 4, or 5. This suggests that on gradual approach to the native state, the entropy available for the protein is depleted and the folding pathway is channeled into few well-defined routes. The corresponding implication is that experiments that probe the late stages of folding only sample the "smoother" part of the free energy landscape.

The natural question related to equilibrium intermediates is how often in total the $\mathbf{I}_{E,i}$ states appear in the slow folding pathways. It turns out that approx 50% of all slow folding trajectories pass through at least one of $\mathbf{I}_{E,i}$. The other half becomes trapped in numerous diverse sets of intermediate states $\mathbf{I}_{K,i}$, which are almost always unique for a given trajectory. The crucial difference between these intermediate states and equilibrium intermediates $\mathbf{I}_{E,i}$ is that $\mathbf{I}_{K,i}$ are *never*

Table 3
The List of Kinetic Intermediates

Intermediate	$E_{p,i}{}^a$	$x_i{}^b$	$P_i{}^c$	$r_i{}^d$	$P_i{}^e$	$R^2_{g,i}{}^f$
$\mathbf{I}_{K,1}$	−5.57	0.70	0.03	0.58	0.20	4.86
$\mathbf{I}_{K,2}$	−5.87	0.65	0.01	0.05	0.70	6.75
$\mathbf{I}_{K,3}$	−8.25	0.63	0.05	0.38	0.35	4.64
$\mathbf{I}_{K,4}$	−5.14	0.84	0.01	0.30	0.25	3.91
$\mathbf{I}_{K,5}$	−8.42	0.67	0.02	0.45	0.08	4.45
$\mathbf{I}_{K,6}$	−9.57	0.68	0.03	0.22	0.38	4.50
$\mathbf{I}_{K,7}$	−9.42	0.59	0.01	0.15	0.25	4.44
$\mathbf{I}_{K,8}$	−4.74	0.76	0.01	0.20	0.35	2.71
$\mathbf{I}_{K,9}$	−6.84	0.79	0.01	0.20	0.35	3.50
$\mathbf{I}_{K,10}$	−4.32	0.73	0.01	0.10	0.15	4.11
$\mathbf{I}_{K,11}$	−9.40	0.61	0.02	0.05	0.60	5.04
$\mathbf{I}_{K,12}$	−5.49	0.73	0.01	0.20	0.10	3.65

$^a E_{p,i}$ is the potential energy of an intermediate i in the units of $\varepsilon_h \approx 1.25$ kcal/mol.
$^b x_i$ is the overlap value for an intermediate i (*40*).
$^c P_i$ is the fraction of slow folding trajectories, sampling an intermediate i.
$^d r_i$ is the ratio of the first passage times to i and to the native state.
$^e p_i = < \Delta\tau_{E(K),i,m}/\tau_{F,m} >$, where $\Delta\tau_{E(K),i,m}$ is the total lifetime of an intermediate i in a trajectory m, $\tau_{F,m}$ is the folding time for a trajectory m, and $< ... >$ implies the average over all slow folding trajectories.
$^f R^2_{g,i}$ is the radius of gyration squared for an intermediate i in the units of a. The native radius of gyration $R^2_{g,N} = 6.69a$ ($a = 3.8$ Å).

visited after the native state is reached. For this reason, $\mathbf{I}_{K,i}$ are termed as kinetic intermediates. To describe their characteristics we have selected twelve $\mathbf{I}_{K,i}$ intermediates, which have the largest average lifetimes $< \Delta\tau_{K,j} >$ (**Table 3**). The average $< \Delta\tau_{K,j} >$ for the twelve $I_{K,i}$ is larger that the average lifetime of equilibrium intermediates $\mathbf{I}_{E,i}$. More importantly, the fraction of slow folding trajectories containing a given $\mathbf{I}_{K,i}$ does not exceed 5% with the majority of kinetic intermediates appearing only in a single trajectory. The structural characteristics of $\mathbf{I}_{K,i}$ are also very different from those of $\mathbf{I}_{E,i}$. For example, kinetic intermediates have high potential energies, low native content, and are more compact than equilibrium intermediates (**Tables 2** and **3**). As a result, β-strands in $\mathbf{I}_{K,i}$ are highly distorted. As a rule, these structures contain just a few native dihedral angles. In contrast, equilibrium intermediates have very few deviations in dihedral angles from the native values. These observations imply that $\mathbf{I}_{K,i}$ are compact, randomly packed structures, which have little in common with the native conformation.

In summary, we found that intermediate states in slow folding trajectories can be divided into two categories. The equilibrium intermediates $\mathbf{I}_{E,i}$

frequently occur in folding pathways and the protein often revisits them after reaching the native state. These native-like stable states appear roughly halfway to the native state. There is evidence of repeated interconversion between various $I_{E,i}$ and the native state. Because equilibrium intermediates are accumulated in appreciable amount during folding, these states are the prime candidates for experimental detection. On the other hand, it is unlikely that any of the kinetic intermediates $I_{K,i}$ could have experimental significance because these states are rarely sampled more than once along folding trajectories. Unless kinetic intermediates appear late in the folding process they cannot be detected experimentally. A clear elucidation of their role will require fast folding experiments that can trap them along the folding pathways.

5. Multiple Folding Routes and Transition States

To understand how proteins and RNA fold it is necessary to describe the structures of all the kinetic species including the elusive transition states *(43)*. In simple chemical reactions, that can often be described using low dimensional reaction coordinates, the bottleneck region separating the reactants and products is relatively well defined, which makes it possible to infer the transition state structures from detailed kinetic studies. In contrast, in the bottleneck regions of the protein folding reactions there is a nearly complete compensation of the enthalpy gain upon forming favorable tertiary contacts and the free energy loss resulting from a decrease in the conformational entropy *(44)*. This makes the barrier to folding relatively low and broad. Consequently, it is fruitful to describe the bottleneck structures in protein folding as a set of conformations that constitutes the transition state ensemble *(12)*.

Currently, the only available experimental tool for mapping out the structures of the TSE is through protein engineering that requires constructing a series of well-designed mutations peppered throughout the protein *(8)*. Denoting wild-type as WT and M as mutant, response to a site mutation is quantified using *(45)*

$$\Phi_F = \frac{\Delta\Delta G^\ddagger}{\Delta\Delta G} = \frac{\Delta G_M^\ddagger - \Delta G_{WT}^\ddagger}{\Delta G_M - \Delta G_{WT}} \simeq \frac{ln\dfrac{k_f^{WT}}{k_f^M}}{ln\dfrac{k_u^{WT}}{k_u^M}\dfrac{k_f^M}{k_f^{WT}}}, \qquad (7)$$

where $\Delta\Delta G^\ddagger$ and $\Delta\Delta G$ are the changes in the free energy of activation and stability respectively, $k_f (k_u)$ is the folding (unfolding) rate. The approximate equality in **Eq. 7** holds good for proteins that fold by two-state kinetics. Protein engineering method combined with Φ_F-value analysis was first introduced by Fersht in the context of enzyme catalysis *(46)* and later formalized to decipher the structures of the TSE in the unimolecular folding reaction *(47)*. This method has been used to map out the TSE structures in a number of proteins *(15–17,48,49)*.

The earliest studies pointed out that $\Phi_F = 0$ and 1 lend themselves to a straightforward interpretation. The meaning of the majority (85%) *(50)* of the small (less than 0.6) Φ_F-values is less clear *(17,51)*. Nonclassical Φ_F-values (those outside the range $0 \le \Phi_F \le 1$), which are observed in proteins such as CI2 *(17)* and SH3 domain *(15)*, can arise from multiple routes to the native state *(51)*. Mutations that alter the fraction of molecules that flow through these pathways can produce nonclassical Φ_F-values. This implies that there is diversity in the structures of the TSE. Simple arguments outlined next show that other scenarios for the origin of nonclassical Φ_F-values are also possible.

5.1. Origins of Nonclassical Φ_F-Values

A few possibilities for the magnitude and sign of Φ_F-values can be outlined based on the free energy profile for two-state folders (**Fig. 7**). We first consider nonclassical Φ_F-values, postponing until **Subheading 5.1.2.** a discussion of the conventional interpretation of **Eq. 7**.

5.1.1. Narrow TSE

If a mutation destabilizes the native conformation and enhances k_f, then Φ_F can become less than zero (**Eq. 7**). Negative Φ_F-values have been found in the well-studied proteins including CI2 (LI32) *(17)* and SH3 (D15A) domain *(15)*. If we assume that there is very little dispersion in the TSE, as is apparently the case in CI2 *(47)* and SH3 domain *(16)*, a possible interpretation of $\Phi_F < 0$ is that non-native interactions are involved in the bottleneck region. Because mutation increases the folding rate, it follows that the TSE, that is presumed to be conformationally restricted, has to be stabilized provided that the free energy of the denatured state ensemble (DSE) is not affected (**Fig. 7**). The additional stabilization of the TSE has to come from attractive non-native interactions (**Eq. 7**). This interpretation finds support in simulations of lattice models with side chains *(13,14)*.

Alternatively, $\Phi_F < 0$, if a stabilizing mutation retards the folding rate (**Fig. 7**). An example of this is KA24 *(17)*, which is in the minicore of CI2, for which $\Delta\Delta G < 0$ while $\Delta\Delta G^{\ddagger} > 0$.

5.1.2. Multiple Transition States

Negative Φ_F-values can arise by an entirely different mechanism. Let us consider a protein that folds by multiple routes (**Fig. 3**). The best-studied example is lysozyme, which folds by KPM *(41)*. In this case a fraction of molecules ($\simeq 25\%$ *[42]*) reaches the native conformation rapidly, while in the remaining fraction only part of the structure (α-domain in lysozyme) is formed. The fraction of unfolded molecules obeying KPM is $P_u(t) \simeq \phi_{WT} e^{-t/\tau_f^{WT}} + (1 - \phi_{WT}) e^{-t/\tau_s^{WT}}$ where τ_f and τ_s are the fast and slow folding time-scales. The mean folding time is $\tau_F = \phi_{\alpha} \tau_f^{\alpha} + (1 - \phi_{\alpha}) \tau_s^{\alpha}$ where α is either *WT* or *M*.

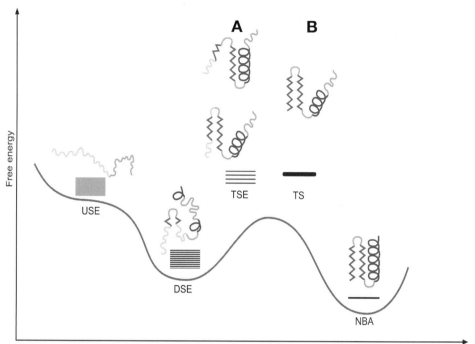

Fig. 7. Ensembles of states, which are populated during folding, are depicted in the free energy vs reaction coordinate diagram. Unfolded state ensemble (USE) is a homogeneous collection of conformations without persistent secondary or tertiary structure. After initiation of folding (temperature jump or denaturant concentration change), USE rapidly transforms into the denatured state ensemble with clearly identifiable local structure usually bearing the signatures of the native state. The set of high free energy conformations comprises the transition state ensemble (TSE), which can be usually grouped into few structurally distinct transition state clusters (**A**). Because of topological restrictions, the TSE may be conformationally restricted (**B**). Native state and proximal native-like conformations form the native basin of attraction, in which all secondary and tertiary interactions are fully formed. Horizontal black bars represent the diversity of structures in the ensemble. DSE, denatured state ensemble; NBA, native basin of attraction; TS, transition state.

We consider two limiting cases:

1. Apparent two-state folding: if folding occurs through two coupled microscopic routes, the native state would be populated in a single step provided that $\tau_f \approx \tau_s$. Because the folding times for these routes are similar, i.e., $\tau_s^{WT} \approx \tau_f^{WT} (1 + \epsilon_1)$ and $\tau_s^{M} \approx \tau_f^{M} (1 + \epsilon_2)$, where $(\epsilon_1/\epsilon_2 \sim 0(1)$ and $\epsilon_1, \epsilon_2 < 1$

$$\frac{\Delta\Delta G\Phi_F}{RT} \approx ln\frac{T_f^M}{T_f^{WT}} + ln\frac{1-\epsilon_2}{1-\epsilon_1}\frac{\phi_M}{\phi_{WT}} . \tag{8}$$

Assuming that the mutation is destabilizing and $\tau_f^M \simeq \tau_F^{WT}$, $\Phi_F > 1$, if $(e \in_1 \phi_{WT} - \in_2 \phi_M) > (e - 1)$ $(lne = 1)$, while $\Phi_F < 0$ if $\phi_M > \phi_{WT}$. Numerical lattice model simulations illustrate these arguments *(51)*. Our arguments show that negative Φ_F-values can also emerge, if a stabilizing mutation decreases the amplitude of the fast phase, i.e., $\phi_M < \phi_{WT}$.

Nonclassical Φ_F values are obtained, if the amplitude of the fast channel is altered on mutation, even though the overall folding still follows two-state kinetics. This suggests that nonclassical Φ_F-values can result, if there are multiple transition states (disperse TSE), the free energies of which are similar. We should emphasize that the number of structurally distinct TSE is small, as has been explicitly shown using lattice models with side chains *(14)* and more realistic models of β-hairpins *(52)*.

2. TSE in proteins exhibiting KPM: in polypeptide chains that fold by KPM, R = $\tau_s^{WT}/\tau_f^{WT} >> 1$ (**Fig. 3**). In lysozyme $R \approx 100$ *(42)*. The energy landscape is riddled with kinetic traps that profoundly alter the folding mechanism. As a result, the structures of the TSE should be determined separately for the pool of molecules that fold by fast and slow tracks. Lattice model calculations using three dimensional 48-mer sequences show that the molecules folding by fast track via nucleation–collapse mechanism have multiple (three or four) distinct transition state structures (Giugliarelli, G., Klimov, D., and Thirumalai, D. unpublished). On the other hand, TSE in the slow track show much less diversity. If the fast and slow tracks are not separately analyzed, then $\Phi_F \approx \pm ln(\tau_s/\tau_f)$, which can be much greater than unity or much less than zero. These arguments suggest that TSE in proteins exhibiting complex kinetics can also be probed using Φ_F-value analysis provided all the possible scenarios are explored.

5.2. "Classical" Φ_F-Values: Role of the Denatured State Ensemble

If $\Phi_F \approx 1$, mutations affect the native structure and the TSE to the same extent implying that residue is structured in the TSE *(45)*. Because we expect that most non-disruptive mutations do not alter the free energies of DSE significantly, the condition for $\Phi_F \approx 1$ is $\Delta G^M - \Delta G^{WT} \approx \Delta G_M^{\ddagger} - \Delta G_{WT}^{\ddagger}$. This interpretation holds, if the entropy associated with the TSE is negligible, i.e., one has a narrow TSE (**Fig. 7**). Extensive studies in CI2 *(17,47)* have suggested that there is essentially one dominant pathway so that observation of $\Phi_F \approx 1$ for A16G indicates near complete structuring of A16 in the TSE. Concepts such as nucleation–collapse mechanism in CI2 *(17)* and the polarized transition state in the SH3 domains *(15)* have emerged based on the finding that few residues have $\Phi_F \approx 1$.

Small Φ_F-values imply that the residues have as much structure in the TSE as they do in the DSE. The difficulty in unambiguously interpreting the meaning of small Φ_F-values is that the nature of the DSE may play a crucial role. To illustrate this point we consider β-hairpin formation in two mutants (M1 and M2) of the 16-mer C-terminal hairpin from the GB1 protein *(52)*. We have identified the TSEs for M1 and M2 using a pattern recognition algorithm *(53)*. The

TSE for M1 is narrow, while there are four classes of TS structures for M2, three of which are displayed in **Fig. 8**. The DSE is taken to be the set of conformations that form within $\delta\tau_F$ ($\delta \approx 0.05$ and τ_F is the folding time) of the initiation of folding. Clustering these conformations shows profound differences in the structures of DSE (**Fig. 8**). The DSE for M1 is twice as structured as that of M2. These observations are reflected in the Φ_F-values calculated for the hydrophobic cluster: $\Phi_F \approx 0.9$ for M1 and $\Phi_F \approx 0.1$ for M2 *(53)*.

Structural characterization of α-spectrin SH3 domain by NMR *(54)*, in which except for L33V all other mutations have classical Φ_F-values, shows that the DSE has many characteristics in common with the TSE in accord with the β-hairpin results (**Fig. 8**). Interpreting fractional Φ_F-values in these systems, for which the TSE is predicted to be narrow, is possible, because the transition structures have to be "topology preserving" *(9)*, i.e., the fold restricts the range of possible conformations in the bottleneck region.

5.3. Φ_F-Values Using Single Molecule Experiments

The simple method based on Φ_F-values has proved to be a powerful probe of the fleeting transition state structures. Recent advances in using temperature *(55)* and denaturant dependent *(56)* Φ_F as global reaction coordinates have revealed additional features of the TSE in proteins, thus expanding the scope of the method. We anticipate that, in the not too distant future, single molecule fluorescence experiments *(57)* can provide Φ_F-values of individual protein molecules. Such experiments will provide near continuous movie of the transition from DSE past the TSE into the NBA (**Fig. 7**). The power of the Φ_F-value analysis can be fully realized if nonclassical values are interpreted using the energy landscape perspective of folding.

6. Conclusions

In this chapter we have focused on the plausible role of equilibrium and kinetic intermediates in protein folding. It is likely that folding of relatively large proteins might be dominated by the formation of an ensemble of intermediates. Both equilibrium and kinetic intermediates are relevant for fully understanding the folding kinetics. Because it is difficult to trap and characterize kinetic intermediates, experiments have focused exclusively on equilibrium intermediates. The best-studied example is the pH 4.0 intermediate **I** of apoMb. The data suggests that the cooperativity of formation of **I** is complex and depends sensitively on mutations as well anion concentration. In this relatively well-understood case, the equilibrium and kinetic intermediates are similar. The complexity of multiple intermediates and their role in folding are illustrated using simple off-lattice models. Using these simulations we have elucidated the differences between equilibrium and kinetic intermediates. We argue that the role of intermediates and the nature of transition state ensembles

DSE   TSE   NBA

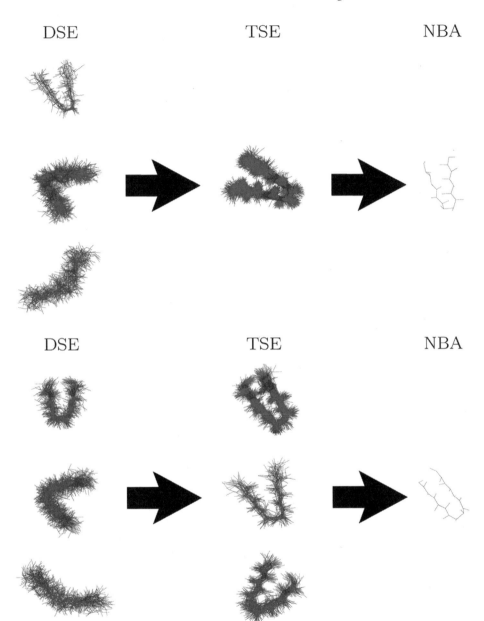

DSE   TSE   NBA

Fig. 8. Structures of denatured state ensemble (DSE) and transition state ensemble (TSE) for the two mutants of the C-terminal 16-mer β-hairpin of GB1 protein. Upper and lower panels correspond to M1 and M2, respectively. The DSE falls into three distinct clusters, which differ in the size and the extent of structure. The DSE of M1 is considerably more structured than M2, reflecting the tight turn in M1. The TSE of M1 is narrow, while that of M2 has multiple transition states.

in proteins that follow multiple routes will require fast folding and single molecule experiments.

Acknowledgments

We are grateful to Robert L. Baldwin for providing us with a table of experimental data used in **ref. 30**. This work was supported in part by the National Science Foundation grant CHE02-09340.

References

1. Dill, K. A. and Chan, H. S. (1997) From Levinthal to pathways to funnels. *Natur. Struct. Biol.* **4,** 10–19.
2. Onuchic, J. N., Luthey-Schulten, Z. A., and Wolynes, P. G. (1997) Theory of protein folding: an energy landscape perspective. *Ann. Rev. Phys. Chem.* **48,** 545–600.
3. Dobson, C. M., Sali, A., and Karplus, M. (1998) Protein folding: A perspective from theory and experiment. *Angew. Chem. Int. Edit.* **37,** 868–893.
4. Thirumalai, D. and Klimov, D. K. (1999) Deciphering the time scales and mechanisms of protein folding using minimal off-lattice models. *Curr. Opin. Struct. Biol.* **9,** 197–207.
5. Eaton, W. A., Munoz, V., Hagen, S. J., et al. (2000) Fast kinetics and mechanisms in protein folding. *Ann. Rev. Biophys. Biomol. Struct.* **29,** 327–359.
6. Daggett, V. and Fersht, A. R. (2003) Is there a unifying mechanism for protein folding? *Trends Biochem. Sci.* **28,** 18–25.
7. Fersht, A. R. (1995) Characterizing transition states in protein folding. *Curr. Opin. Struct. Biol.* **5,** 79–84.
8. Thirumalai, D. (1995) From minimal models to real proteins: time scales for protein folding. *J. Physiq. I* **5,** 1457–1467.
9. Guo, Z. and Thirumalai, D. (1995) Kinetics of protein folding: nucleation mechanism, time scales, and pathways. *Biopolymers* **36,** 83–103.
10. Fersht, A. R. (1995) Optimization of rates of protein folding: the nucleation-condensation mechanism and its implications. *Proc. Natl. Acad. Sci. USA* **92,** 10,869–10,873.
11. Klimov, D. K. and Thirumalai, D. (1998) Lattice models for proteins reveal multiple folding nuclei for nucleation-collapse mechanism. *J. Mol. Biol.* **282,** 471–492.
12. Onuchic, J. N., Socci, N. D., Luthey-Schulten, Z., and Wolynes, P. G. (1996) Protein folding funnels: The nature of the transition state ensemble. *Fold. Des.* **1,** 441–450.
13. Li, L., Mirny, L. A., and Shakhnovich, E. I. (2000) Kinetics, thermodynamics and evolution of non-native interactions in protein folding nucleus. *Nat. Struct. Biol.* **7,** 336–342.
14. Klimov, D. K. and Thirumalai, D. (2001) Multiple protein folding nuclei and the transition state ensemble in two state proteins. *Prot. Struct. Funct. Genet.* **43,** 465–475.
15. Riddle, D. S., Grantcharova, V. P., Santiago, J. V., Alm, E., Ruczinski, I., and Baker, D. (1999) Experiment and theory highlight role of native topology in SH3 folding. *Natur. Struct. Biol.* **6,** 1016–1024.

16. Martinez, J. C. and Serrano, L. (1999) The folding transition states between SH3 domains is conformationally restricted and evolutionary conserved. *Nat. Struct. Biol.* **6,** 1010–1016.
17. Itzhaki, L. S., Otzen, D. E., and Fersht, A. R. (1995) The nature of the transition state of chymotrypsin inhibitor 2 analyzed by protein engineering methods: evidence for a nucleation-condensation mechanism for protein folding. *J. Mol. Biol.* **254,** 260–288.
18. Baker, D. (2000) A surprising simplicity of protein folding. *Nature* **405,** 39–42.
19. Koga, N. and Takada, S. (2001) Roles of native topology and chain-length scaling in protein folding: a simulation study with a Go-like model. *J. Mol. Biol.* **313,** 171–180.
20. Ivankov, D. N., Garbuzynskiy, S. O., Alm, E., Plaxco, K. W., Baker, D., and Finkelstein, A. V. (2003) Contact order revisited: Influence of protein size on the folding rate. *Prot. Sci.* **12,** 2057–2062.
21. Li, M. S., Klimov, D. K., and Thirumalai, D. (2003) Thermal denaturation and folding rates of single domain proteins: size matters. *Polymer* **45,** 573–579.
22. Schuler, B., Lipman, E. A., and Eaton, W. A. (2002) Probing the free-energy surface for protein folding with single-molecule fluorescence spectroscopy. *Nature* **49,** 743–747.
23. Dyson, H. J. and Wright, P. (1998) Equilibrium NMR studies of unfolded and partially folded proteins. *Natur. Struct. Biol.* **5,** 499–503.
24. Kim, P. and Baldwin, R. L. (1990) Intermediates in the folding reactions of small proteins. *Annu. Rev. Biochem.* **59,** 631–660.
25. Hughson, F. M., Wright, P. E., and Baldwin, R. L. (1990) Structural characterization of a partly folded apomyoglobin intermediate. *Science* **249,** 1544–1548.
26. Jennings, P. A. and Wright, P. E. (1993) Formation of a molten globule intermediate early in the kinetic folding pathway of apomyoglobin. *Science* **262,** 892–896.
27. Uzawa, T., Akiyama, S., Kimura, T., et al. (2004) Collapse and search dynamics of apomyoglobin folding revealed by submillisecond observations of alpha-helical content and compactness. *Proc. Natl. Acad. Sci. USA* **101,** 1171–1176.
28. Cardenas, A. E. and Elber, R. (2003) Kinetics of cytochrome C folding: atomically detailed simulations. *Proteins: Struct. Funct. Genet.* **51,** 245–257.
29. Thirumalai, D. and Woodson, S. A. (1996) Kinetics of folding of proteins and RNA. *Acc. Chem. Res* **29,** 433–439.
30. Luo, Y., Kay, M. S., and Baldwin, R. L. (1997) Cooperativity of folding of the apomyoglobin pH 4 intermediate studied by glycine and proline mutations. *Nat. Struct. Biol.* **4,** 925–930.
31. Serrano, L., Kellis, J. T., Jr., Cann, P., Matouschek, A., and Fersht, A. R. (1992) The folding of an enzyme. II. Substructure of barnase and the contribution of different interactions to protein stability. *J. Mol. Biol.* **224,** 783–804.
32. Klimov, D. K. and Thirumalai, D. (1998) Cooperativity in protein folding: from lattice models with side chains to real proteins. *Fold. Des.* **3,** 127–139.
33. Santoro, M. M. and Bolen, D. W. (1988) Unfolding free energy changes determined by the linear extrapolation method. 1. Unfolding of phenylmethanesulfonyl α-chymotrypsin using different denaturants. *Biochemistry* **27,** 8063–8068.

34. Breslauer, K. J. (1995) Extracting thermodynamic data from equilibrium melting curves for oligonucleotide order-disorder transitions. *Methods Enzymol.* **259,** 221–242.

35. Allen, D. L. and Pielak, G. J. (1998) Baseline lengths and automated fitting of denaturant data. *Prot. Sci.* **7,** 1262, 1263.

36. Cutler, R. L., Pielak, G. J., Mauk, A. G., and Smith, M. (1987) Replacement of cysteine-107 of *Saccharomyces cerevisiae* iso-1- cytochrome c with threonine: Improved stability of the mutant protein. *Prot. Eng.* **1,** 95–99.

37. Luo, Y. and Baldwin, R. L. (1998) Trifluoroethanol stabilizes the pH 4 folding intermediate of sperm whale apomyoglobin. *J. Mol. Biol.* **279,** 49–57.

38. Thirumalai, D., Klimov, D. K., and Woodson, S. A. (1997) Kinetic partitioning mechanism as unifying theme in the folding of biomolecules. *Theor. Chem. Acct.* **1,** 23–30.

39. Camacho, C. and Thirumalai, D. (1993) Minimum energy compact structures in random sequences of heteropolymers. *Phys. Rev. Lett.* **71,** 2505–2508.

40. Veitshans, T., Klimov, D. K., and Thirumalai, D. (1997) Protein folding kinetics: Time scales, pathways, and energy landscapes in terms of sequence dependent properties. *Fold. Des.* **2,** 1–22.

41. Kiefhaber, T. (1995) Kinetic traps in lysozyme folding. *Proc. Natl. Acad. Sci. USA* **92,** 9029–9033.

42. Matagne, A., Radford, S. E., and Dobson, C. M. (1997) Fast and slow tracks in lysozyme folding: insight into the role of domains in the folding process. *J. Mol. Biol.* **267,** 1068–1074.

43. Fersht, A. R. (1998) *Structure and Mechanism in Protein Science: A Guide to Enzyme Catalysis and Protein Folding.* W.H. Freeman and Co., New York, NY.

44. Wolynes, P. G., Onuchic, J. K., and Thirumalai, D. (1995) Navigating the folding routes. *Science* **267,** 1619, 1620.

45. Fersht, A. R., Matouschek, A., and Serrano, L. (1992) The folding of an enzyme. I. Theory of protein engineering analysis of stability and pathway of protein folding. *J. Mol. Biol.* **224,** 771–782.

46. Fersht, A. R., Leatherbarrow, R. L., and Wells, T. N. C. (1987) Structure-activity relationships in engineered proteins: analysis of use of binding energy by linear free energy relationships. *Biochemsitry* **26,** 6030–6038.

47. Fersht, A. R., Itzhaki, L. S., ElMasry, L. F., Matthews, J. M., and Otzen, D. E. (1994) Single versus parallel pathways of protein folding and fractional formation of structure in the transition state. *Proc. Natl. Acad. Sci. USA* **91,** 10,426–10,429.

48. Fulton, K. F., Main, E. R. G., Daggett, V., and Jackson, S. E. (2000) Mapping the interactions present in the transition state for unfolding/folding FKBP12. *J. Mol. Biol.* **291,** 445–461.

49. Hamill, S. J., Stewart, A., and Clarke, J. (2000) The folding of immunoglobulin-like greek key protein is defined by a common-core nucleus and regions constrained by topology. *J. Mol. Biol.* **297,** 165–178.

50. Goldenberg, D. P. (1999) Finding the right fold. *Nat. Struct. Biol.* **6,** 987–990.

51. Ozkan, S. B., Bahar, I., and Dill, K. A. (2001) Transition states and the meaning of ϕ-values in protein folding kinetics. *Natur. Struct. Biol.* **8,** 765–769.

52. Klimov, D. K. and Thirumalai, D. (2000) Mechanisms and kinetics of β-hairpin formation. *Proc. Natl. Acad. Sci. USA* **97,** 2544–2549.
53. Klimov, D. K. and Thirumalai, D. (2002) Stiffness of the distal loop restricts the structural heterogeneity of the transition state ensemble in SH3 domains. *J. Mol. Biol.* **315,** 721–737.
54. Kortemme, T., Kelly, M. J., Kay, L. E., Forman-Kay, J., and Serrano, L. (2000) Similarities between the spectrin SH3 domain denatured state and its folding transition state. *J. Mol. Biol.* **297,** 1217–1229.
55. Crane, J. C., Koepf, E. K., Kelly, J. W., and Gruebele, M. (2000) Mapping the transition state of the WW domain β-sheet. *J. Mol. Biol.* **298,** 283–292.
56. Oliveberg, M. (2001) Characterization of the transition states for protein folding: Towards a new level of mechanistic details in protein engineering analysis. *Curr. Opin. Struct. Biol.* **11,** 94–100.
57. Kelley, A. M., Michalet, X., and Weiss, S. (2001)Chemical physics: single-molecule spectroscopy comes of age. *Science* **292,** 1671, 1672.

16

Thinking the Impossible

How to Solve the Protein Folding Problem With and Without Homologous Structures and More

Rita Casadio, Piero Fariselli, Pier Luigi Martelli, and Gianluca Tasco

Summary

Structure prediction of proteins is a difficult task as well as prediction of protein–protein interaction. When no homologous sequence with known structure is available for the target protein, search of distantly related proteins to the target may be done automatically (fold recognition/ threading). However, there are difficult proteins for which still modeling on the basis of a putative scaffold is nearly impossible. In the following, we describe that for some specific examples, human expertise was able to derive alignments to proteins of similar function with the aid of machine learning-based methods specifically suited for predicting structural features. The manually curate search of putative templates was successful in generating low-resolution three-dimensional (3D) models in at least two cases: the human tissue transglutaminase and the alcohol dehydrogenase from *Sulfolobus solfataricus*. This is based on the structural comparison of the model with the 3D protein structure that became available after prediction. For protein–protein interaction, a knowledge-based method can give predictions of putative interaction patches on the protein surface; this feature may help in adding additional weight to specific nodes in nets of interacting proteins.

Key Words: Protein folding; protein structure prediction; building by homology, threading; *ab initio* methods; remote homologs; validation; 3D modeling of membrane proteins; 3D modeling of globular proteins; prediction of protein interacting patches.

1. Synopsis

One of the most difficult problems in molecular and structural biology is the protein folding problem. Even today, in spite of the enormous computational power that different Grid projects have made available, scientists are still missing a good comprehensive target function, whose minima can give the interatomic

From: *Methods in Molecular Biology, vol. 350: Protein Folding Protocols*
Edited by: Y. Bai and R. Nussinov © Humana Press Inc., Totowa, NJ

distances necessary to reconstruct the protein three-dimensional (3D) structure starting from the residue sequence. However thanks to several efforts, today we know that at least for some 30,000 sequences nature, or at least crystallographers, solved the problem; this is approximately the number of proteins known with atomic resolution and collected in the Protein Data Bank (PDB) of structures solved with atomic resolution *(1)*. The good news is that in the last 10 yr the protein folding problem has been addressed from a new perspective. Instead of trying to optimize target functions, a knowledge-based approach has been taken into account. By considering known 3D structures, the question to be asked was: how and what can we learn from experiments where sequences are found already folded into their presumably native and functional fold? In this chapter, we will, therefore, answer to a few relevant questions: when is it possible to fold a protein with a high probability of success? Under which constraints can we think of computing a good/decent/bad protein 3D model? And more…can we think of predicting where and if our "pet" protein will ever interact with other proteins without knowing about very complex protein–protein interaction networks?

2. Is it Possible to Fold a Protein Sequence?

Nowadays, if you have a "pet" protein and still this is not known with atomic resolution, the first step that you may want to try is a database search *(2)*. By aligning your sequence toward the database of known structures, you may fall in the cases depicted in **Fig. 1**.

1. Your sequence is significantly similar to a sequence of known structure.
2. Your sequence has a low or no significant similarity with any of the proteins known at atomic resolution.

The threshold value of sequence homology identity, with respect to which you may end in either case, is historically in the range of 30 to 40% *(3)*.

In the first case, you may easily or less easily get a model for your protein based on *building by comparison*, a well-exploited technique according to which sequence alignment allows adopting the 3D coordinates of the template protein for the target *(4)*. Your protein is modeled by adopting as scaffold the template whose sequence has the highest level of identity with the target.

In the second case the situation gets more complicated and, therefore, more difficult. The basic idea is now to find among the existing folds in the database those that can best fit the target sequence or portions of it. Several methods have been developed to optimize this search, as previously reported, and some of them are available in the web *(5)*.

When you have not found any homolog sequences to the target, then you either get in touch with a crystallographer or a nuclear magnetic resonance spectroscopy expert, depending on the size of your protein, or you are left with the most risky

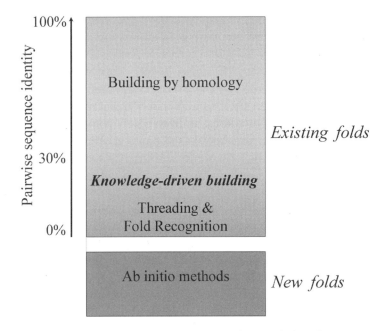

Fig. 1. Schematic diagram showing the methods for predicting the structure of a protein from its amino acid sequence when a possible fold for the novel chain is known or not. For comparative modeling (upper part) the accuracy of the model depends significantly on the sequence identity between the target sequence and the known structure on which the model is based.

choice: to try *ab initio* methods that, with respect to the results, are endowed with all the uncertainty that a computational method, although formally correct, may give rise to when it comes to predicting sequence folding. In other words, if you have good reasons to believe that the target protein is very similar to a given template, you can compute a 3D structure of the protein with a reliability that correlates with the level of sequence identity of the target with the template; the higher the sequence identity, the better the 3D model of the target protein *(4)*.

The efficacy of the predictive methods is routinely scored in the Community Wide Experiment on the Critical Assessment of Techniques for Protein Structure Prediction (CASP), a semi-annual community-wide blind test for a fair comparison of the different existing protein structure prediction techniques implemented in various laboratories around the world. In CASP some newly solved 3D structures of proteins are withheld and only their sequences are made public. The structure of those proteins will not be published until a given deadline, which gives CASP participants the chance to submit their predictions for the proteins. The different CASP editions (six editions every 2 yr since 1994, http://predictioncenter.llnl. gov/), have indicated that the patterns outlined above give satisfactory results in the majority of the cases if the level of sequence identity is high (>40%) (building by homology); the methods, however, may give satisfactory results with sequence identity less than 40% when *fold recognition/threading* are properly applied to sequences that are distantly related to others. In either case, solving the problem of predicting the protein structure is a problem of performing the correct alignment toward a related or distantly related protein or set of proteins *(5,6)*.

CASP4 and CASP5 indicated that, among the *ab initio* procedures, the best results were obtained with a fragment-based method called Rosetta, that relies on a scoring function assigning a score to 3D structures and guiding the search from the extended chain to a protein-like fold *(7)*.

3. Fold Recognition/Threading

When a novel protein shares a sequence identity lower than 25–30% with 3D resolved proteins, a pairwise sequence alignment with standard methods no longer assures to catch structural similarity *(3)*. However, proteins sharing low sequence identity can share the same structure, either because they are distantly related from the evolutionary point of view (remote homologous sequences), or because, though evolutionary unrelated, their structures converged for functional reasons.

On these bases, fold recognition methods have been developed in order to find in the database of 3D structures the fold (if it exists) that can be assigned to a novel sequence *(8)*. The general strategy is to add information to the target sequence before searching for a suitable template. This information can come either from the comparison with all other sequences in the database that allows compiling and compare the sequence profiles, or from the prediction of structural features starting from the amino acid sequence *(9–11)*. Implementations of fold recognition methods are available and protein structure prediction can be automatically obtained. Among the best performing methods you may consider: 3DPSSM *(12)*, TOPITS *(13)*, GenTHREADER *(14)*, and 3DSHOTGUN *(15)*.

4. Our Solutions to the Protein Folding Problem for Difficult Cases

Fold recognition methods, available in the web, may fail when a very difficult sequence is at hand. Difficult sequences are all those proteins that for several reasons cannot be modeled in an automatic fashion because they do not fall in any of the categories previously mentioned. In these cases, constraints different from those implemented in the web methods may also help in getting a better alignment with template and/or, as in the case of remote homologs, in selecting the template, which would be impossible with a standard routine approach. These cases often are brought about by experimentalists along with specific questions related to some functional features that need to be reconciled with the molecular properties of the protein.

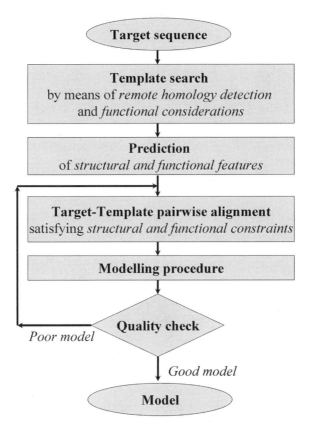

Fig. 2. Flow chart of a knowledge-based comparative model building. Functional information and predictions of structural features are used for selecting the template and for constraining the alignment between the target and the template.

In solving some of these specific cases, including both membrane and globular proteins, we thought to focus on human expertise-driven strategies that generally help in choosing the template. This consists in collecting all the possible information on protein function, possible ligands and effectors, and detection of remote homologs that may help in searching the database for specific templates that are functionally related to the target sequence. Once the target(s) has been selected, the problem can be treated again as an alignment problem and different predicted features may help in determining the most satisfactory alignment between the target and the template sequences. After this expert-driven alignment procedure, modeling can be similar to that adopted in *building by homology*. The overall approach is shown in **Fig. 2**. A routine ensures that different alignments are tried in order to optimize the structural requirement of a good 3D model. The comparative model building is performed with the MODELLER program *(16)*. For a given alignment, several model structures are then built and evaluated with the PROCHECK or PROSAII suite of programs *(17,18)*. Only the best-evaluated model is retained after the analysis. Close contacts are removed with the DEBUMP program of the WHATIF package *(19)*.

4.1. Machine Learning and Protein Structure Predictions

When it is necessary to build models for proteins of unknown spatial structure, which have very little homology with other proteins of known structure, nonstandard techniques need to be developed and the tools for protein structure predictions may, therefore, help in protein modeling.

In most of our examples, the search of the optimal alignment is accomplished by considering protein structural features computed with predictive methods (**Fig. 3**). Computational tools can bridge the gap between sequence and protein 3D structure based on the notion that information is to be retrieved from the databases and that knowledge-based methods can help in approaching a solution of the protein folding problem. To this aim our group has implemented machine learning-based predictors capable of performing with some success in different tasks, including predictions of the secondary structure of globular and membrane proteins, of the topology of membrane proteins, and of the presence of signal peptides. Moreover, we have developed methods for predicting contact maps in proteins and the probability of finding a cysteine in a disulfide bridge, tools which can contribute to the goal of predicting the 3D structure starting from the sequence (the so called "*ab initio*" prediction) (www.biocomp.unibo.it).

What is interesting and peculiar to our and other methods present in the web, including the ones predicting the protein transmembrane topology, is that all the machine learning approaches are trained on nonredundant sets of protein structures derived from the database of proteins solved with atomic resolution and containing the specific feature(s) to be predicted. All our predictors take advantage of evolution information derived from the structural alignments of homologous (evolutionary-related) proteins and taken from the sequence and structure databases. The choice of the machine learning approach is problem-specific: our predictive methods are based on neural networks, hidden Markov models, or hidden neural networks, depending on the addressed problem (www.biocomp.unibo.it) *(20)*.

4.2. Some Test Cases Including Membrane and Globular Proteins

4.2.1. Membrane Proteins

A homology search may be particularly unfortunate when it comes to a membrane protein sequence, since we know very few templates for membrane

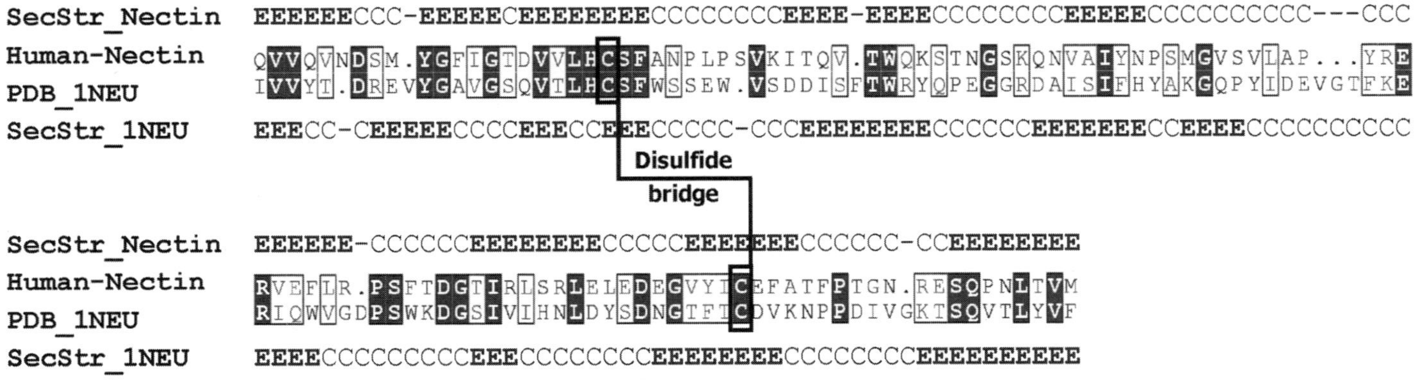

Fig. 3. Alignment between the sequences of the human nectin and of the myelin adhesion molecule, P, whose structure is deposited in the PDB database with the code 1NEU. Dark-filled boxes highlight identical residues, whereas white boxes indicate similar residues. The secondary structure of nectin has been predicted using SECPRED (*36*). The bonding state of the two cysteine residues of nectin has been predicted using CYSPRED (*37*). The secondary structure and the cysteine bonding state for 1NEU have been computed with the DSSP program (*38*). The alignment between the target and the template has been forced in order to conserve sequence identity, secondary structure, and disulfide-bonded cysteine residues.

proteins that exploit a small subset of the functions that membrane proteins perform. Unfortunately, unless our chain is very similar to one of the membrane proteins in the database, we will not find templates suited for adopting a building-by-homology procedure.

Prompted by the necessity to model this important and abundant class of proteins, we developed specific predictors for the topology prediction of membrane proteins, starting from their sequence: HMMB2TMR, a HMM-based predictor and ENSEMBLE, an ensemble predictor (consisting of one neural network and two HMMs), respectively, for all β-membrane proteins and all α-membrane proteins *(20)*. Two models computed with the expert-driven approach depicted in **Fig. 2**, are shown in **Fig. 3**.

The voltage-dependent anion channels (VDAC) of *Saccharomyces cerevisiae* and of *Drosophila melanogaster* were modeled on the outer membrane proteins of prokaryotes, the only one available with atomic resolution. This was possible considering that for β-barrel membrane proteins in prokaryotes, the barrel architecture is rather conserved, changing only the number of strands in the barrel. After predicting the number of strands with HMMB2TMR, the alignment of the target to different templates with the same number of strands was computed by overlapping the topology. The models were then refined using existing data in the literature *(21)* or with site-directed mutagenesis *(22)*.

For all α-membrane proteins, the situation is more difficult because different proteins have different architectures and different number of transmembrane α-helices. In this case, it is important that the function is known. We modeled the oxoglutarate carrier of bovine mitochondria using the recently solved 3D structure of the ADP/ATP carrier as a template. Also in this case the model was refined considering experimental and site-directed mutagenesis data *(23)*.

4.2.2. Globular Proteins

Concerning globular proteins, several models have been computed (**Fig. 4**), each one sharing very little sequence identity with the proteins whose structures were available at the time we performed the modeling procedure.

The first test case is the modeling of the 3D of human integrin β_3 with the specific aim of designing peptidomimetic molecules suited to act as inhibitors of its action mechanism. The model of an experimentally characterized metal ion binding site in the protein, know to interact with RGD peptides, was computed adopting as scaffold a similar metal binding site in an α-integrin fragment, known at atomic resolution *(24)*.

In the case of the human tissue transglutaminase, despite only a 34% sequence identity with the crystallized human factor XIII used as a template, the strong functional similarity assured a good alignment, computed by constraining the key residues essential for the activity. Our model was in agreement

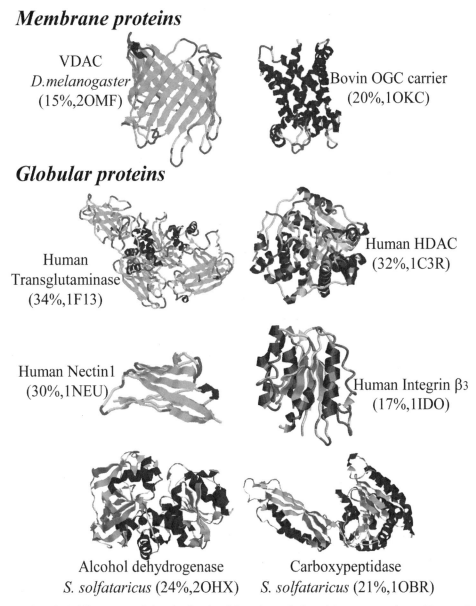

Membrane proteins

VDAC
D.melanogaster
(15%,2OMF)

Bovin OGC carrier
(20%,1OKC)

Globular proteins

Human
Transglutaminase
(34%,1F13)

Human HDAC
(32%,1C3R)

Human Nectin1
(30%,1NEU)

Human Integrin β3
(17%,1IDO)

Alcohol dehydrogenase
S. solfataricus (24%,2OHX)

Carboxypeptidase
S. solfataricus (21%,1OBR)

Fig. 4. Different models obtained with a knowledge-driven procedure. For each model, the PDB code of the template structure and the percent sequence identity with the target are indicated in parentheses. *See* **Subheadings 4.2.1.–4.2.2.** for more details.

with the data obtained by small-angle neutron and X-ray scattering, and was adopted to simulate the conformational changes induced by calcium ion binding and related to the regulation of protein activity *(25)*.

Similar considerations drove the recent modeling of the human histone deacetylase 1 (HDAC1) on the basis of a HDAC homolog from *Aquifex aeolicus*. Although distantly related (32% sequence identity), the functional similarity allowed us to conserve four residues involved in zinc ion coordination in the alignment between the target and the template. The model obtained by this procedure, particularly accurate in the active site zone, was then adopted for docking to the protein a newly discovered ligand affecting its activity, the 9-hydroxystearic acid *(26)*.

A more complicated case study was the modeling of the extracellular domain of the human nectin on a myelin membrane adhesion molecule P0, 30% identical in sequence, and only partially similar in function. In this case the alignment between the target and the template was constrained imposing the superimposition of the predicted secondary structure of the two proteins. Moreover, two cysteine residues in the nectin were predicted to form a disulfide bridge and their positions were aligned with the two disulfide-bound cysteines of the myelin adhesion molecule (**Fig. 3**). This model was good enough to suggest a molecular interpretation of the effect of different mutations on the binding between the nectin and the glycoprotein G of the herpes simplex virus *(27)*.

Finally, the computing of the two last models reported in **Fig. 4**, required an extensive application of both functional information from experiments and prediction of structural features. Alcohol dehydrogenase (ADH) from *Sulfolobus solfataricus* was known to bind two zinc ions into two different sites and to be functional in a tetrameric form. Unfortunately in the databases of structurally resolved proteins only dimeric two-zinc containing or tetrameric one-zinc containing ADHs were present. Therefore, the model was built taking into account the two features and adopting two different templates at the same time, in order to model both the second zinc binding site and the tetrameric interaction patches (both templates sharing less than 25% sequence identity with *Sulfolobus* ADH). The final alignment between the target and the templates was aided by the prediction of the secondary structure *(28)*.

Also in the case of the carboxypeptidase from *S. solfataricus* (CPSso) the interplay between functional information, the prediction methods and the experimental validation was relevant for obtaining a reliable model. From experimental results it was known that CPSso contains one zinc atom per monomer. In the structural database two kinds of zinc-containing carboxypeptidases could be found, having a similar fold around the active site, but having completely different overall folds and quaternary structures. Two templates were chosen: carboxypeptidase from *Thermoactinomyces vulgaris*, 16% identical to CPSso, and carboxypeptidase G2 from *Pseudomonas spirillum*, 21% identical. Two models were built, constraining the alignments with the conservation of the key residues and superimposing the predicted secondary structures of the target with that of the templates. The differences in the zinc ion binding residues between the models allowed devising experiments for deciding which model was right. The site-directed mutagenesis determined that it was that based on carboxypeptidase G2. Moreover, small angle X-ray diffraction experiments confirmed the compatibility between the quaternary structure of the protein and the shape of the tetrameric form of the model obtained on the basis of carboxypeptidase G2 crystallographic symmetries *(29)*.

5. Validation

After having computed a 3D model for a protein, the most appropriate question is: how good is it? This is a very difficult question to answer because the truth is not known. In principle and in several cases, including the ones we have previously described, models are to be taken as low-resolution models of the protein, so to say that we are responsible of the backbone folding of the protein and not of the side chain positions of the residues. The correct prediction of side chain rotamers is, however, still a problem even in the case of building by homology, when sequence similarity is rather high. Basically, our models were good for fitting experimental data and helped in interpreting them at the molecular level. Often, site-directed mutagenesis experiments, inspired by the models, were also sufficient to prove some structural features. So far, in only three cases, we could test our expertise in generating the models toward the real structures that became available after we published our results (**Fig. 5**).

In the case of the metal binding fragment of human integrin β_3, the superimposition with the real whole structure of the protein indicates that, although the overall fold was somewhat correct, the correct structural alignment was rather poor. In spite of this, the model scaffold was good enough to assist the design of effective peptidomimetic molecules *(24,30)*.

When the 3D structure of the human tissue transglutaminase became available, the alignment of the model with the real structure was much better, especially considering the length of the modeled protein (680 residues) *(25,31)*.

The 3D structure of the *S. solfataricus* ADH was solved at atomic level (0.185-nm resolution) shortly after the publication of the model. The superimposition between the model and the experimental structure gives a root mean square deviation of about 0.25 nm, a quite satisfactory result, considering the extent of human expertise involved in the building procedure *(28,32)*.

6. More...

As a consequence of wide-scale projects aiming to characterize all the metabolic networks going on in a cell during its life cycle, we know that proteins interact which each others and with nucleic acids. Several data on protein–protein interactions are presently available for different organisms; however, the number and type of interactions still depend on the experimental technique *(33)*.

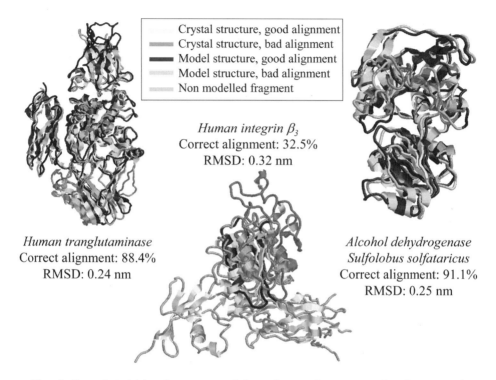

Fig. 5. Superimposition between models and crystal structures for three proteins. The crystal structures considered for the comparison are the human tranglutaminase 1KV3 *(31)*, the integrin β_3 extracted from the α_V–β_3 complex 1L5G *(30)*, and the alcohol dehydrogenase from *Sulfolobus solfataricus* 1JVB *(32)*. The portions in which the model and the structure are coincident (correct alignment) and/or badly aligned are according to the gray scale. The coverage of this portion is indicated for each model as the percentage of correct alignment. The percentage of correct alignment and the root mean square deviations are computed with the CE program *(39)*.

A bottom-up approach in validating some of the nodes in the nets of protein–protein interactions is to be able to address the problem starting from the molecular detailed structure of the protein. To this aim we developed a predictor suited to detect interacting patches on the protein surface **(Fig. 6)**. The predictor is based on neural networks, trained on a set of nonredundant interacting heterodymers known with atomic resolution and a filter to leave out spurious assignments. The predictor is available on the web together with a database of the PDB structures predicted according to our method *(34,35)*.

7. Conclusions and Perspectives

A recent survey of a decade of CASP experiments lists the major bottlenecks that knowledge-based methods have to overcome in the next years: close

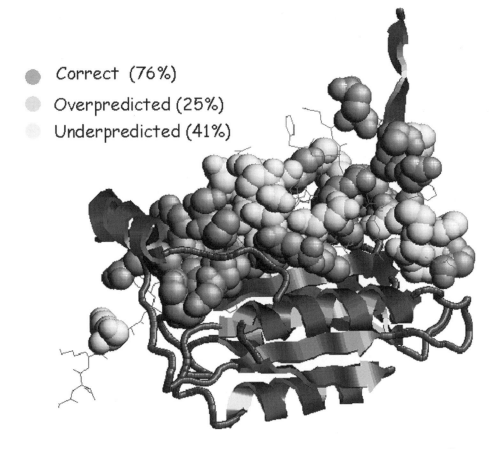

Fig. 6. Predicted interaction patches in chain B of the inhibited interleukin-1β converting enzyme (1IBC). Prediction was performed with a modified version of ISPRED, a neural network-based predictor available at www.biocomp.unibo.it *(34)*.

evolutionary relationship models approaching experimental accuracy, improved alignments, refinement of remote evolutionary relationship models, and reliable discrimination among possible template-free models *(5)*. In the meantime, with our expert-driven approach we emphasize the role that human expertise still has in partially solving these bottlenecks that are crucial in finding general solutions to the problem with automatic procedures.

With our validated test cases we highlight that modeling difficult proteins may be unsatisfactory in the case of fragments *(24)*, more satisfactory even for very large proteins when several pieces of information are taken into account, including those derived from top scoring predictors *(25,28)*. Certainly, in the two last cases, what mainly helped us in producing correct models was the choice of templates on the basis of function similarity of ligand binding sites and the alignment constrained by prediction of structural features. Work is in progress in our and other laboratories to assess the role of protein interaction patch predictions in validating experimental networks of protein–protein interaction *(33)*.

Acknowledgments

This work was supported by the following grants: a COFIN2003 of the Ministero della Istruzione dell'Universita' e della Ricerca (MIUR), a PNR 2001-2003 (FIRB art.8) projects on Postgenomics delivered to RC and a grant delivered by the European Union's VI Framework Programme for the BioSapiens Network of Excellence project. PF acknowledges a MIUR grant on Proteases.

References

1. Sternberg, M. (1997) *Protein Structure Prediction: A Practical Approach,* Oxford University Press, Oxford, UK.
2. Lesk, A. M. (2005) *Introduction to Bioinformatics, 2nd ed.* Oxford University Press, Oxford, UK.
3. Rost, B. (1999) Twilight zone of protein sequence alignments. *Protein Eng.* **12,** 85–94.
4. Sanchez, R., Pieper, U., Melo, F., et al. (2000) Protein structure modeling for structural genomics. *Nat. Struct. Biol.* **7 Suppl,** 986–990.
5. Moult, J. (2005) A decade of CASP: progress, bottlenecks and prognosis in protein structure prediction. *Curr. Opin. Struct. Biol.* **15,** 285–289.
6. Bourne, P. E. (2003) CASP and CAFASP experiments and their findings. *Methods Biochem. Anal.* **44,** 501–507.
7. Bonneau, R., Strauss, C. E., Rohl, C. A., et al. (2002) De novo prediction of three-dimensional structures for major protein families. *J. Mol. Biol.* **322,** 65–78.
8. Friedberg, I., Jaroszewski, L. Y., and Godzik, A. (2004) The interplay of fold recognition and experimental structure determination in structural genomics. *Curr. Opin. Struct. Biol.* **14,** 307–312.
9. Capriotti, E., Fariselli, P., Rossi, I., and Casadio, R. (2004) A Shannon entropy-based filter detects high-quality profile-profile alignments in searches for remote homologues. *Proteins* **54,** 351–360.
10. Prybylski, D. and Rost. B. (2004) Improving fold recognition without folds. *J. Mol. Biol.* **341,** 255–269.
11. Tramontano, A. (2005) *The Ten Most Wanted Solutions in Protein Bioinformatics,* CRC Press, London, UK.
12. Kelley, L. A., McCallum, R. M., and Sternberg, M. J. (2000) Enhanced genome annotation using structural profiles in the program 3D-PSSM. *J. Mol. Biol.* **299,** 499–520.
13. Rost, B. (1995) TOPITS: threading one-dimensional predictions into three-dimensional structures. *Proc. Int. Conf. Intell. Syst. Mol. Biol.* **3,** 314–321.
14. Mc Guffin, L. J. and Jones, D. T. (2003) Improvement of the GenTHREADER method for genomic fold recognition. *Bioinformatics* **19,** 874–881.

15. Fischer, D. (2003) 3D-SHOTGUN: a novel, cooperative, fold-recognition meta-predictor. *Proteins* **51,** 434–441.
16. Sali, A. and Blundell, T. L. (1993) Comparative protein modelling by satisfaction of spatial restraints. *J. Mol. Biol.* **234,** 779–781.
17. Laskowski, R. A., MacArthur, M. W., Moss, D. S., and Thornton, J. M. (1993) PROCHECK: a program to check the stereochemical quality of protein structures. *J. Appl. Cryst.* **26,** 283–291.
18. Sippl, M. J. (1993) Recognition of errors in three-dimensional structures of proteins. *Proteins* **17,** 355–362.
19. Vriend, G. (1990) WHAT IF: a molecular modeling and drug design program. *J. Mol. Graph.* **8,** 52–56.
20. Casadio, R., Fariselli, P., and Martelli, P. L. (2003) In silico prediction of the structure of membrane proteins: is it feasible? *Brief Bioinf.* **4,** 341–348.
21. Casadio, R., Jacoboni, I., Messina, A., and De Pinto, V. (2002) A 3D model of the voltage-dependent anion channel. *FEBS Lett.* **520,** 1–7.
22. Aiello, R., Messina, A., Schiffler, B., et al. (2004) Functional characterization of a second porin isoform in drosophila melanogster: DmPorin2 forms voltage-independent cation selective pores. *J. Biol. Chem.* **279,** 25,364–25,373.
23. Morozzo della Rocca, B., Miniero, D. V., Tasco, G., et al. (2005) Substrate-induced conformational changes of the mitochondrial oxoglutarate carrier: a spectroscopic and molecular modelling study. *Mol. Membr. Biol.* **22,** 443–445.
24. Casadio, R., Compiani, M., Facchiano, A., et al. (2002) Protein structure prediction and biomolecular recognition: from protein sequence to peptidomimetic design with the human beta 3 integrin *SAR QSAR Envir Res* **13,** 473–486.
25. Casadio, R., Polverini, E., Mariani, P., et al. (1999) The structural basis for the regulation of tissue transglutaminase by calcium ions. *Eur. J. Biochem.* **262,** 672–679.
26. Calonghi, N., Cappadone, C., Pagnotta, E., et al. (2005) Histone deacetylase 1: a target of 9-hydroxystearic acid in the inhibition of cell growth in human colon cancer. *J. Lipid Res.* **46,** 1569–1603.
27. Menotti, L., Casadio, R., Bertucci, C., Lopez, M., and Campadelli-Fiume, G. (2002) Substitution in the murine nectin1 receptor of a single conserved amino acid at a position distal from the Herpes Simplex virus gD binding site confers high-affinity binding to gD. *J. Virol.* **76,** 5463–5471.
28. Casadio, R., Martelli, P. L., Giordano, A., Rossi, M., and Raia, C. A. (2002) A low-resolution 3D model of the tetrameric alcohol dehydrogenase from Sulfolobus solfataricus. *Protein Eng.* **15,** 215–223.
29. Occhipinti, E., Martelli, P. L., Spinozzi, F., et al. (2003) 3D structure of Sulfolobus solfataricus carboxypeptidase developed by molecular modeling is confirmed by site-directed mutagenesis and small angle X-ray scattering. *Biophys. J.* **85,** 1165–1175.
30. Xiong, J. P., Stehle, T., Zhang, R., et al. (2002) Crystal structure of the extracellular segment of Integrin alpha V-beta3 in complex with an Arg-Gly-Asp ligand. *Science* **296,** 151–155.

31. Liu, S., Ceriona, R. A., and Clardy, J. (2002) Structural basis for the guanine nucleotide-binding activity of tissue transglutaminase and its regulation of transamidation activity. *Proc. Natl. Acad. Sci. USA* **99,** 2743–2747.

32. Esposito, L., Sica, F., Raia, C. A., et al. (2002) Crystal structure of alcohol dehydrogenase from the hyperthermophilic archaeon Sulfolobus Solfataricus at 1.85 Angstrom resolution *J. Mol. Biol.* **318,** 463–477.

33. Aloy, P., Pichaut, M., and Russel, R. B. (2005) Protein complexes: structure prediction challenges for the 21st century. *Curr. Opin. Struct. Biol.* **15,** 15–22.

34. Fariselli, P., Pazos, F., Valencia, A., and Casadio, R. (2002) Prediction of protein–protein interaction sites in heterocomplexes with neural networks. *Eur. J. Biochem.* **269,** 1356–1361.

35. Fariselli, P., Zauli, A., Rossi, I., Finelli, M., Martelli, P. L., and Casadio, R. (2003) A neural network method to improve prediction of protein–protein interaction sites in heterocomplexes. *Proc 13th IEEE Workshop on Neural Networks for Signal Processing (NNSP'03)* **13,** 33–41.

36. Jacoboni, I., Martelli, P. L., Fariselli, P., Compiani, M., and Casadio, R. (2000) Predictions of protein segments with the same aminoacid sequence and different secondary structure: a benchmark for predictive methods. *Proteins* **41,** 535–544.

37. Martelli, P. L., Fariselli, P., Malaguti, L., and Casadio, R. (2002) Prediction of the disulfide bonding state of cysteines in proteins with hidden neural networks. *Protein Eng.* **15,** 951–953.

38. Kabsch, W. and Sander, C. (1983) Dictionary of protein secondary structure: pattern recognition of hydrogen-bonded and geometrical features. *Biopolymers* **22,** 2577–2637.

39. Shindyalov, I. N. and Bourne, P. E. (1998) Protein structure alignment by incremental combinatorial extension (CE) of the optimal path. *Protein Eng.* **11,** 739–747.

Index

321